Advancing Medicine through Nanotechnology and Nanomechanics Applications

Keka Talukdar
Nadiha High School, India

Mayank Bhushan
Pondicherry University, India

Anil Shantappa Malipatil
Guru Nanak Dev Engineering College - Bidar, India

A volume in the Advances in Medical Technologies and Clinical Practice (AMTCP) Book Series

IGI GLOBAL
DISSEMINATOR OF KNOWLEDGE

www.igi-global.com

Published in the United States of America by
IGI Global
Medical Information Science Reference (an imprint of IGI Global)
701 E. Chocolate Avenue
Hershey PA 17033
Tel: 717-533-8845
Fax: 717-533-8661
E-mail: cust@igi-global.com
Web site: http://www.igi-global.com

Library of Congress Cataloging-in-Publication Data

Names: Talukdar, Keka, 1975- editor. | Bhushan, Mayank, 1983- editor. | Malipatil, Anil Shantappa, 1984- editor.
Title: Advancing medicine through nanotechnology and nanomechanics applications / Keka Talukdar, Mayank Bhushan, and Anil Shantappa Malipatil, editors.
Other titles: Advances in medical technologies and clinical practice book series. 2327-9354
Description: Hershey, PA : Medical Information Science Reference, [2017] | Series: Advances in medical technologies and clinical practice book series | Includes bibliographical references and index.
Identifiers: LCCN 2016032785| ISBN 9781522510437 (h/c) | ISBN 9781522510444 (ebook)
Subjects: | MESH: Nanomedicine | Nanostructures
Classification: LCC R857.N34 | NLM QT 36.5 | DDC 610.28--dc23 LC record available at https://lccn.loc.gov/2016032785

This book is published in the IGI Global book series Advances in Medical Technologies and Clinical Practice (AMTCP) (ISSN: 2327-9354; eISSN: 2327-9370)

British Cataloguing in Publication Data
A Cataloguing in Publication record for this book is available from the British Library.

All work contributed to this book is new, previously-unpublished material. The views expressed in this book are those of the authors, but not necessarily of the publisher.

Advances in Medical Technologies and Clinical Practice (AMTCP) Book Series

ISSN: 2327-9354
EISSN: 2327-9370

Mission

Medical technological innovation continues to provide avenues of research for faster and safer diagnosis and treatments for patients. Practitioners must stay up to date with these latest advancements to provide the best care for nursing and clinical practices.

The **Advances in Medical Technologies and Clinical Practice (AMTCP) Book Series** brings together the most recent research on the latest technology used in areas of nursing informatics, clinical technology, biomedicine, diagnostic technologies, and more. Researchers, students, and practitioners in this field will benefit from this fundamental coverage on the use of technology in clinical practices.

Coverage

- E-Health
- Medical Informatics
- Biomechanics
- Clinical Studies
- Nursing Informatics
- Biomedical Applications
- Clinical Nutrition
- Diagnostic Technologies
- Clinical High-Performance Computing
- Telemedicine

IGI Global is currently accepting manuscripts for publication within this series. To submit a proposal for a volume in this series, please contact our Acquisition Editors at Acquisitions@igi-global.com or visit: http://www.igi-global.com/publish/.

The Advances in Medical Technologies and Clinical Practice (AMTCP) Book Series (ISSN 2327-9354) is published by IGI Global, 701 E. Chocolate Avenue, Hershey, PA 17033-1240, USA, www.igi-global.com. This series is composed of titles available for purchase individually; each title is edited to be contextually exclusive from any other title within the series. For pricing and ordering information please visit http://www.igi-global.com/book-series/advances-medical-technologies-clinical-practice/73682. Postmaster: Send all address changes to above address.

Titles in this Series

For a list of additional titles in this series, please visit: www.igi-global.com

Pattern and Data Analysis in Healthcare Settings
Vivek Tiwari (Maulana Azad National Institute of Technology, India) Basant Tiwari (Devi Ahilya University, India) Ramjeevan Singh Thakur (Maulana Azad National Institute of Technology, India) and Shailendra Gupta (AISECT University, India)
Medical Information Science Reference • copyright 2017 • 358pp • H/C (ISBN: 9781522505365) • US $215.00 (our price)

Novel Approaches for Drug Delivery
Raj K. Keservani (Rajiv Gandhi Proudyogiki Vishwavidyalaya, India) Anil K. Sharma (Delhi Institute of Pharmaceutical Sciences and Research, India) and Rajesh Kumar Kesharwani (National Institute of Technology, Warangal, India)
Medical Information Science Reference • copyright 2017 • 515pp • H/C (ISBN: 9781522507512) • US $210.00 (our price)

Emerging Research in the Analysis and Modeling of Gene Regulatory Networks
Ivan V. Ivanov (Texas A&M University, USA) Xiaoning Qian (Texas A&M University, USA) and Ranadip Pal (Texas Tech University, USA)
Medical Information Science Reference • copyright 2016 • 418pp • H/C (ISBN: 9781522503538) • US $195.00 (our price)

Applied Case Studies and Solutions in Molecular Docking-Based Drug Design
Siavoush Dastmalchi (Tabriz University of Medical Sciences, Iran) Maryam Hamzeh-Mivehroud (Tabriz University of Medical Sciences, Iran) and Babak Sokouti (Tabriz University of Medical Sciences, Iran)
Medical Information Science Reference • copyright 2016 • 367pp • H/C (ISBN: 9781522503620) • US $225.00 (our price)

Methods and Algorithms for Molecular Docking-Based Drug Design and Discovery
Siavoush Dastmalchi (Tabriz University of Medical Sciences, Iran) Maryam Hamzeh-Mivehroud (Tabriz University of Medical Sciences, Iran) and Babak Sokouti (Tabriz University of Medical Sciences, Iran)
Medical Information Science Reference • copyright 2016 • 456pp • H/C (ISBN: 9781522501152) • US $235.00 (our price)

Advancing Pharmaceutical Processes and Tools for Improved Health Outcomes
Tagelsir Mohamed Gasmelseid (International University of Africa, Sudan)
Medical Information Science Reference • copyright 2016 • 388pp • H/C (ISBN: 9781522502487) • US $185.00 (our price)

Editorial Advisory Board

Table of Contents

Detailed Table of Contents

Ion channels are naturally occurring pores found in all living organisms. These may be of different types, activated by many stimuli, but their addition or alteration may cause serious physiological issues. They allow some ions to pass through them and are resistant for some ions. So they are ion selective. For the pharmaceutical industry, ion channels are now bird's eye view as many drugs, intended to cure a disease, target ion channels located in a specific part of our body. Quantum dots, nanoparticles, carbon nanomaterials change the channel properties in a considerable amount. A huge potential of targeting the ion channels for changing the treatment procedure completely is now opening a new side of nanomechanics, nanofluidics, biophysics and medicine. The following paragraphs will describe how experimental as well as theoretical studies can bring a lot of change in the field of medicine. The fatal disease like cancer and male infertility; these two important issues are addressed here. The dependent of such diseases on ion channels and the way of treating them are emphasized.

This study shows the importance of electron microscopy in the analysis of the interaction of microorganisms (Staphylococcus aureus) with polymeric (polyurethane) dental prostheses. Starting from the biofilm formation and the biodestruction of the plastic material resulting in the production of polyurethane nanoparticles, the focus is on the bacterial secretion of membrane vesicles (in the range of 20-50 nm) loaded with plastic nanoparticles (from 2-3 to 10 nm) and on the toxicological threat that these delivery devices represent when interacting with host cells. The nanoparticles deliverance led by the bacterial infections dynamics opens new ways to the possibility of delivering drugs to selected cells.

The use of micro- and nanoparticles is rapidly advancing and has been most commonly used in medical and biological research that offers an encouraging scope in broad range of disciplines. Manipulation of the biomaterials to their micro- and nano-scale renders their properties and behavior different from that of the same material in the mass scale and make them more reactive than large particles. The removal of tooth structure and its restoration with synthetic material to solve the problems caused by dental caries, trauma and fracture is a practice nearly as old as dentistry. Efforts are made to create micro- and nanomaterials that can revolutionize these ancestral therapies and dental procedures. The use of these materials had shown some promising applications in caries control, endodontic therapy, regenerative dentistry, periodontology and oral biofilm management. This review aims to discuss the recent advances and future potential of polymer-based micro- and nanoparticles in dentistry.

The principle of mass detection using nano biosensors is based on the fact that the resonant frequency is very much sensitive to the mass of the bio-molecule, as with mass changes stiffness varies. The change of the attached mass on the CNT causes a shift to the resonant frequency. The key issue of mass detection is in quantifying the shift in the resonant frequency due to the mass of the attached molecule.This study, explores the vibration responses of the cantilever single and double-walled carbon nanotube with various attached microbes on the tip with an aim of developing a sensor. The biological objects studied include Alanine with Amino terminal residue, Deoxyadeonosine with free residue, Coronaviridae, Bartonella bacilliformis etc.. This sensor will be utilized to facilitate the identification of bacteria or virus that may be attached to CNT.

The idea of formulating brain permeable nanoparticles stems from the need to treat various neurological disorders like Parkinson's disease, Alzheimer's disease, schizophrenia, depression and brain tumors. Neuropeptides, antibiotics, anticancer drugs and many CNS active drugs cannot cross blood brain barrier (BBB). Studies have revealed that when these drugs are loaded on to nanoparticles they not only cross BBB, but also exhibit decreased side effects. The drug can be dissolved, dispersed, encapsulated inside the nanoparticle or attached on to surface of nanoparticles. In 1995, dalargin was the first drug to be delivered across blood brain barrier (BBB) using polysorbate 80 coated nanoparticles. The size of nanoparticles is usually between 10-1000nm. For crossing BBB it should be less than 300 nm.

Aeromonas hydrophila is a heterotrophic, gram negative bacterium which is primary or secondary cause of ulcers, fin rot, tail rot, and hemorrhagic septicaemia in fish. The treatments for this infection are only restricted to some antibiotics. So, novel materials are being searched for combating with bacterial infections and the resulting consequences. In this chapter, Ag2O/CuO nanocomposites were synthesized chemically and characterized by X-ray diffraction (XRD), scanning electron micrograph (SEM) and transmission electron microscope (TEM). The antimicrobial activities of Ag2O/CuO nanoparticles (NPs), was checked by both well diffusion and turbidometric (spectrophotometric) method. Synthesized nanoparticles exhibited their antimicrobial efficacy in both the standard inhibitory assays; these results thus provide a scope for further research on the application of Ag2O/CuO nanoparticles as disinfectant and/or antibiotic in the fishery industry.

Human body has a hierarchy of structures. There are many organs which are affected much earlier by a disease when it is detected. Most modern sensors can detect anomalies when its concentration in the body fluid reaches to millimolar range. But more sensitive biosensors should detect disease from much lower concentration like femtomolar range. So, extremely sensitive biosensors are needed for early detection of fatal diseases at their early stage. It should detect the target molecule from a very low concentration of analyte. Also, molecules which we often need to detect are too small in size. So Nanotechnology and biotechnology should shake hand to detect nanosized particles from an extremely low concentration solution. Hence we are in a real need of a biosensor. Here we are interested in charged biomolecules and will discuss the performance of Field- Effect Transistor based biosensosrs by computational method.

The reactivity of acid base cements forming hydroxyapatite (HA) such as, β-tricalcium phosphate (β-TCP) and dicalcium phosphate dehydrate (DCPD). Amorphous calcium phosphates, prepared by precipitation from supersaturated solutions, can also react to form apatitic cements since they are thermodynamic unstable with respect to HA and have a setting reaction more independent of particle size. Calcium phosphate cement containing an antibiotic can be used for filling bone defects and to ensure local antibiotherapy. Therefore, in the present chapter proposal, cement paste was prepared by combining cement liquids comprised of Na2HPO4 with cement powders. Gentamicin sulfate was also loaded on the cements and its in vitro release was evaluated over a period of time. The cement setting times were compared before and after drug addition. According to results, the initial and final setting times of samples increase after drug addition.

The limitation of orthopedic fractures and large bone defects treatments has brought the focus on fabricating bone grafts that could enhance ostegenesis and vascularization in-vitro. Developing biomimetic materials such as mineralized nanofibers that can provide three-dimensional templates of the natural bone extracellular-matrix is one of the most promising alternative for bone regeneration. Understanding the interactions between the structure of the scaffolds and cells and therefore the control cellular pathways are critical for developing functional bone grafts. In order to enhance bone regeneration, the engineered scaffold needs to mimic the characteristics of composite bone ECM. This chapter reviews the fabrication of and fabrication techniques for fabricating biomimetic bone tissue engineering scaffolds. In addition, the chapter covers design criteria for developing the scaffolds and examples of enhanced osteogenic differentiation outcomes by fabricating biomimetic scaffolds.

Foreword

WHAT IS NANOMECHANICS?

Nanomechanics is a research field that studies material behaviors at nanoscale level.

In recent years, nanomechanics has become an indispensable part of theoretical foundation for many engineering fields and emerging technologies such as nanotechnology as well as biomedical and bioenvironmental technologies. Because of its multidisciplinary characteristics, the term nanomechanics has 'multidisciplinary interpretations', and it has been used with different meanings in different contexts. In the area of applied mechanics, nanomechanics is often referred to as a hierarchical mechanics and mathematics paradigm that is mainly used to study the effective material properties of composite materials. A major objective of this kind of study is to find the statistical average material properties of the heterogeneous material through various homogenization methods. In condensed solid state physics and statistical mechanics, this process is called the Coarse Graining. One of the fundamental challenges of the contemporary statistical physics is how to construct accurate coarse grain models.

Traditionally, the standard nanomechanics methodology in engineering applications treats a composite material as a generic continuum model with a two-level paradigm: nanoscopic structure and macroscopic structure. The material proper- ties at nanoscale are usually given as a priori, and the task is to find the material behaviors at nanoscale, which are also called as the effective or overall material properties. From this perspective, a material point at the macro-level may be viewed as a nanoscopic material ensemble. In principle, the constitutive relations at macro level should be able to be derived from the ensemble average of nano- objects that are governed by the nanoscale physical laws, which can be quantum mechanics, lattice dynamics, nanoscale plasticity, or elasticity, etc. Two subtle points worth further clarification:

(1) the constitutive relations or material behaviors at macroscale may be very much different from their counterparts at nanoscale, so the task of nanomechanics is to find the unknown macroscale constitutive laws.

INTRODUCTION TO NANOMECHANICS

The conventional two-level paradigm of nanomechanics is a special mathematical homogenization model that is usually not associated with any fixed length scale. When studying material properties of a metal, 1 mm may be viewed as macroscale, and the length scale at nanolevel may range from 1nm to 100 nm; whereas studying the deformation of a dam, the macroscale may be up to 103 m.

In engineering applications, the conventional nanomechanics deals with practical engineering problems of a broad spectrum: effective material properties of composite/synthetic materials, such as cementitious materials, geotechnical materials, etc.; constitutive modeling of bio-materials, such as bone, muscle, blood flow, etc.; phase transformations; defects in solids, such as dislocation motion and crack growth; and environmental problems, such as air pollution, ground water flow and chemical transport, etc.

A Contemporary condensed matter physics and applied mechanics in general agree that the physics at molecular or atomic level can be described by the quantum mechanics or related approximation theories, e.g. density functional theory; the physics between the nm scale to sub-μm scale is governed by nanomechanics though presently we are mainly relying on the molecular dynamics simulation; from μm scale to or sub-mm length scale, nanomechanics and related mesoscale mechanics are playing more important roles; and the macroscopic phenomenological theory is generally valid at the length scale mm level or up.

In this book, we shall focus on several areas of nanomechanics and nanomechanics. Different from traditional nanomechanics, a salient feature of nanomechanics is its multiscale and multiphysics characteristics. It has some features presented in quantum mechanics, or quantum statistical mechanics, manifesting the statistical effects at atomic or sub-atomic level; on the other hand, it also shares many features of continuum mechanics, because a nanostructure could contain millions of atoms.

The impetus for contemporary nanomechanics is primarily due to the emergence of nanoscience and biomedical technology. It appears that traditional physics alone is not sufficient to deal with many engineering problems that are emerging from nanotechnologies and nanoengineering. There is a call for nanomechanics and nanocomputational mechanics to serve as the infrastructure of these developing engineering fields. For instance, much attention has been focused recently on material properties of thin films; manufacturing devices and components of a nanoelectromechanical system (NEMS), such as nano sized sensors, motors; mechanics of nanotubes and nanowires; and nanobiophysics/biochemistry systems, e.g. protein/DNA interaction in biomolecular simulation, etc.

Babak Sadeghi
Tonekabon Branch, IAU, Iran

Preface

Nanotechnology is the most significant field of research in the present day scenario. It is a multidisciplinary subject encompassing various branches of science and technology. But the technological expertise is considered to be properly exploited when it is worthy to use for the benefit of mankind. In the journey of living healthy life or saving lives of people, nanotechnology has gifted us efficient drug delivery systems, found a way of scafolding for nerve regeneration, has given more specific imaging technology, presented materials with therapeutic efficacy. Not only that, it has a lot of potential for gene therapy and cell repair by nanorobots. Personalized medicine is now not a dream only. Nanotechnology has the potential to take the challenge for treating disease like cancer. In this context the knowledge of nanomechanics is also necessary while dealing with engineered nanostructure and nanodevices. When we talk about nanomaterials for sensor application, or tissue engineering we have to deal with material properties while dynamics of molecular system is necessary for preparing nano electro mechanical systems.

Nanoscience has offered us nanoparticles and nanomaterials, nanocomposites, quantum dots, carbon nanotubes and the process of manipulating them is conferred by nanotechnology. Moreover, nanomachines can add a further superior hand that assembles miniaturized machines and nanobots for the refinement of clinical medicine. High expectations are developed that, in the coming few decades we will be able to combat against fatal diseases by much improved techniques and tools with the boon of nanotechnology. Though at the present stage, it will be very much optimistic idea to take the effect of nanomedicine in the clinical trial in a large scale, but some medical products that use nanotechnology are commercially available and are in use in clinical trial. To bring a revolutionary breakthrough in the world of medicine or for the personalized medicine, it needs many investigations, both theoretical and experimental, before coming to reality.

However, the challenge may be critical, molecular imaging, improved diagnostic technique, targeted drug or gene delivery, but these are not a far reaching goal and that is going to be established by many ongoing researches throughout the world. We expect in vivo applications of medicine through nanotechnology.

In the current status of nanomedicine, some potential applications are catching the attention of the present day researchers and scientists. Nanopolymer scaffolds mimicking the natural pattern can substitute teeth or bone, supersensitive sensors can detect disease causing biomolecules in very short time and from an extremely low concentration fluid, nanotweezer can take image and manipulate nanosized object inside body during surgery, nanoparticles may serve as therapeutic agents, and many more. Nanotechnology and nanomechanics is proving its excellence in all sectors of medical field and hence the topic will be relevant at least for the coming two decades.

Hence this book, written by eminent researchers, is a useful one in an appropriate time. Authors of the book have presented a number of major issues related to the progress of medical science through the knowledge of nanoscience and development of nanotechnology. Many of them are working in their laboratory for the state-of-art development of nanomedicine. It encompasses the detection of disease, therapeutic uses of nanoparticles, nanocomposites, fibers, modeling and simulation of nanobiosensor, carbon nanotube in sensor application etc. The subject matter which is nowadays a matter of intense study like tissue engineering is included here. Though all the topics falling under the title could not be addressed in a single book, but we think that the book will draw one's attention as it covers very relevant matters falling under nanomedicine.

Professionals and researchers working in the field of nanomedicine, healthcare, nanotechnology and pharmacology will surely find interest in the topics of the book. Moreover, the book will provide insights on the basic science factors behind all research outcomes. So a large number of M.Sc., M.Tech and Ph.D. students will find the book very useful for their knowledge and future research. Medical students, engineering students and the persons with technological background will also find interest in this publication. Common persons who want to enhance their knowledge about nanomedicine may also read this book. It consists of eleven chapters, each of them is introduced here;

Chapter 1 is a review chapter illustrating the role of ion channels in future medicine. It briefly points out the ion channels responsible for cancer and the way to fight against them from Nanotechnological and Nanomechanical point of view. Male infertility, which was not so opened up until now, is primarily discussed in the chapter where ion channels are responsible to regulate sperm motility. Mathematical modeling, necessary for this purpose is described briefly.

In chapter 2, a new way of cell specific drug delivery is reported by the authors and established by SEM and TEM study. Interaction of polyurethane oral prostheses with microorganisms is revealed in the work and bacterial secretion of membrane vesicles loaded with plastic nanoparticles is on spotlight. Toxicological effect of such devices while interacting with host cells are deeply observed and thus it is an optimistic step from in-vitro to in-vivo when drug delivery is concerned.

Chapter 3 deals with an extensive review of the structure of the materials and the fabrication techniques used to develop mineralized nanofibrous biomimetic scaffolds for bone tissue engineering. Different fabrication procedures of scaffolds for bone tissue engineering are discussed. The limitation of using synthetic biodegradable polymers as biomimetic scaffolds is that they cannot promote cell adhesion, proliferation, or differentiation to the extent that a natural polymer is able to because synthetic polymers often lack biological cues. Surface modification procedures, required for this purpose are described. The great potential of such scaffolds for clinical use will evoke a clear idea in the mind of the readers and they can learn the mechanism of their fabrication in the laboratory.

In chapter 4, the authors have exposed a great applicability for single walled and double walled carbon nanotubes for sensing biological molecules including virus and bacteria. This study, explores the vibration responses of the cantilever double-walled carbon nanotube (DWCNT) with various attached microbes on the tip with an aim of developing a sensor. Depending on the frequency shift it senses the presence of microbe attached to the tip.

Antibiotics, anticancer drugs and many CNS active drugs cannot cross blood brain barrier (BBB) and hence many neurological disorders like Parkinson's disease, Alzheimer's disease, schizophrenia, depression and brain tumors are limited by their crossing efficiency of BBB. Chapter 5 shows how the drugs, when loaded on to nanoparticles, not only cross BBB, but these have decreased side effects. The review gives description of various processes of drug deliveries to brain and compares how nanotechnology based research is far more helpful in targeted drug delivery to brain causing fewer side effects.

Chapter 6 is an original research article describing the process of chemical synthesization of Ag_2O/CuO nanocomposites and its characterization by XRD, SEM and TEM. The nanocomposite has microbial activities against ulcers, fin rot, tail rot, and hemorrhagic septicaemia in fish. The antimicrobial activites are confirmed by well diffusion and turbidometric (spectrophotometric) method. Thus this chapter provides a scope for further research on the application of Ag2O/CuO nanoparticles as disinfectant and/or antibiotic in the fishery industry.

Chapter 7 is a computational work on FET-based nanobiosensor. The theory behind settling time, sensitivity, noise ratio is described elaborately. The sensor crosses the Nernst limit of voltage sensitivity in dual gate configuration. The study focuses recent idea for supersensitive nanobiosensor.

Calcium phosphate is a substitute of bone and has excellent biocompatibility and bioactive properties. It can be used as bone filler or bone augmentation. In Chapter 8, the procedure of preparation of gentamicin sulfate loaded Calcium phosphate bone cement consisting of β-TCP and DCPD phases is described and the result of the investigation of the cement injectability in a constant injection velocity is discussed in vitro. Here injectibility is the ability of cement to move through a syringe–catheter device. The treatment of orthopedic fracture and large bone defect by bone grafting is now limited for ostegenesis and vascularization in-vitro.

Chapter 9 covers design criteria for developing the scaffolds with biomimetric materials like mineralized nanofibres. It cites examples of enhanced osteogenic differentiation outcomes by fabricating biomimetic scaffolds for bone regeneration.

Chapter 10 deals with the synthesis of various types of magnetic nanoparticles using different approaches depending of their ability to generate either single core of multcore magnetic nanoparticles. The various biomedical applications of magnetic nanoparticles like Magnetic Resonance Imaging (MRI), drug delivery etc along with possible limitations and challenges for their extended applications in medicine are also discussed.

Due to the promising characteristics of nanoparticles and nanomaterials, nanodevices for drug delivery is finding many new ways to decrease side effects in the patients' body and mind. There is a possibility of using polymeric nano herbal drugs as smart nano brain drugs to take care of mental health too. The review of chapter 11 gives an explicit detail of polymeric herbal nano drugs, their preparation, uses and advantages over other ones. It also gives some direction of treating some challenging diseases in the same way.

So the current book is a well-written collection of many major research topics of interest. It is an aggregation of reviews and original research articles which are written by some of the experts of the specific field. Reading this book will make some basic concept of the subject and will create a pathway for the future researchers for a more developed world of medicine with the help of nanotechnology and nanomechanics.

Chapter 1
Ion Channels, Nanomechanics, and Nanomedicine

Keka Talukdar
Nadiha High School, India

ABSTRACT

Ion channels are naturally occurring pores found in all living organisms. These may be of different types, activated by many stimuli, but their addition or alteration may cause serious physiological issues. They allow some ions to pass through them and are resistant for some ions. So they are ion selective. For the pharmaceutical industry, ion channels are now bird's eye view as many drugs, intended to cure a disease, target ion channels located in a specific part of our body. Quantum dots, nanoparticles, carbon nanomaterials change the channel properties in a considerable amount. A huge potential of targeting the ion channels for changing the treatment procedure completely is now opening a new side of nanomechanics, nanofluidics, biophysics and medicine. The following paragraphs will describe how experimental as well as theoretical studies can bring a lot of change in the field of medicine. The fatal disease like cancer and male infertility; these two important issues are addressed here. The dependent of such diseases on ion channels and the way of treating them are emphasized.

DOI: 10.4018/978-1-5225-1043-7.ch001

INTRODUCTION

Ion channels allow the flow of ions through cell. They are integral membrane protein assemblies and activated by different stimuli. The activation procedure depends on the nature of the channel and some are activated by second messengers, some by voltage or temperature or sometimes by mechanical stress. These factors open the channel gate and changes the amount of ions entering or leaving the cell. The research on ion channel came into light by a series of work of Hodgkin and Huxley (1952a, 1952b). They measured the electrical current in the channel. They represented the membrane as an electrical circuit and measured the flow of current. There are two ways of production of current; one is the charging of the membrane capacity and other is the ionic movement through the resistors parallel to the capacitance. Sodium and potassium currents have different contribution to the total current. The movement of the isolated giant axon of Loligo got a clear reason which is the change of sodium and potassium permeability of the channel to electrical stimuli. The presence of the action potential was also established. Using only voltage clamping, the first current from individual ion channels was measured by Hladky and Haydon (1972). Gramicidin ion channel was taken by these researchers as it is made up of small peptides placed in the lipid bilayer. It is a simple channel to study the structure function relationship of the ion channels. Agre and Mckinnon (2003) were awarded the Noble prize for discovering ion channels in cell membrane. For the discovery of water channel and structure and mechanistic studies of ion channels. In the noble lecture of Mackinnon (2004) he explained the structural principles of the gating and regulation of potassium ion channel which governs the electrical signaling in the nervous system with many other physiological processes. Ion channels play pivotal functions in biological cells such as it produces action potential where ion channels are present, e.g. nerve and muscle. the regulation of hormone release from endocrine cells (Hille, 2001). Alteration of deletion of parts of ion channel induces a number of diseases, like disorder of the nervous system and cardiac system.

BACKGROUND

Considerable information about ion channels are gathered up till now that added important means of further progress in the field. The C-type inactivated K+ channels serves as water pathway (Hoomann, NJahnke, & Horner, 2013). Though it allows water transportation through it, it prevents ions. water flow is another important action observed by these channels. Homology models of the pore loop domain of voltage-gated potassium channels kv1.1-kv1.6 are generated by Liu and Lin (2004). Polyamines are marked as the gating molecules for inward-rectifier K+ channels

(Oliver, Baukrowitz & Fakler, 2000). As there are many reviews that talk about the role of diverse ion channels in different types of cancer, the following paragraphs highlight the relation and ion channel focusing on the process of treatment. Male fertility is a such a big issue in current age, so stress is given on that problem which are in proper light only before a few years. Significant role of different ion channels in male fertility and infertility is discussed in detail with Ca^{+2} signal mechanism.

ION CHANNELS AND CANCER

In channels are identified as a reason of quick spreading cancer cells. The voltage dependent anion channels, known as VDAC family accelerate the process of cell death during the development of the disease in patient's body. The chnnels are VDAC1, VDAC2, VDAC3 (McCommis & Baines, 2012; Shoshan-Barmatz & Golan, 2012; Shoshan-Barmatz & Mizrachi, 2012). Mitochondrial permeability transition (MPT) is the last cell death process related to all cancer types (Brenner & Grimm, 2006). The death is accelerated by stress created in cells and the presence of toxic agents. Unlike other immune cells, ion channels of cancer cells are hyperpolarised and thus help in growth of cancer cells and metastasis. Ion channels play multiple role to maintain the Ca^{+2} level in the cell which is the final reason of carcinoma in cell. The flux of Ca^{+2} ions is maintained in turn by the normal flow of K^+ and Cl^- ions. Any change in Ca^{2+} homeostasis occurs due to malfunctioning of the channels. So to keep the intercellular level concentration of Ca^{2+} ions is the target of cancer preventive drugs. Different types of channels are found which is responsible for regulating Ca^{+2} signaling, e.g, voltage-activated, receptor-activated etc. Some are activated second messenger.

Mitochondrial and other organellar channels are responsible for this growth. The Ca^{+2} signaling in the inter cellular region governs the process the ion flow and of ATP production in mitochondria, and hence by modulating the proper ion channel, direct impact can be made to handle the fast spreading of cancer cells. As the ion channels in the inner mitochondrial membrane play the main role in apoptosis, hence the related ion channels are of great interest to challenge such a live taking disease. The review of Lienza et al. (Lienza, 2013) is an explicit one to know the relation of mitochondrial ion channels and cancer.

The channels which are to be deeply related to cancer are K_V, K_Ca, K_{IR}, K_{2p} which are potassium channels. Sodium channel N_{aV}, calcium channel C_{aV} and store-operated calcium channels (SOC) are other two channels responsible for cell proliferation, increase in cell volume, invasion, migration, chemoresistance, apoptosis etc. Other channels are TRP channels, Na^+/H^+ exchanger, calcium channels, sodium pumps etc.

The important role of transient receptor potential channels (TRP) TRPV5 and TRPV6 are recently reported by Saotome, Singh, Yelshanskaya and Sobolevsky (2016). They proposed a Ca^{+2} permeation mechanism through the cell. According to them in the transmembrane domain like K+ channel, Ca^{2+} selectivity can be found by knowing the coordination of Ca^{2+} by a ring of asparic acid derived side chains in the selectivity filter (Pardo, 2004; Kunzelmann, 2005; Ousingsawat, 2007; Hoffmann, 2009; Afrasiabi, 2010; Bödding, 2009).

TRPM7 is a divalent cation-permeable channel which regulates cellular homeostasis of ions with an intrinsic kinase. It acts as a cellular sensor and signal transducer by its ion channels and kinase activity. It is a transporter of cations, mainly Mg^{+2} and sometimes Ca^{+2}. TRPM7 has a novel role in preventing senescence. When TRPM7 is inactive, cell size is increased or multiple nuclei are observed indicating rapid cancerous cell division. Invasion and metastasis of pancreatic cancer is an immediate result of inactive TRPM7 channel (S. Yee, Chan, D. Yee and K.Yee, 2012). TRPM8 is also responsible for malignancy in lung (Du et al., 2014), breast (Chodon et al., 2010), urinary bludder (Mukerji et al., 2006), prostate (Zhang & Barritt, 2006) and pancreas (Yee, 2016).

Ways to Control Cell Apoptosis and Cancer

Attacking of the body cells by cancer and the disease development involve several steps. 1. Alteration or deletion of genes or proteins related to cell division, 2. apoptosis and 3. differention. At the later stage angiogenesis followed by rapid proliferation of cancer cell occurs. At last the cell metastasize other tissues, either in vicinity or distant cells. Mutations of some particular proteins like *ras* protein and p53 tumor-supressor protein can induce neoplastic transformation in cells. (McCormick, 1991). In different ways ion channels prevent cell proliferation. Voltage gated K^+ channels, Cl^- channels perform two-fold action; maintains action potential throughout the cell circle and controls the swelling of the cell during proliferation. (Lang et al., 2000). This in turn, regulates apoptosis. Cystolic process increases Ca^{+2} level and controls apoptosis and proliferation. Pharmaceutical science is now in its way to inhibit these channels to inhibit cell proliferation for various cancer cell lines (Lehen'kyi, Shapovalov, Skryma. & Prevarskaya, 2011) In the extensive review of Li and Xiong (Li & Xiong, 2011) the involvement of various potassium channels are described.

In different ways drugs can interact with ion channels. They can either directly block the channel pore and inhibit the channel or interact with the agonist binding site so that binding is prevented. There are different conformational states of the channels. Using proper drugs which interact with the allosteric site, channel modification can be obtained. By knowing properly, the composition of the channel membrane, suitable drugs can be used to repair them. Drugs can also be used to enhance the possibility of forming complexes by the reaction of ion channel with different membrane proteins and hence to alter channel function (Andersen, 2008; Arcangeli & Becchetti, 2010).

But while doing this, it should be handled with care such the the drugs must be specific for a specific ion channels and they don't do any harm to the other normal tissues. Nanomechanics can take a serious part in this respect. Nanocomputers, nanorobots may serve the purpose to target specific drug attack. (Arcangeli & Becchetti, 2010).

Synthetic Ion Channels

If any ion channel gets deleted or alter or if it is dysfunctioning, for this synthetic ion channels can be developed that helps ion flow in between lipid bilayer. Amphiphiles are used to form pore. The hydraphiles was prepared in the laboratory by Gokel and Negin (2013). They are made of two parts; one is head group made from crown and the central element is polar. Gramicidin A β-helix is the ideal structure of ion channels from where motivation of preparing synthetic ion channels are developed. Ion selectivity can also be achieved by using suitable compound as a basic building block of the channel (Koert, 2005). These channels mimic the structure of the natural ion channels and prepared to function for the same. Potential uses are in biosensing, or substitute of original ion channel.

ION CHANNELS AND MALE INFERTILITY

After successful penetration, the spermatozoa maturate in the female reproductive tract. For their maturation and for the required physiological responses of the sperm for fertilization require the proper functioning of some of the ion channels. Hyperactivated motility, chemotaxis and acrosome reaction are the main physiological processes before fertilization. Mammalian spermatozoa are normally quiescent which become motile when pass through epididymis. During fertilization the motility becomes critical. Motility is hyperactivated during capacitation (Aaustin, 1951; Chang, 1951). During hyperactivation, highly asymmetrical flagellar bending is observed which imparts a great swimming force to the sperm. Also capacitation process includes the increase of ability to undergo acrosome reaction. Capacitation results in the increase of intracellular Ca^{+2} concentrations, controlling of sperm membrane potential and intracellular pH. all these function are maintained by several ion channels. Hyperactivation is required for the required motility that imparts the sperm a force necessary to penetrate the cumulus and zona pellucid surrounding the egg (Suarez, 2003, 2006). A specific Calcium permeability is required for the ion channels that help the process. These are voltage dependent Ca^{2+} channels (CatSpers and Ca_Vs), cyclic nucleotide-gated channels. Another class of calcium permeant channel is transient receptor potential (TRP) channels (Darszon et al., 2005; 2006).

Acrosome Reaction and Molecular Events Attached

During fertilization a sperm has to fuse with the plasma membrane and then penetrate the female egg. But penetrating through the egg's hard shell is a problem for the sperm. So when the sperm approaches the egg, the sperm shells undergo a process, known as acrosome reaction. At the zona pellucida (ZP) of the egg, the membrane of the acrosome fuses with the plasma membrane of the oocyte and expose the content of acrosome. Surface antigens and numerous enzymes are exposed in this process which is required for breaking of the egg's hard coating and for fertilization. A series of intracellular and membrane physicochemical changes occur in the female reproductive tract, called capacitation. This process is assiciated by a decrease in the rigidity of plasma membrane. The rate of flowing of cholesterol through the cell increases at that time. The phospholipid content of membrane is changed and hyperpolarization of the plasma membrane are derived. The whole itracellular and metabolic factors are considerably changed (Patrat, Serres & Jouannet, 2000).

Role of Intracellular Ca^{2+} in Regulating the Sperm Activity in the Female Reproductive Tract

A numerous number of spermetozoa enter into the female reproductive tract. But only few can propagate through the environment filled with viscous fluid of the reproductive tract and reach the site of fertilization. Three factors are responsible to prevent the sperms to enter the site of fertilization. These are Immune response of the female, viscosity of the cervical mucus and the thick and sticky oviductul epithelium. The oviductul epithelium, moreover, is covered with mucus (Kirichok &Lishko, 2011). For fertilization, several processes are to be completed. First of all, the sperms should acquire motility. Not only that, hyperactivated motility is essential to overcome the viscous mucus and sticky epithelium of the fallopian tube. Two more processes, viz, chemotaxis and acrosomal exocytosis are also important to attain hyperactivated state.

Motility is easy to observe in sperm but not well understood. As it is now quite impossible to observe the behavior of individual motile sperm by experiment, the factors responsible for sperm motility are still not clear (Darszon et al., 2006). Mammalian sperm exhibits two general forms of motility, activated and hyperactivated. The first process is observed in the freshly ejaculated sperm which is associated with low amplitude flagellar beat. Stimulated by bicarbonate (HCO^-_3) and Ca^{2+}, hyperactivated motility is resulted. Though these ions do not directly take part, but regulate the enzyme adenylyl cyclase which in turn activates cAMP on protein kinase (PKA), necessary for the exaggerated large amplitude flagellar bending. Ca2+ is required for the initiation and maintenance of hyperactivated motility by directly regulating components of axonemal machinery. In stagnant human sperm progesterone regulates long lasting oscillations in [Ca2+]I is generated by creation of a progesterone gradient. Source of Ca^{2+} is mainly the redundant nuclear envelope (RNE) which is contained in the sperm head. The change in motility and the increase in [ca2+]i depend on the external Ca^{2+}. There are number of voltage gated calcium channels in mammalian sperm (Ca_V). Other channels that influence the motility are cyclic nucleotide-gated channel family (CNG). Definitely, the role of the novel class of ion channels, namely CatSpers is evident now which are confined in testes.

Cav and CatSper Ion Channels

A new class of ion channels, known as CatSper is found in flgellum. These are specific for specific sperms. They are discovered in 2001 as a new family of calcium selective ion channel. Before that it was believed that voltage gated Ca^{2+} channel; Ca_v and voltage gated cation channels (VGCC) were responsible for the increase of Ca2+ level (Florman, Arnoult, Kazam, Li & O'Toole, 1998). At least ten genes are present in Cav channel. These channels may be of two types; high voltage activated and low voltage activated. Both types of channels are found in mammalian sperm (Darszon et al., 2006). However, the notion about the role of Ca_V channels in the elevation of $[Ca^{2+}]_i$ was found to be wrong as male mice deficient in $Ca_V2.2$, $Ca_V2.3$ and $Ca_V3.1$ were found to be fertile (Choi et al., 2007). Though Ca_V3 currents have been recorded in mouse testicular sperm, no Ca_V currents have been detected in epididymal sperm. These experimental results raise a different possibility of $[Ca^{2+]}_i$ elevation in sperm.

In 2001, sperm specific subunits of a new family of ion channels were discovered. These cation channels of spermatozoa are commonly known as CatSpers. At present the CatSper family is comprised of 4 members (CatSpers 1-4) (Ren et al, 2001; Quill, Ren, Clapham & Garbers, 2001; Lobley, Pierron, Reynolds, Allen & Michalovich, 2003; Jin et al, 2005; Qi et al., 2007). Seven CatSper subunits corresponding to the heteromeric CatSper channel have been identified. In CatSper channels assemblage of tetramer of α like subunitsis identified which produces current. In mammals, these channels resemble the structure of voltage-activated calcium channels, Ca_V. CatSper1, CatSper2, CatSper3 and CatSper4 are restricted to testis and localized to the principle piece of sperm tail. Also recently identified putative 2TM auxiliary CatSperβ (Liu, Xia, Cho, Clapham & Ren, 2007) protein and two putative 1TM associated CatSperγ (Chung, Navarro, Krapivinsky, Krapivinsky &

Clapham, 2011) and CatSperδ (Wang, Liu, Cho & Ren, 2009) proteins are found in the same part of the sperm. CatSper1 and 2 both are essential for the hyperactivation of the spermatozoa. Hyperactivation is the main process for fertilization. Sequence identifiers have obtained 22-27% similarity between the members of CatSper family as far as ion transfer domain is concerned (Lobley et al., 2003).

In all kinds of cellular motion, the Ca^{+2} ion concentration is a governing factor. In normal swimming sperm does not require the increase of Ca^{+2} ions. But the influx of Ca^{2+} ions is necessary during capacitation, chemotaxis and hyperactivated motility and also to initiate acrosome reaction. Ca+2 level is increased by a non –genomic mechanism via steroid hormone progesterone in human sperm (Lishko, Botchkina & Kirichok, 2011; Strunker et al., 2011). Progesterone is released by cumulus cells surrounding the oocyte. A number of Ca^{2+} dependent physiological processes are mediated by progesterone. Human CatSper is activated by the increase of intracellular pH and extracellular progesterone. Human $I_{CatSper}$ is potently blocked by an inhibitor of T-type voltage-gated Ca^{2+} channels (Lishko et al., 2012).

One recent patch-clamp experiment (Brenker et al., 2012) has shown that different odorants, control the Ca^{+2} influx and motility directly without involving G-protein-coupled receptors and cAMP-signalling pathways in mammalian sperm. During the journey through the female genital tract, the membrane-permeable analogous of cyclic nucleotides activates CatSper channel through extracellular site. This channel complex serves as a polymodal sensor for different chemicals which are responsible for male fertility.

Very recent experiment reveals that CatSper ion channel agonists do not induce robust hyperactivation in human sperm (Alsamari et al., 2013a). Rather 4-Aminopyridin causes hyperactivation by mobilizing stored Ca^{2+}. On the other hand, activated CatSper facilitates the penetration of viscous medium but not the mobilization of the stored Ca^{2+}. Also activation of CatSper does not help the penetration of the zona pellucida. Progesterone or pH value serve as stimulants for increase of CatSper current, $I_{CatSper}$ only.

Medical Point of View

Until now there is only a very little progress regarding male infertility. The only thing under consideration is semen parameter. But even if this parameter has a normal value, male infertility is reported. The only way to treat sperm dysfunction is given by reproductive technology, which include a range of invasive treatments. Although pharmacological treatments which are non-invasive, are hampered for lack of our knowledge. But in current time, the better understanding of the role of some extremely important ion channels are coming out. These findings point toward the generation and regulation of calcium signals that control sperm hyperactivated

Figure 1. The structure of ion channel engineered from calcium-release activated channel with pdb entry 4HKR. Pore is shown in the middle portion. The image is colored according to residue type.

motility which is found in mice. The main regulator is Ca^{2+}. The two sources of Ca^{2+} required for hyperactivation are 1) the entry of Ca^{2+} through pH dependent CatSper channels in the plasma membrane of the principal piece of flagellum and 2) mobilization of stored Ca^{2+} in the neck or midpiece region (Alsamari et al., 2013b). Regulation of the mobilization of stored Ca^{2+} in human sperm by nitric oxide is reported (Alsamari et al., 2013b) through protein (possibly RyRs) S-nitrosylation. In vitro studies show that human sperm function may be improved by improving the mobilization of stored Ca^{2+} and which again can be done by exposing human sperm to NO donors. Sperm treatment with NO greatly enhances the progesterone effect on $[Ca^{2+}]_i$ (de Oliveira & Mendes, 2008). Experimental studies also point towards the possibility of modulation of hyperactivation by release of Ca^{2+} from inner cell. This helps to reverse the direction of the sharp flagellar bending. So the sperms get a reverse way of propagation. Treatments with 4-AP and thimesoral in mice (Chang & Suarez, 2011) in this basic idea is reported. But until the complete picture of all the factors leading to male infertility is obtained, successful treatment of the same is not possible.

Mathematical Modeling for Ca2+ Signaling Mechanism

The dependence of the triggering of hyperactivation is evident from the experimental studies, but the in vivo mechanism is not known. The detailed knowledge about the working principle of the channel, i.e. their storing and clearing mechanisms

Figure 2. The transmembrane helix for the calcium release activated channel ORAI (PDB entry 4HKR): extracellular and cytoplasmic end of the pore is shown with C and N terminals

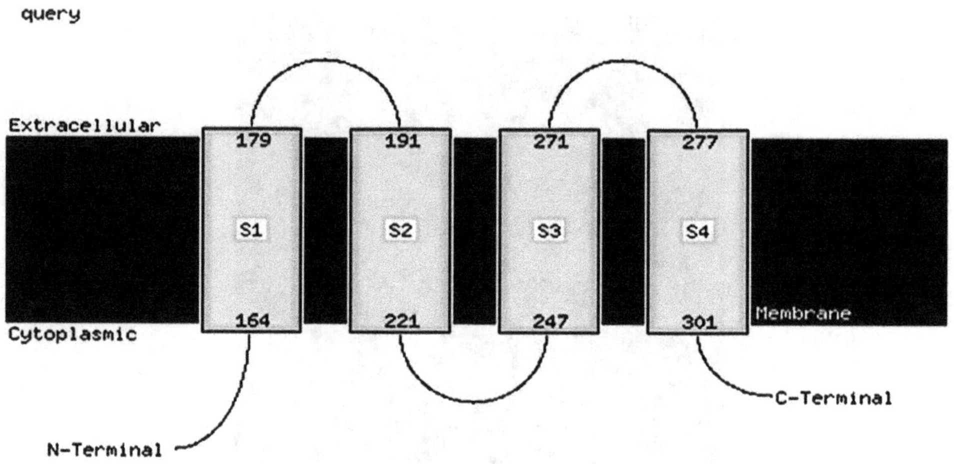

are required for successful clinical remedy. Developing of mathematical models is necessary in this respect to understand the non-linear interactions in the Ca^{2+} signaling pathway.

However, different theoretical investigations regarding structure-function relationship of ion channels have been performed. That includes many cell types like pituitary gland (Blum, Reed, Janovick &Conn, 2000)) vertebrate olfactory receptor neurons (Dougherty, Wright & Yew, 2005), muscle (Wang et al., 2008) and Xenopus oocyte (Atri, Amundson, Clapham & Sneyd, 1993). Mathematical model of Olson et al. (Olson, Suarez & Fauci, 2010) gives the Ca^{2+} flux due to channels, stores and also the diffusion of ions along the length of the flagellum. The variation of diffusivity due to microtubules, fibers, mitochondria, acrosome, nucleus and Jensen's ring was also included in this model.

The author of this chapter also performed several ion channel simulation work with ion like Calcium release activated calcium channel, nerve ion channel. It was observed that voltage and ligand binding play crucial role in changing the structure of the channel. Ligand binding sometimes opens the pore of the channel. Figure 1 shows a engineered ion channel pore (Talukdar & Shantappa, 2016a, 2016b). Figure 1 gives the picture of an engineered ion channel. The channel is engineered to form pore. Figure 2 is the transmebbrane helix structure of the same channel. Figure 3 gives the ligand binding view of a channel with the protein pdb 1BL8.

Figure 3. The structure change of the calcium releases activated channel (PDB entry 4HKR) on ligand binding. The binding site is shown by cyan color. The structure is colored by residue name with open source software VMD.

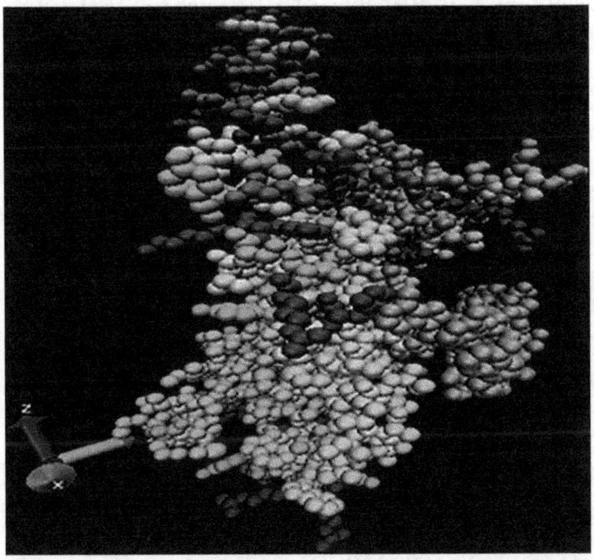

CONCLUSION

Studying ion channels for the treatment of many fatal diseases like cancer, tumor and to accept the challenge of infertility in men is two important field of research. Potassium channels, Calcium channels, Sodium channels are responsible for the alteration of ion channels and hence the apoptosis. The knowledge of nanomechanics to prepare synthetic ion channels is one of the process to prepare artificial means for ion flow. New generation drugs are intended to direct attack ion channels such that various cancer cell line can be treated with less or no side effect. Newly discovered family of iobn channels, known as CatSper channels, which govern the calcium ion release, influence quite considerably to male fertility which was until now not so opened up. The flow of ions through channels needs the help of nanomechanics and nanofluidics. A new generation of medicine targeting ion channels are developing to take a big challenge to fight cancer.

REFERENCES

Afrasiabi, E., Hietamäki, M., Viitanen, T., Sukumaran, P., Bergelin, N., & Törnquist, K. (2010). Expression and significance of HERG (KCNH2) potassium channels in the regulation of MDAMB-435S melanoma cell proliferation and migration. *Cellular Signalling*, *22*(1), 57–64. doi:10.1016/j.cellsig.2009.09.010 PMID:19765650

Alasmari, W., Costello, S., Correia, J., Oxenham, S. K., Morris, J., Fernandes, L., & Barratt, C. L. et al. (2013). Ca2+ signals generated by CatSper and Ca2+ stores regulate different behaviors in human sperm. *The Journal of Biological Chemistry*, *288*(9), 6248–6258. doi:10.1074/jbc.M112.439356 PMID:23344959

Alsmari, W., Barratt, C. L., Publicover, S. J., Whalley, K. M., Foster, E., Kay, V., & Oxenham, S. K. et al. (2013). The clinical significance of calcium-signaling pathways mediation human sperm hyperactivation. *Human Reproduction (Oxford, England)*, *28*(4), 866–876. doi:10.1093/humrep/des467 PMID:23406974

Andersen, O. S. (2008). Perspectives on how to drug an ion channel. *The Journal of General Physiology*, *131*(5), 395–397. doi:10.1085/jgp.200810012 PMID:18411334

Arcangeli, A., & Becchetti, A. (2010). New Trends in Cancer Therapy. *Targeting Ion Channels and Transporters*, *3*, 1202–1224.

Atri, A., Amundson, J., Clapham, D. E., & Sneyd, J. (1993). A single-pool model for intracellular calcium oscillations and waves in the Xenopus laevis oocyte. *Biophysical Journal*, *65*(4), 1727–1739. doi:10.1016/S0006-3495(93)81191-3 PMID:8274661

Austin, C. R. (1951). Observations on the penetration of the sperm in the mammalian egg. *Australian Journal of Scientific Research, Series B: Biological Sciences*, *4*, 581–596. PMID:14895481

Blum, J. J., Reed, M. C., Janovick, J. A., & Conn, P. M. (2000). A mathematical model quantifying GnRH- induced LH secretion from gonadotropes. *American Journal of Physiology. Endocrinology and Metabolism*, *278*, 263–272. PMID:10662710

Bödding, M. (2007). TRP proteins and cancer. *Cellular Signalling*, *19*(3), 617–624. doi:10.1016/j.cellsig.2006.08.012 PMID:17029734

Brenker, C., Goodwin, N., Weyand, I., Kashikar, N. D., Naruse, M., Krähling, M., & Strünker, T. et al. (2012). The CatSper channel: A polymodal chemosensor in human sperm. *The EMBO Journal*, *31*(7), 1654–1665. doi:10.1038/emboj.2012.30 PMID:22354039

Brenner, C., & Moulin, M. (2012). Physiological roles of the permeability transition pore. *Circulation Research, 111*(9), 1237–1247. doi:10.1161/CIRCRESAHA.112.265942 PMID:23065346

Chang, H., & Suarez, S. S. (2011). Two distinct Ca2+ signaling pathways modulate sperm flagellar beating patterns in mice. *Biology of Reproduction, 85*(2), 296–305. doi:10.1095/biolreprod.110.089789 PMID:21389347

Chang, M. C. (1951). Fertilizing capacity of spermatozoa deposited into the fallopian tubes. *Nature, 168*(4277), 697–698. doi:10.1038/168697b0 PMID:14882325

Chodon, D., Guilbert, A., Dhennin-Duthille, I., Gautier, M., Telliez, M. S., Sevestre, H., & Ouadid-Ahidouch, H. (2010). Estrogen regulation of TRPM8 expression in breast cancer cells. *BMC Cancer, 10*(1), 212. doi:10.1186/1471-2407-10-212 PMID:20482834

Choi, S., Na, H. S., Kim, J., Lee, J., Lee, S., Kim, D., & Shin, H. S. et al. (2007). Attenua.ted pain responses in mice lacking CaV32 T-type. channels. *Genes Brain & Behavior, 6*(5), 425–431. doi:10.1111/j.1601-183X.2006.00268.x PMID:16939637

Chung, J. J., Navarro, B., Krapivinsky, G., Krapivinsky, L., & Clapham, D. E. (2011). A novel gene required for male fertility and functional CATSPER channel formation in spermatozoa. *Nature Communications, 2*(1), 153. doi:10.1038/ncomms1153 PMID:21224844

Darszon, A., Acevedo, J. J., Galindo, B. E., Hernandez-Gonzalez, E. O., Nishigaki, T., Trevino, C. L., & Beltran, C. et al. (2006). Sperm channel diversity and functional multiplicity. *Reproduction (Cambridge, England), 131*(6), 977–988. doi:10.1530/rep.1.00612 PMID:16735537

Darszon, A., Martínez, L. P., Acevedo, J. J., Cruz, H. A., & Treviño, C. L. (2006). T-type Ca2+ channels in sperm function. *Cell Calcium, 40*(2), 241–252. doi:10.1016/j.ceca.2006.04.028 PMID:16797697

Darszon, A., Nishiqaki, T., Wood, C., Trevino, C. L., Felix, R., & Beltran, C. (2005). Calcium channels and Ca2+ fluctuations in sperm physiology. *International Review of Cytology, 243*, 79–172. doi:10.1016/S0074-7696(05)43002-8 PMID:15797459

de Oliveira, M., & Mendes, G. S. (2008). Modulation of calcium signaling in human sperm by nitric oxide [Ph.D. Thesis]. University of Birmingham.

Dougherty, D. P., Wright, G. A., & Yew, A. C. (2005). Computational model of the cAMP-mediated sensory response and calcium-dependent adaptation in vertebrate olfactory receptor neurons. *Proceedings of the National Academy of Sciences of the United States of America, 102*(30), 10415–10420. doi:10.1073/pnas.0504099102 PMID:16027364

Du, G. J., Li, J.-H., Liu, W.-J., Liu, Y.-H., Zhao, B., Li, H.-R., & Duan, Y.-J. et al. (2014). The combination of TRPM8 and TRPA1 expression causes an invasive phenotype in lung cancer. *Tumour Biology, 35*(2), 1251–1261. doi:10.1007/s13277-013-1167-3 PMID:24037916

Florman, H. M., Arnoult, C., Kazam, I. G., Li, C., & O'Toole, C. M. (1998). A perspective on the control of mammalian fertilization by egg-activated ion channels in sperm: A tale of two channels. *Biology of Reproduction, 59*(1), 12–16. doi:10.1095/biolreprod59.1.12 PMID:9674987

Gokel, G. W., & Negin, S. (2013). Synthetic Ion Channels: From Pores to Biological Applications. *Accounts of Chemical Research, 46*(12), 2824–2833. doi:10.1021/ar400026x PMID:23738778

Hille, B. (2001). *Ion channels of excitable membranes* (3rd ed.). Sunderland, Massachusetts, USA: Sinauer Associates, Inc.

Hladky, S. B., & Hadlay, D. A. (1972). Ion transfer across lipid membranes in the presence of gramicidin A. I. Studies of the unit conductance channel. *Biochimica et Biophysica Acta, 274*(2), 294–312. doi:10.1016/0005-2736(72)90178-2 PMID:5048999

Hodgkin, A. L. & Huxley, A. F. (1952b). The components of membrane conductance in the giant axon of Loligo. *Journal of Physiology*, 116(4), 473-496.

Hodgkn, A. L., & Huxley, A. F. (1952a). Currents carried by sodium and potassium ions through the membrane of the giant axon of Loligo. *The Journal of Physiology, 116*(4), 449–472. doi:10.1113/jphysiol.1952.sp004717 PMID:14946713

Hoffmann, E. K., Lambert, I. H., & Pedersen, S. F. (2009). Physiology of cell volume regulation in vertebrates. *Physiological Reviews, 89*(1), 193–277. doi:10.1152/physrev.00037.2007 PMID:19126758

Hoomann, T., Jahnke, N., Horner, A., Keller, S., & Pohl, P. (2013). Filter gate closure inhibits ion but not water transport through potassium channels. *Proceedings of the National Academy of Sciences of the United States of America, 10*(26), 10842–10847. doi:10.1073/pnas.1304714110 PMID:23754382

Jin, J. L., O'Doherty, A. M., Wang, S., Zheng, H., Sanders, K. M., & Yan, W. (2005). CatSper3 and CatSper4 encode two cation channel-like proteins exclusively expressed in the testis. *Biology of Reproduction*, *73*(6), 1235–1242. doi:10.1095/biolreprod.105.045468 PMID:16107607

Kirichok, Y., & Lishko, P. V. (2011). Rediscovering sperm ion channels with the patch-clamp technique. *Molecular Human Reproduction*, *17*(8), 478–499. doi:10.1093/molehr/gar044 PMID:21642646

Koert, U. (2005). Synthetic ion channels: Functional analysis and structural studies. *Physical Chemistry Chemical Physics*, *7*(7), 1501–1506. doi:10.1039/b418561g PMID:19787974

Kunzelmann, K. (2005). Ion channels and cancer. *The Journal of Membrane Biology*, *205*(3), 159–173. doi:10.1007/s00232-005-0781-4 PMID:16362504

Lang, F., Ritter, M., Gamper, N., Huber, S., Fillon, S., Tanneur, V., & Gulbins, E. et al. (2000). Cell volume in the regulation of cell proliferation and apoptotic cell death. *Cellular Physiology and Biochemistry*, *10*(5-6), 417–428. doi:10.1159/000016367 PMID:11125224

Leanza, L., Biasutto, L., Managò, A., Gulbins, E. Zoratti, M. and Szabò, I. (2013). Intracellular ion channels and cancer. *Frontiers in Physiology*, 4, 227(1-7)

Lehen'kyi, V., Shapovalov, C., Skryma, R., & Prevarskaya, N. (2011). Ion channnels and transporters in cancer. 5. Ion channels in control of cancer and cell apoptosis. *American Journal of Physiology. Cell Physiology*, *30*(6), C1281–C1289. doi:10.1152/ajpcell.00249.2011 PMID:21940667

Lishko, P., Kirichok, Y., Ren, D., Navarro, B., Chung, J.J. & Clapham, D.E. (2012), The control of male fertility by spermatozoan ion channels. *Annual Review of Physiology*, 74, 1-10.23.

Lishko, P. V., Botchkina, I. L., & Kirichok, Y. (2011). Progesterone activates the principal Ca2+ channel of human sperm. *Letter to Nature*, *471*(7338), 387–391. doi:10.1038/nature09767 PMID:21412339

Liu, H. L., & Lin, J. C. (2004). A set of homology models of pore loop domain of six eukaryotic voltage- gated potassium channels Kv1.1-Kv. Proteins: Structure. *Function and Ioinformatics*, *55*, 558–567. doi:10.1002/prot.20065

Liu, J., Xia, J., Cho, K. H., Clapham, D. E., & Ren, D. (2007). CatSperbeta, a novel transmembrane protein in the CatSper channel complex. *The Journal of Biological Chemistry*, *282*(26), 18945–18952. doi:10.1074/jbc.M701083200 PMID:17478420

Lobley, A., Pierron, V., Reynolds, L., Allen, L., & Michalovich, D. (2003). Identification of human and mouse CatSper3 and CatSper4 genes: Characterisation of a common interaction domain and evidence for expression in testis. *Reproductive Biology and Endocrinology*, *1*(1), 53. doi:10.1186/1477-7827-1-53 PMID:12932298

McCommis, K. S., & Baines, C. P. (2012). The role of VDAC in cell death: friend or foe. *Biochimica et Biophysica Acta*, *1818*(6), 1444–1450. doi:10.1016/j.bbamem.2011.10.025 PMID:22062421

McCormick, F. (1991). GTP-binding proteins as oncogenes in human tumors. *Environmental Health Perspectives*, *93*, 17–18. PMID:1773789

Mukerji, G., Yiangou, Y., Corocran, S. L., Selmer, I., Smith, G. D., Benham, C. D., & Anand, R. et al. (2006). Cool and menthol receptor TRPM8 in human urinary bladder disorders and clinical correlations. *BMC Urology*, *6*(1), 6. doi:10.1186/1471-2490-6-6 PMID:16519806

Navarro, B., Kirichok, Y., Chung, J.-J., & Clapham, D. E. (2008). Ion channels that control fertility in mammalian spermatozoa. *The International Journal of Developmental Biology*, *52*(5-6), 607–613. doi:10.1387/ijdb.072554bn PMID:18649274

Oliver, D., Baukrowitz, T., & Fakler, B. (2000). Polyamines as gating molecules of inward-rectifier K^+ channels. *European Journal of Biochemistry*, *267*(19), 5824–5829. doi:10.1046/j.1432-1327.2000.01669.x PMID:10998040

Olson, S. D., Suarez, S. S., & Fauci, L. J. (2010). A model of CatSper channel mediated calcium dynamics in mammalian spermatozoa. *Bulletin of Mathematical Biology*, *72*(8), 1925–1946. doi:10.1007/s11538-010-9516-5 PMID:20169416

Ousingsawat, J., Spitzer, M., Puntheeranurak, S., & Terracciano, L. (2007). Expression of voltage-gated potassium channels in human and mouse colonic carcinoma. *Clinical Cancer Research*, *13*(3), 824–831. doi:10.1158/1078-0432.CCR-06-1940 PMID:17289873

Oxenham (2013). The clinical significance of calcium-signaling pathways mediation human sperm hyperactivation. *Human Reproduction*, 28(4), 866-876.

Pardo, L. A. (2004). Voltage-gated potassium channels in cell proliferation. *Physiology (Bethesda, MD)*, *19*(5), 285–292. doi:10.1152/physiol.00011.2004 PMID:15381757

Patrat, C., Serres, C., & Jouannet, P. (2000). The acrosome reaction in human spermatozoa. *Biology of the Cell*, *92*(3-4), 255–266. doi:10.1016/S0248-4900(00)01072-8 PMID:11043413

Qi, H., Moran, M. M., Navarro, B., Chong, J. A., Krapivinsky, G., Krapivinsky, L., & Clapham, D. E. et al. (2007). All four CatSper ion channel proteins are required for male fertility and sperm cell hyperactivated motility. *Proceedings of the National Academy of Sciences of the United States of America*, *104*(4), 1219–1223. doi:10.1073/pnas.0610286104 PMID:17227845

Quill, T. A., Ren, D., Clapham, D. E., & Garbers, D. L. (2001). A voltage-gated ion channel expressed specifically in spermatozoa. *Proceedings of the National Academy of Sciences of the United States of America*, *98*(22), 12527–12531. doi:10.1073/pnas.221454998 PMID:11675491

Ren, D., Navarro, B., Perez, G., Jackson, A. C., Hsu, S., Shi, Q., & Clapham, D. E. et al. (2001). A sperm ion channel required for sperm motility and male fertility. *Nature*, *413*(6856), 603–609. doi:10.1038/35098027 PMID:11595941

Saotome, K., Singh, A. K., Yelshanskaya, M. V., & Sobolevsky, A. I. (2016). Crystal structure of the epithelial calcium channel TRPV6. *Nature*, *534*(7608), 506–511. doi:10.1038/nature17975 PMID:27296226

Shoshan-Barmatz, V., & Golan, M. (2012). MitochondrialVDAC1: Function in cell life and death and a target for cancer therapy. *Current Medicinal Chemistry*, *19*(5), 714–735. doi:10.2174/092986712798992110 PMID:22204343

Shoshan-Barmatz, V., & Mizrachi, D. (2012). VDAC1: From structure to cancer therapy. *Frontiers in Oncology*, *29*, 164. PMID:23233904

Strunker, T., Goodwin, N., Brenker, C., Kashikar, N. D., Weyand, I., Seifert, R., & Kaupp, U. B. (2011). The CatSper channel mediates progesterone-induced Ca2+ influx in human sperm. *Nature*, *471*(7338), 382–386. doi:10.1038/nature09769 PMID:21412338

Suarez, S. S., & Ho, H. C. (2003). Hyperactivation of mammalian sperm. *Cellular and Molecular Biology*, *49*, 351–356. PMID:12887087

Suarez, S. S., & Pacey, A. A. (2006). Sperm transport in the female reproductive tract. *Human Reproduction Update*, *12*(1), 23–37. doi:10.1093/humupd/dmi047 PMID:16272225

Talukdar, K., & Shantappa, A. (2016a). Molecular dynamics study of nerve ion channel. *Materials Today: Proceedings*, *3*(6), 1828–1834. doi:10.1016/j.matpr.2016.04.081

Talukdar, K., & Shantappa, A. (2016b). Computational study of a calcium release-activated calcium channel. *AIP Conference Proceedings*, *1731*, 050130. doi:10.1063/1.4947784

Wang, H., Liu, J., Cho, K. H., & Ren, D. (2009). A novel, single, transmembrane protein CATSPERG is associated with CATSPER1 channel protein. *Biology of Reproduction*, *81*(3), 539–544. doi:10.1095/biolreprod.109.077107 PMID:19516020

Wang, I., Politi, A., Tania, N., Bai, Y., Sanderson, M. J., & Sneyd, J. (2008). A mathematical model of airway and pulmonary arteriole smooth muscle. *Biophysical Journal*, *94*(6), 2053–2064. doi:10.1529/biophysj.107.113977 PMID:18065464

Yee, N. S. (2016). Chapter four-TRPM8 Ion channels as potential cancer biomarker and target in pancreatic cancer. *Advances in Protein Chemistry and Structural Biology*, *104*, 127–155. doi:10.1016/bs.apcsb.2016.01.001 PMID:27038374

Yee, N. S., Chan, A. S., Yee, J. D., & Yee, R. K. (2012). TRPM7 and TRPM8 Ion Channels in Pancreatic Adenocarcinoma: Potential Roles as Cancer Biomarkers and Targets. *Scientifica*.

Zhang, L., & Barritt, G. J. (2006). TRPM8 in prostate cancer cells: A potential diagnostic and prognostic marker with a secretory function? *Endocrine-Related Cancer*, *13*(1), 27–38. doi:10.1677/erc.1.01093 PMID:16601277

Chapter 2
Analysis of the Bacterial Vesicles' Enhanced Toxicological Threat Via Electron Microscopy

Roberta Curia
University of Milano-Bicocca, Italy

Marziale Milani
University of Milano-Bicocca, Italy

Lyubov V Didenko
Gamaleya Research Institute for Epidemiology and Microbiology, Russia

George A Avtandilov
Moscow State University of Medicine and Dentistry, Russia

Natalia V Shevlyagina
Gamaleya Research Institute for Epidemiology and Microbiology, Russia

Francesco Tatti
FEI Company, Italy

ABSTRACT

*This study shows the importance of electron microscopy in the analysis of the interaction of microorganisms (*Staphylococcus aureus*) with polymeric (polyurethane) dental prostheses. Starting from the biofilm formation and the biodestruction of the plastic material resulting in the production of polyurethane nanoparticles, the focus is on the bacterial secretion of membrane vesicles (in the range of 20-50 nm) loaded with plastic nanoparticles (from 2-3 to 10 nm) and on the toxicological threat that these delivery devices represent when interacting with host cells. The nanoparticles deliverance led by the bacterial infections dynamics opens new ways to the possibility of delivering drugs to selected cells.*

DOI: 10.4018/978-1-5225-1043-7.ch002

BACTERIAL BIOFILM, PLASTIC BIODESTRUCTION, AND NANOPARTICLES FORMATION

Staphylococcus aureus is a Gram-positive bacterium permanently present in the human body as a commensal, and often as a pathogen as well (Lowy, 1998). When polyurethane prosthetic devices come into contact with *S. aureus*, its ability to colonize medical implants, adhering to the polymeric surface and forming a biofilm, results in the development of a community with dynamics able to highly affect the material's stability. The biofilm formation is a four-steps process (Figure 1) where the bacterial initial adhesion onto the polymer is followed by the formation of microcolonies, by the maturation phase with the appearance of the exopolysaccharide matrix and eventually by the detachment of nomadic cells (Didenko et al., 2012; Didenko et al., 2013; Curia et al., 2014; Ghannoum & O'Toole, 2004).

The action of a bacterial biofilm, characterized by high cell density and limitation of nutrients (Fux, Costerton, Stewart, & Stoodley, 2005), provokes damages on the prosthesis and causes the detachment of debris (micro- and nano-particles) from the bulk material (biodestruction) (Didenko et al., 2012; Curia et al., 2014) that will be well documented by scanning and transmission electron microscopy techniques (Figures 2, 3, and 4).

Figure 1. Scheme of the formation of a coccal biofilm and polyurethane biodestruction
1. Initial adhesion of S. aureus cells to the polyurethane surface;
2. Formation of microcolonies;
3. Biofilm maturation, appearance of the exopolysaccharide matrix, initial biodestruction of polyurethane and absorption of polymeric nanoparticles by bacterial cells;
4. Detachment of nomadic cells loaded with polyurethane nanoparticles from the mature biofilm.

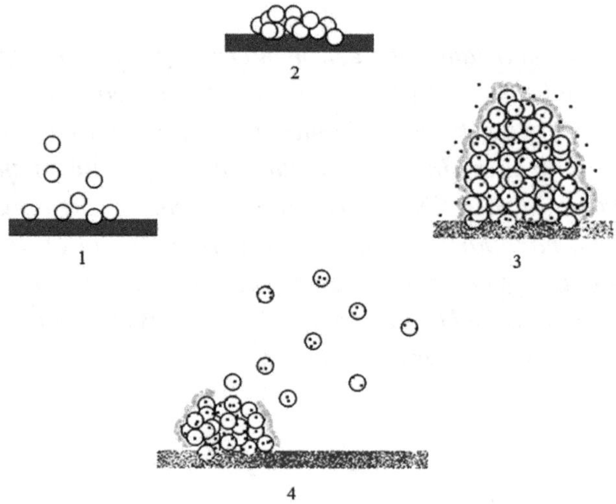

Figure 2. SEM (Scanning Electron Microscope) image of a S. aureus biofilm on a polyurethane surface after a 45-days-long incubation. It is evident the presence of the exopolysaccharide matrix. Polyurethane nanoparticles are detectable throughout the whole biofilm embedded in the matrix (circles).

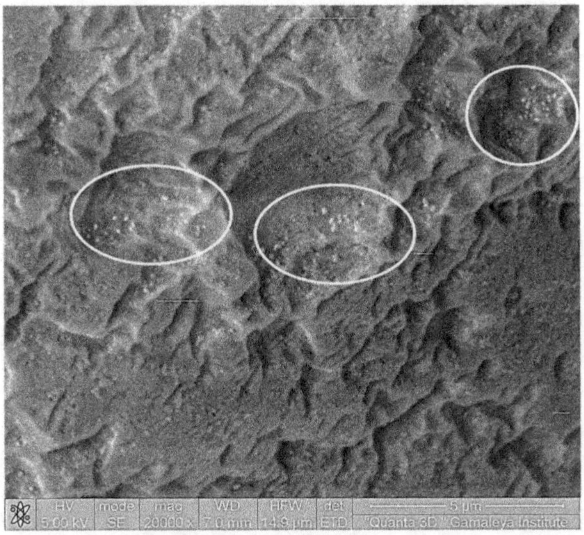

Figure 3. STEM Dark Field image of a portion of a S. aureus cell. Bright polyurethane nanoparticles are visible in the external medium, on the cell wall and inside the bacterium. A trail of vesicles (arrows) from the periphery to the center of the cell is detectable. The cell wall is undergoing a modification phase, and on the left it is visible that components of the peptidoglycan cell wall are dispersing in the extracellular environment. Vesicles exiting the bacterial cell are visible by the scaling of the cell wall (circle).

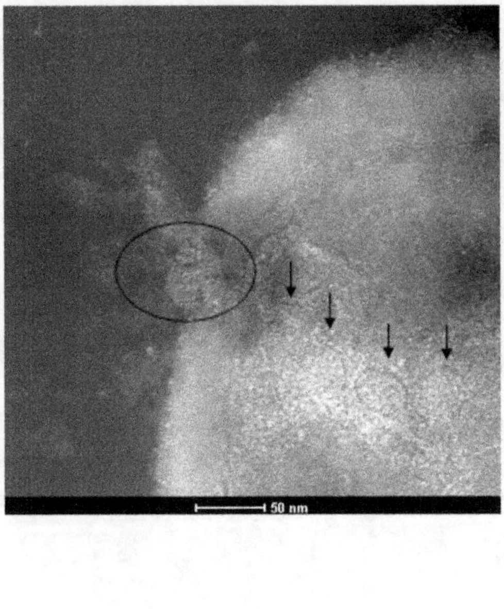

Figure 4. magnification of a membranous vesicle inside a S. aureus cell. In this TEM Bright Field image, electron dense polyurethane nanoparticles (2-3 nm) appear darker than the surrounding components. Vesicle diameter measures 30 nm. Horizontal Field Width (HFW) = 65 nm. The thickness of the vesicle's plasma membrane is of about 4-5 nm.

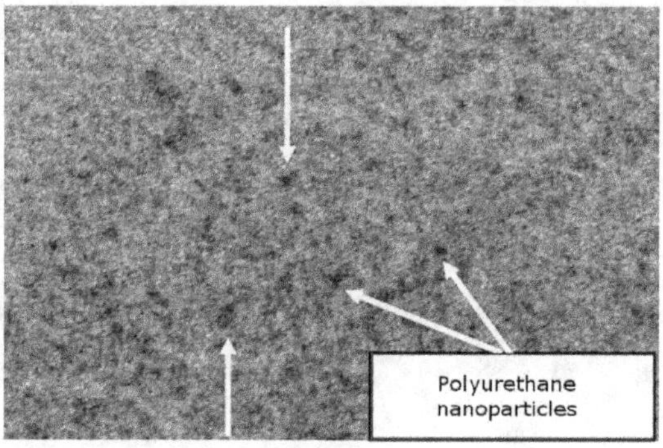

Polyurethane derives from the condensation of polyisocyanate and polyol, with intramolecular urethane bonds. The biodegradation of polyurethane is carried out by bacteria that attack the urethane bond, thus operating a polyurethanolytic activity (Howard, 2011; Shah, Hasan, Hameed, & Ahmed, 2008) on the polymer and as a consequence micro- and nano-sized polyurethane particles are produced and embedded in the exopolysaccharide matrix (Figure 2) (Didenko et al., 2012; Curia et al., 2014). The polyurethanolytic action of bacteria is carried out by two kinds of polyurethases, a membrane bound polyurethanesterase and an extracellular polyurethanesterase, that cooperate in order to make it possible for bacteria to hydrolyze polyester chains (Shah et al., 2008).

The final product of the microbial attack on the urethane bond is a new kind of nanoparticle, non-engineered, thus with properties (Revell, 2006) different from nanoparticles used in the research and industrial fields.

Since the neo-formed polyurethane particles are entrapped in the exopolysaccharide matrix, when cells (planktonic or nomadic cells) move away from the whole mature biofilm, polyurethane particles are dragged away with the breaking off bacterial cells.

NANOPARTICLES UPTAKE PROCESS BY *S. AUREUS*

Protein Corona

Electron microscopy points out that polyurethane micro- and nano- particles are embedded in the exopolysaccharide matrix, but only the nanosized ones are actually internalized within the bacteria. TEM (Transmission Electron Microscope) and STEM (Scanning Transmission Electron Microscope) images show that nanoparticles are internalized into *S. aureus* and that they are surrounded by membrane vesicles (Figures 3 and 4), thus electron microscopy allows the extraction of information about the uptake process.

The polyurethane nanoparticles' pathway from outside the cell to the interior of the bacterium can be influenced by the presence of a further structure that surrounds nanoparticles in a biological medium: the protein corona. Hypothetically nanoparticles could be found at a basic state (as they are), but in biological fluids, proteins, according to their characteristics such as type, size and flexibility, compete for the nanoparticles' surfaces on the basis of their category, dimension, shape and hydrophobicity, forming a protein cover (Sahneh, Scoglio, & Riviere, 2013). Hence the interaction of polymeric nanoparticles with living systems via colloidal forces and dynamic biophysicochemical relations leads to the formation of the protein corona, a structure that shields the nanoparticle redefining the charge distribution of the nanomaterial (Cedervall et al., 2007; Nel et al., 2009). Colloidal forces and dynamic biophysicochemical relations can be seen as complex combinations of elementary electromagnetic interactions. Two major features to be kept into account are the time dependence of all processes, and the cellular and intracellular presence of endogenous electromagnetic fields. In Cifra, Fields, and Farhadi (2011) the authors review theories and experiments on how cells can generate and detect electromagnetic fields, and their possible mediation of cellular interactions.

Since polyurethane nanoparticles under examination are non-engineered and derive directly from the biodestructive action of *S. aureus* on polymers, it is difficult to predict their effective charge. The analysis of the composition and the spatial arrangement of the protein corona could be a good method for the determination of the net (positive, negative or neutral) charge of the polyurethane nanoparticles, but it would be difficult because of the fast dynamics of the composition of the corona complex (Sahneh et al., 2013) and of the number of the different types of molecules involved (Tenzer et al., 2013).

The protein corona affects the uptake of nanoparticles since alterations of fundamental electromagnetic properties modify what the cell sees (Lynch, Salvati, & Dawson, 2009). Hence *S. aureus* cells internalize nanoparticles shielded by the protein cover that acts as a camouflage, hiding the actual nature of the cargo.

Endocytosis

Endocytosis includes all the absorption methods (e.g. clathrin mediated endocytosis, pinocytosis, phagocytosis) differing from one another for several factors, among others the vesicles' size or the origin of the cargo (Šamaj et al., 2004). This process, widely studied in the eukaryotic domain, is also present in prokaryotes. In Figures 3 and 4 it is shown that endocytic membranous vesicles envelop polyurethane nanoparticles within *S. aureus*; hence it is possible to affirm that nanoparticles (possibly shielded from the protein corona) are internalized within the bacterial cell via endocytosis.

The presence of the protein cover influences the uptake process either enhancing or inhibiting the absorption: since the cell interfaces with proteins belonging to the shield and not with the nanoparticle itself, *S. aureus* is not able to detect the real cargo which is being internalized. In this way the uptake process is enhanced and bacteria internalize even potentially toxic materials. On the other hand, due to similar charge or different size and stability, proteins covering nanoparticles could not be affine to the bacterial uptake structures, thus inhibiting the internalization.

MEMBRANE VESICLES' SPATIAL ORGANIZATION

Cytoskeleton

The disposition of membranous vesicles loaded with polyurethane nanoparticles visible in Figure 3 suggests that a sort of pre-established spatially organized pathway exists from the periphery of *S. aureus* to its interior (Didenko, Shevlyagina, Curia, Erega, & Milani, 2014; Milani et al., 2016). The intracellular mesh along which membranous vesicles move into the bacterial cell is the cytoskeleton, not only important for the vesicular trafficking, but also for supporting the maintenance of cell shape providing mechanical strength, for cell division, and chromosome segregation (Michie & Löwe, 2006). The correlation of cytoskeleton and endocytosis is supported by several authors (May & Machesky, 2001; Engqvist-Goldstein & Drubin, 2003; Šamaj et al., 2004): during endocytosis cytoskeletal structures such as actin filaments and microtubules are actively involved, and the dynamic interactions are necessary for the endocytic machinery to work (Šamaj et al., 2004).

Even though the prokaryotic cytoskeleton has been studied for fewer years in comparison with the studies of eukaryotic cells, it has been confirmed the presence of a bacterial cytoskeleton (van den Ent, Amos, & Löwe, 2001; Møller-Jensen & Löwe, 2005). Prokaryotic cytoskeletal structures, homologues of actin (MreB, ParM and FtsA) and tubulin (FtsZ), are similar to those of eukaryotes not only structurally but functionally as well, revealing their evolutionary relationship (Carballido-López

& Errington, 2003; Amos, van den Ent, & Löwe, 2004; Löwe, van den Ent, & Amos, 2004; Michie & Löwe, 2006). Then it is natural to correlate the cytoskeleton and the functions carried out in the eukaryotic domain with the homologue cytoskeletal structures and the tasks performed in the prokaryotic one.

Cytoskeletal structures not only have a role in the vesicles formation for the cargo internalization, but are extremely important in promoting the movement of the vesicles through the cell. In the initial stage of the uptake process there are changes in the cell membrane; as regards the eukaryotic domain several studies have analyzed these modifications at membrane level, focusing on the membrane shape, the invagination depth and the width and height of the curvature, measuring all the phases from the U-shaped invagination, till the neck constriction and scission of the vesicle (Avinoam, Schorb, Beese, Briggs, & Kaksonen, 2015). The endocytic pathway is a key process that define how cells interact with the environment, but also how they cope with their inner cargoes. Endocytosis and vesicular trafficking are strictly connected: actin provides the energy for the internalization process whenever the cells uptake cargoes or the membrane displays a high rigidity, furthermore actin filaments and microtubules along with motor proteins are required for the transport of the uptaken vesicles. Molecular motors are proteins that convert chemical energy (hydrolysis of nucleotide triphosphates) into mechanical work and movement. Motor proteins associated to cytoskeletal structures are myosins, dyneins and kinesins; these proteins act moving along actin filaments or microtubules using the cytoskeleton as tracks (Iyer, Romanowicz, & Laudon, 2004; Schliwa & Woehlke, 2003). All three types of cytoskeletal motors are involved in the transport of organelles and vesicles, with microtubules backing off short and long-range transport, whereas actin mostly supports short-range movements (Granger, McNee, Allan, & Woodman, 2014; Kamal & Goldstein, 2000; Schliwa & Woehlke, 2003). The vesicular trafficking within the cell is highly coordinated and vesicles moving along microtubules can stop their movement and switch direction, moreover the switching between cytoskeletal structures is also possible, in fact factors as Rho and GTPases promote the switch of the vesicles from actin filaments to microtubules and vice versa (Brown, 1999; Döhner, Nagel, & Sodeik, 2005; Granger et al., 2014; Schliwa & Woehlke,2003).

The evolutionary correlation of prokaryotes with eukaryotes is confirmed, but in comparison with the most studied eukaryotic cells, there is a lack of knowledge when it comes to bacterial strategy of vesicular transport. Electron Microscopy images of *S. aureus* (Figures 3 and 4) can help filling the gap elucidating steps of the vesicles' transport by the analysis of their spatial disposition, where polyurethane nanoparticles act as built-in markers. From the images it is also possible to determine the size (range of 20-50 nm) and the number of the vesicles.

In the membranous vesicle is it possible to observe a multiplicity of polyurethane nanoparticles.

Vesicles Classification

From the previous paragraph it is evident that cells (both prokaryotic and eukaryotic) are highly organized and that the machinery for the vesicular trafficking has a functional role in maintaining the spatial distribution of the cell and supporting the transport of vesicles, but it is important to underline that cytoskeletal structures also promote the vesicles maturation (Granger et al., 2014).

A classification of different types of vesicles and their functions can be done. The classical endocytic pathway starts with Early Endosomes (the vesicular compartment that first receives the incoming cargo) that, moving along microtubules towards the center of the cell, mature and convert into Late Endosomes. They are the linking point between the recycling pathway (Late Endosomes retrograde at Early Endosomes) and the degradation cycle (Late Endosomes undergo a further maturation into lysosomes) (Huotari & Helenius, 2011). Early and Late Endosomes finally can exit the cell.

In eukaryotic cells, independently from which state of maturation vesicles are in, these spherical membranous structures (which size ranges from 30 nm to 1 μm) can contain different types of cargo (biomolecules, transmembrane receptors, genetic information as well as waste material or drugs) that can be delivered to the extracellular environment. Secreted vesicles can have several functions: they can serve as intercellular signaling for immune suppression or for mediate cell killing at a distance, or can act as potential decoys to distract the immune system; they have a role in inflammation and can present antigens; they can transfer signaling components or genetic information; in some cases they are able to pass the Blood Brain Barrier and can be used for clinical applications (therapy, prognosis, or as biomarkers) (van der Pol, Böing, Harrison, Sturk, & Nieuwland, 2012).

Analyzing Electron Microscopy images it is not possible to make the same classification for the inner prokaryotic vesicles, but it can be affirmed that if a vesicle loaded with polyurethane nanoparticles is found outside *S. aureus* (Figure 3), that very vesicle was previously inside the bacterium. In this case polyurethane nanoparticles act as markers, testifying that vesicles which have absorbed the polymer were once within the bacterium.

The secretion pathway is active not only in eukaryotes, but also in prokaryotes and Archea, confirming that the intercellular communication via vesicles has been conserved throughout evolution (Deatherage & Cookson, 2012; van der Pol, Boing, Harrison, Sturk, & Nieuwland, 2012).

Bacterial Outer Membrane Vesicles

Cell-derived vesicles in the prokaryotic domain are called outer membrane vesicles or just membrane vesicles. It was thought that the production of membranous cell-derived structures was a peculiarity of Gram-negative bacteria, but in the last years, studies revealed that even Gram-positive bacteria are able to release membrane vesicles (Brown, Wolf, Prados-Rosales, & Casadevall, 2015; Curia, Erega, & Milani, 2015; Deatherage & Cookson, 2012; Lee, 2012; Gurung et al., 2011; Lee et al., 2009). The formation and secretion of extracellular vesicles are evolutionally conserved processes and the secreted vesicles preserve similar functions and mechanisms of release (Brown et al., 2015).

Gram-negative bacteria produce and secrete outer membrane vesicles which diameter ranges from 20 nm to 200 nm, whereas vesicles secreted by Gram-positive bacteria are smaller, with a diameter that goes from 20 nm to 100 nm. Even though organisms synthesize and regulate vesicles differently, membrane vesicles secreted by Gram-positive and Gram-negative bacteria seem to have similar functions: transfer of several types of cargo (proteins, phospholipids, signaling molecules, virulence factors, nucleic acids and genetic information) without the activation of the immune system, involvement in cell signaling and the formation of biofilm, ability to sense nutrients, association with cytotoxicity and killing of other bacteria, invasion of eukaryotic host cells and consequent cause of infection and disease, and antibiotic resistance (Brown et al., 2015; van der Pol et al., 2012).

The assumption that membrane vesicles were secreted only by Gram-negative bacteria was probably related to the absence of any barrier between the bacterium and the extracellular medium (outer membrane vesicles derive from the outer membrane of Gram-negative bacteria), in fact in the periplasmic space that separates the inner and the outer membrane there is only a thin layer of peptidoglycan; the presence of a thick peptidoglycan cell wall, surrounding a single lipid membrane, in Gram-positive bacteria was thought to represent an unsurpassable obstacle for the release of vesicles (Brown et al., 2015). So far it is not known how the process works in Gram-positive bacteria, but three hypotheses of how membrane vesicles can traverse the thick peptidoglycan layer after the release from the plasma membrane have been advanced: vesicles could be forced out the cell wall by the turgor pressure and pass through pores; the action of enzymes could modify the cell wall, loosening it and increasing the pores' size; vesicles could exit the cell through protein channels (Brown et al., 2015). The idea that cell enzymes contained within vesicles have a role in the vesicular transport could be widespread to *S. aureus* membrane vesicles, in fact they carry enzymes able to degrade the peptidoglycan cell wall, making it possible for the vesicles to pass thanks to modifications in the cell wall, but nowadays further studies are necessary to understand all the details of how vesicles cross the cell wall of Gram-positive bacteria (Brown et al., 2015).

The secretion of membrane vesicles could appear as an expensive process in terms of energy for bacteria, but cell-derived vesicles can bring advantages to microorganisms: vesicles could be responsible for the *S. aureus*' high resistance against antibiotics acting as a decoy for antibodies, making them actually non effective against the whole microorganism (Deatherage & Cookson, 2012; Lee, 2012; Gurung et al., 2011). Therefore, membrane vesicles could be a mean through which bacteria misdirect the immune system of the host (Deatherage & Cookson, 2012).

From Electron Microscopy images (Figure 3) it seems that the hypothesis of peptidoglycan degrading enzymes could be the reason of the secretion of *S. aureus* membrane vesicles. In fact, it is possible to detect a budding area where the cell wall appears partially degraded, and by this zone are located vesicles astride the plasma membrane, exiting the bacterium, and vesicles outside the bacterial cell, both containing polyurethane nanoparticles that act as markers. Hence, Electron Microscopy with its ability to take snapshots at fixed times, combined with literature can contribute to the comprehension of processes occurring at nanoscale. In particular, the existence of transported cargoes surrounded by vesicles from outside to inside the cell (endocytosis), and the actual transfer of material within vesicles from within the cell to the extracellular medium (secretion of vesicles) clarifies that vesicular trafficking in bacteria is a two-directions process, and that the transport of vesicles is supported by cytoskeletal structures.

Toxicological Threat of Polyurethane

Whenever nanoparticles are concerned it is relevant to assess the risks that the material used can represent for the health of patients. In the current study polyurethane dental prostheses (provided by Dentalur Russia, Russian Federation) with different kinds of surfaces were used. First of all, it is necessary to analyze the biological effect of exposure to bulk polyurethane. Toluene diisocyanate is the principal industrial isocyanate implied as precursor in the production of polyurethanes (www.cdc.gov/niosh/pdfs/7311022b.pdf). Literature reports toxicological effects of inhaled free toluene diisocyanate in the form of irritation of the pulmonary tract, respiratory illnesses, bronchitis and in some cases hypersensitive and allergic responses. Polyurethanes are produced by the catalyzed reaction between toluene diisocyanates and hydroxyl groups found in polyether and polyester polyols (R-NCO + R'OH → R-NHCO-OR') (Howard, 2011; www.cdc.gov/niosh/pdfs/7311022b.pdf).

Analysis of the Bacterial Vesicles' Enhanced Toxicological Threat

Topic of this work is the interaction of a culture of *S. aureus* isolated from a patient with a periodontal disease with plastic surfaces. The bacterial action on polyurethane affects the urethane bond, thus creating a polyurethanolytic action, and creating free toluene diisocyanates which are highly reactive with organic compounds containing hydrogen atoms attached to oxygen, nitrogen or sulfur, such as water or proteins. Since polyurethanes are commonly used in food packaging, Plastic Directives have been drawn up to establish the maximum quantity for each compound accepted per unit. Toluene diisocyanate are highly toxic and regulations have estimated the maximum quantity of isocyanates compounds to 1 mg kg^{-1}, furthermore when toluene diisocyanates reacts with water, they yield an unstable form of carbamic acid that dissociates in a primary aromatic amine and CO_2. Primary amines are more toxic and not more than 10 µg kg^{-1} should migrate into food (Aznar et al., 2011). Since oral prostheses are in a constantly humid environment and often in contact with food, these evaluations of possibly toxic effects should be taken into account in order to properly assess the hazardous effects of polyurethane prostheses.

Another important issue is raised by the ability of toluene diisocyanates to combine with a variety of hydroxyl sites on the biological molecules yielding several reactive compounds with the tendency to polymerize, resulting in the denaturation of proteins. This field needs to be investigated with further studies to better understand what happens at a molecular scale and which abnormalities can occur.

Electron Microscopy allowed researchers to detect polyurethane nanoparticles within bacterial cells (Figures 2, 3 and 4). Toxicological threats presented by the same material (polyurethane) at nanoscale (1 nm is extremely small and amounts to about 10 atoms across (Revell, 2006)), could be underestimated, unexpected and unpredictable. Nanoparticles are characterized by a large surface area in comparison to a tiny volume; atoms exposed to the surface make nanomaterials chemically highly reactive and prone to form agglomerates, guided by van der Waals and electrostatic forces and surface tension, and to associate to proteins (Revell, 2006) (recruitment of biomolecules for the formation of the protein corona).

Nanoparticles detected in Figures 2, 3 and 4 (which size ranges from 2-3 to 10 nm) have marked altered properties in regard to the bulk material and it must be kept in mind that they can be widely disseminated in an organism even trespassing tissues and organs. Moreover, polyurethane nanoparticles enclosed in vesicles within *S. aureus* derive from the bacterial action over a prosthetic device, hence the nanomaterial is not engineered and can be totally unforeseeable.

To assess toxicological threats and risks for the health from the prolonged use of polyurethane prostheses, their wear and the consequences carried out by the bacterial biodestruction, are therefore needed further studies in regard to the bioaccumulation of xenobiotic nanoparticles, their distribution and their hazardousness, taking into account the highly reactive bonds that can be easily set up in a biological medium.

From *in Vitro* to *in Vivo*: Membrane Vesicles as Cargo Deliverers

The details visible in the acquired Electron Microscopy images (Figures 2, 3 and 4) have a great potential to clear up some of the key passages regarding the deterioration of polyurethane prostheses, and the uptake and release of nanoparticles enclosed in vesicles. This information together with knowledge from literature helped researchers to better comprehend the picture and all the structures involved (such as the presence of the protein corona and the role of cytoskeletal structures in the vesicular transport); making connections and analogies between the prokaryotes and the eukaryotes clarified how evolutionally conserved processes evolved differently for diverse organisms.

The potential of combining information from Electron Microscopy images and other investigation techniques brought researchers to think of which outcomes a similar situation could have *in vivo*.

In vitro experiments shed light on the generation of nanoparticles derived from the *S. aureus* biodestructive action on polyurethane oral prostheses, and the subsequent internalization of nanomaterial within microbes. Human beings are a natural reservoir for *S. aureus* (Lowy, 1998) and these bacterial cells are normally present as commensals for instance of the nares and the skin. When *S. aureus* cells come into contact with tissues and the bloodstream, infections take place. The high pathogenicity of *S. aureus* is emphasized by its ability to evade the immune system and the capability to develop antibiotic resistance (Foster, 2005; Rooijakkers, van Kessel, & van Strijp, 2005; Gordon & Lowy, 2008).

When a patient wearing a polyurethane dental prosthesis is affected by a periodontal disease or a gingivitis, *S. aureus* colonies usually present on the oral mucosa can start the formation of a biofilm. This condition, which could be aggravated by poor hygiene (Ghannoum & O'Toole, 2004), can lead to the situation seen in the *in vitro* experiments: an increased number of bacterial cells which deteriorate the plastic material of the prosthetic device and internalize nanosized polyurethane particles. It is known that planktonic cells can detach themselves from a mature biofilm (Ghannoum & O'Toole, 2004), but in this specific case nomadic cells would be loaded with plastic nanoparticles included in vesicles. What would happen?

First of all, there would be the encounter of a bacterial matrix (the exopolysaccharide matrix of biofilms) with mammalian tissues and cells (oral mucosa). As seen for adhesins in the Gram-negative fimbrial structures, even adhesins found in the extracellular matrix of Gram-positive bacteria are able to interact with the mammalian matrix components (Branda, Vik, Friedman, & Kolter, 2005), thus furthering the adhesion of bacterial cells to the mammalian tissue.

Subsequently there would be the detachment of planktonic cells loaded with polyurethane nanoparticles enclosed in membrane vesicles. *S. aureus* cells could gain access to the human body and, anchoring to the cytoskeletal structures of the host (Agerer, Michel, Ohlsen, & Hauck, 2003), could be spread to areas even further from the access point. As it has already been said, *S. aureus* cells during infections are able to evade the immune system and one of the way through which bacteria manage to do so, is taking advantage of the immune cells themselves, for instance hiding in phagosomes (Foster, 2005; Lowy, 2000).

At this point bacterial cells disseminated in several areas of the body could undergo four different outcomes (Table 1):

- *S. aureus* reproduces itself;
- *S. aureus* releases its cargo via exocytosis;
- *S. aureus* secretes vesicles containing polyurethane nanoparticles;
- *S. aureus* dies.

The reproduction of the bacterial cell would increase the number of cells containing membrane vesicles loaded with polymeric nanoparticles.

Table 1. Scheme (to be read from left to right) of the four possible outcomes of an infection of a host mammalian cell carried out by S. aureus cells which absorbed polyurethane nanoparticles. It is notable that whatever the destiny of the bacterial cell is, the spread of polyurethane nanoparticles into the host cell (and subsequently into the whole organism) occurs widely.

S. aureus loaded with polyurethane nanoparticles infects a host mammalian cell.	*S. aureus* reproduces itself.	→	The number of bacterial cells inside the host increases; new born cells contain polyurethane nanoparticles.
	S. aureus releases its cargo via exocytosis.	→	Polyurethane nanoparticles, shielded by the protein corona, freely circulate in the host cell.
	S. aureus secretes membrane vesicles loaded with nanoparticles.	→	Membrane vesicles evade the immune system of the host; guided by infection dynamics membrane vesicles deliver virulence factors together with their cargo to specific targets, spreading nanoparticles in the host.
	S. aureus dies.	→	The bacterial components and polyurethane nanoparticles are spread inside the host cell. Free polyurethane nanoparticles or particles inside membrane vesicles spread into the host cell.

The release of the bacterium's internal cargo would cause the discharge of an indefinite number of nanoparticles (in fact each vesicle can contain one or more nanoparticles) surrounded by the protein corona. The protein cover, shielding its cargo, can fail the host cell (which would recognize only the external biomolecules) acting as a Trojan Horse which actually is loaded with potentially toxic material. In this way the defensive systems of the host cell would not be activated and a threatening cargo would be freely circulating.

The ability of *S. aureus* to secrete cell-derived vesicles is conserved even *in vivo* (Gurung et al., 2011; Lee, 2012) and it increases the bacterial pathogenicity, in fact membrane vesicles can become cytotoxic delivering virulence factors (Brown et al., 2015; Deatherage & Cookson, 2012; Lee, 2012), and induce apoptosis in the host (Gurung et al., 2011). Membrane vesicles are therefore important vehicles of bacterial molecules to host cells (Gurung et al., 2011) and in this case, being loaded with polyurethane nanoparticles, membrane vesicles could secrete their cargo, further concurring to the spread of xenobiotic material.

In case of natural or antibiotic-induced death of *S. aureus* cells all the bacterial components and the cargo would be dispersed in the host cell. Membrane vesicles (if not lysed) would still be a threat for the host cell, but if during the bacterial death membrane vesicles were damaged and lysed they would not represent a danger for the host anymore (Gurung et al., 2011); in any case membrane vesicles could release their noxious cargo via exocytosis.

Concluding, bacteria following their infection schemes release membrane vesicles that contribute to worsen the bacterial infection and are an aggravating for the nanoparticles delivery (Curia, Milani, Didenko, & Shevlyagina, 2013). In an *in vivo* scenario it becomes more important to establish the toxicity of the bulk xenobiotic material, and essential to understand how it behaves at a nanoscale, studying all the interactions that polyurethane nanoparticles could have in a biological medium with bacterial and host mammalian cells. It is necessary to do further investigations to track the pathways that nanoparticles could have in bacteria and from bacteria to host cells, concentrating on the distribution of vesicles, the amount of their xenobiotic content, the mode of release and the accumulation ratio.

Better comprehending the bacteria-host cell interactions and those between bacterial membrane vesicles and host cells could bring to the future use of bacterial cell-derived vesicles as targeted drug delivery vehicles.

Perspectives

In this study we describe a new type of secretion pathway, through which bacteria secrete proteins, cellular components, virulence factors, and a new kind of cargo (polyurethane nanoparticles generated by microorganisms), raising issues about the toxicological risks and the dosimetry of nanoparticles in bacterial membrane vesicles and host cells.

In the same time this study opens new ways to the nanomedicine in the comprehension of the mechanisms of the bacterial secretion of membrane vesicles and their interactions with host cells. Further studies of the delivery of xenobiotic material guided by the dynamics of bacterial infections would be useful since the nanoparticles deliverance led by the infection dynamics opens new ways to the possibility of delivering drugs to selected cells.

Although microscopy cannot cover all the relevant parameters, it is a promising and effective tool in this type of investigation: it can reach the magnification of interest to see membrane vesicles and their cargos, without biochemical preparations that could cover morphology and identity of the protagonists, being biochemistry an indirect analytical tool.

Electron Microscopy Images

Samples Preparation Method

The culture of *S. aureus* was isolated from a patient with a periodontal disease; Dentalur Russia (Russian Federation) provided the samples of polyurethane with different types of surfaces (smooth, rough, surfaces resulted from sawing and splitting operations). *S. aureus* cells were incubated in a Luria Bertani broth with polyurethane with different roughness from 1 to 45 days at uniform temperature of 37° C. A smooth polyurethane slice in broth, and *S. aureus* cells in broth without polymeric materials were the control samples.

A part of the bacterial pellet, obtained after centrifugation at 6000 rpm for 10 minutes, was placed on a thin silicon substrate for the SEM observation, the remaining part underwent preparation for the TEM/STEM analysis. All the phases of samples preparation were carried out at room temperature. Samples for both the Electron Microscopy modalities were primary fixed with aldehyde, postfixed with osmium and stained with lead citrate. At this point samples used in the SEM observation were dried, mounted on an aluminum stub and coated with a gold conductive layer with a thickness of 5 nm in a SPI-Module Sputter/Carbon Coater System (SPI Inc., USA). Instead samples used in the TEM/STEM observation were embedded in agar and cut with an ultramicrotome. Standard drying operation led to structural changes of both the cells and the biofilm, so traditional chemical-physical methods of drying were not applied (Milani, Drobne, & Tatti, 2007). In order to gain a better view of the exopolysaccharide matrix of the biofilm samples were contrasted with copper sulphate and alcian blue.

Electron Microscopes

In this investigation different electron microscopes and techniques (TEM, STEM and SEM) were used. The analysis via the combined use of TEM, STEM and SEM allowed researchers to freely span along the sample, having at disposal the possibility to analyze samples from diverse angles and with a wide range of magnification.

TEM and STEM images were acquired with a Tecnai F20 X-TWIN microscope (FEI Company, USA) equipped with a 200 kV FEG column and a CCD detector.

Electron Microscopy images are obtained depending on how electrons hit the sample, interact with it and leave it. The use of different detectors able to collect the transmitted electrons according to their scattering angle, gave images with better contrast and resolution and with more morphological details, actually improving the investigation of the samples.

In Bright Field (BF) images a detector collected all the electrons passed directly through the sample, blocking the scattered ones. When electrons hit an ultrathin sample, they scatter into high angles or backwards, with the scattering increasing whenever the material investigated has a high atomic number Z. Dark Field (DF) images were acquired collecting electrons scattered after the interactions with the sample; applying an external concentric detector it was possible to acquire High Angle Annular Dark Field (HAADF) images collecting electrons scattered at angles higher than 50 mrad (Krumeich, n. d.; Utsunomiya & Ewing, 2003).

In Bright Field images thick areas appeared darker than the less dense parts. Details not properly visible in the Bright Field images were detectable in the Dark Field ones which in comparison were less intense, but high-contrasted. In Dark Field images the denser and heavier areas with a high atomic number Z scatter stronger, hence they appeared brighter (Krumeich, n.d); this allowed investigators to get information from zones that in Bright Field images appeared to be background noise.

SEM images were acquired with a FEI QUANTA 200 3D FIB/SEM (Focused Ion Beam/Scanning Electron Microscope). This microscope is a Dual Beam equipped with both an electron and an ion column. Images were obtained via the electron column, with a tungsten wire as a source, in both high and low vacuum, mostly at 5 kV electron beam acceleration. The electron beam acceleration ranges from 1 to 30 kV, the choice of working at 5 kV was justified by the fact that with this acceleration the penetration of the beam is limited, thus the sub-superficial parts of the sample give a lower signal contribution allowing to acquire better images (Milani et al., 2007). Moreover, at this acceleration the operator has less problems concerning the charge of the sample. Finally, the advantage of FIB/SEM investigation is the possibility to observe samples which underwent no preparative steps at all as well as sample prepared for the TEM observation embedded in resin.

Electron Microscopies are effective instruments, absolutely essential in biological studies for a better comprehension of processes occurring at a micro- and nano-scale. The potentiality of each electron microscope can be exalted performing investigations using different machines at the same time. Observing the same sample in microscopes with different distinctiveness, and collecting images with diverse characteristics make it possible to do a sample's cross analysis which will result more reliable; this is absolutely necessary for researchers whose aim is to explain and clarify the variety of aspects of the problem.

Despite the extremely precise images provided, via electron microscopy one can investigate only features inside a tiny sample of one or more cells. This limit can be overcome by implementing the researches with studies conducted in different laboratories using various techniques.

In Memoriam of Lyubov Vassilievna Didenko

On December 8th 2015 the life of the outstanding scientist Lyubov V. Didenko ended.

Born on July 22nd 1958, her scientific career began while studying at the medical university; Lyubov V. Didenko received PhD and Doctor of Medicine degree for scientific contribution to the understanding of the pathogenesis of bacteremia in infectious and non-infectious diseases.

She then served at the Department of Infectional Diseases at Botkin Hospital, Moscow (Russian Federation).

Figure 5. Lyubov Vassilievna Didenko (1958-2015)

In the last 20 years L. V. Didenko was the head of the Laboratory of Microorgamisms' anatomy at the Gamaleya Research Institute for Epidemiology and Microbiology of Moscow (Russian Federation).

Throughout her interesting scientific activities, she conducted researches in many directions:

- Interaction of different types of microorganisms, protozoa in environment and clinical settings;
- Ultrastructure of *Helicobacter pylori in vivo* and *in vitro*;
- Process of biofilm formation of *Salmonella typhimurium*, *Burkholderia cepacia*, *Legionella spp*, *Escherichia coli*, *Staphylococcus aureus*, *Brevibacillus laterosporus*;
- Structural features of biofilms in dangerous water systems and on medical supplies;
- Research of generalization of infection process;
- Detection of corona virus in experimental pneumocystic pneumonia;
- Infectious aspects of kidney and prostate stone formation;
- The study of biodegradation of dental prostheses under bacterial colonization.

Lyubov V. Didenko gave relevant contributions to Electron Microscopy of microorganisms' anatomy both in the Transmission and the Scanning Electron Microscopies (TEM and SEM). Moreover she implemented this area of research devoting energies to correlate Focused Ion Beam (FIB) with other microscopies and preparation techniques.

She was not only a talented scientist, but also a true friend to many people, helpful and sincere. She was a woman of culture, able to entertain the listener to the field of her activity, framing the sometimes super technical elements in complex stories resembling theatrical pieces.

We will remember Lyubov V. Didenko forever.

REFERENCES

Agerer, F., Michel, A., Ohlsen, K., & Hauck, C. R. (2003). Integrin-mediated Invasion of *Staphylococcus aureus* into Human Cells Requires Src Family Protein-tyrosine Kinases. *The Journal of Biological Chemistry*, *278*(43), 42524–42531. doi:10.1074/jbc.M302096200 PMID:12893831

Amos, L. A., van den Ent, F., & Löwe, J. (2004). Structural/functional homology between the bacterial and eukaryotic cytoskeletons. *Current Opinion in Cell Biology, 16*(1), 24–31. doi:10.1016/j.ceb.2003.11.005 PMID:15037301

Avinoam, O., Schorb, M., Beese, C. J., Briggs, J. A. G., & Kaksonen, M. (2015). Endocytic sites mature by continuous bending and remodeling of the clathrin coat. *Science, 348*(6241), 1369–1372. doi:10.1126/science.aaa9555 PMID:26089517

Aznar, M., Vera, P., Canellas, E., Nerín, C., Mercea, P., & Störmer, A. (2011). Composition of the adhesives used in food packaging multilayer materials and migration studies from packaging to food. *Journal of Materials Chemistry, 21*(12), 4358–4370. doi:10.1039/c0jm04136j

Branda, S. S., Vik, A., Friedman, L., & Kolter, R. (2005). Biofilms: The matrix revisited. *Trends in Microbiology, 13*(1), 20–26. doi:10.1016/j.tim.2004.11.006 PMID:15639628

Brown, L., Wolf, J. M., Prados-Rosales, R., & Casadevall, A. (2015). Through the wall: Extracellular vesicles in Gram-positive bacteria, mycobacteria and fungi. *Nature Reviews. Microbiology, 13*, 620–630. PMID:26324094

Brown, S. S. (1999). Cooperation between microtubule- and actin-based motor proteins. *Annual Review of Cell and Developmental Biology, 15*(1), 63–80. doi:10.1146/annurev.cellbio.15.1.63 PMID:10611957

Carballido-López, R., & Errington, J. (2003). A dynamic bacterial cytoskeleton. *Trends in Cell Biology, 13*(11), 577–583. doi:10.1016/j.tcb.2003.09.005 PMID:14573351

Cedervall, T., Lynch, I., Lindman, S., Berggård, T., Thulin, E., Nilsson, H., & Linse, S. et al. (2007). Understanding the nanoparticle–protein corona using methods to quantify exchange rates and affinities of proteins for nanoparticles. *Proceedings of the National Academy of Sciences of the United States of America, 104*(7), 2050–2055. doi:10.1073/pnas.0608582104 PMID:17267609

Center for Disease Control and Prevention. (n. d.) Biological effects of exposure. Retrieved from www.cdc.gov/niosh/pdfs/7311022b.pdf

Cifra, M., Fields, J. Z., & Farhadi, A. (2011). Electromagnetic cellular interactions. *Progress in Biophysics and Molecular Biology, 105*(3), 223–246. doi:10.1016/j.pbiomolbio.2010.07.003 PMID:20674588

Curia, R., Erega, A., & Milani, M. (2015). Study of the Formation, Secretion and Trafficking of Membrane Vesicles Loaded with Polyurethane Nanoparticles in Staphylococcus aureus via Electron Microscopy. *HSOA Journal of Nanotechnology: Nanomedicine & Nanobiotechnology*, 2.

Curia, R., Milani, M., Didenko, L. V., Avtandilov, G. A., Shevlyagina, N. V., & Smirnova, T. A. (2014). Beyond the biodestruction of polyurethane: *S. aureus* uptake of nanoparticles is a challenge for toxicology. In A. Méndez-Vilas (Ed.), *Microscopy: advances in scientific research and education* (pp. 16–23). Spain: Formatex Research Center.

Curia, R., Milani, M., Didenko, L. V., & Shevlyagina, N. V. (2013). Electron microscopy broadens toxicology's horizons: The role of nanoparticles vehiculated by bacteria. *Current Topics in Toxicology*, *9*, 93–98.

Deatherage, B. L., & Cookson, B. T. (2012). Membrane Vesicle Release in Bacteria, Eukaryotes, and Archaea: A Conserved yet Underappreciated Aspect of Microbial Life. *Infection and Immunity*, *80*(6), 1948–1957. doi:10.1128/IAI.06014-11 PMID:22409932

Didenko, L., Avtandilov, G., Shevlyagina, N., Shustrova, N., Smirnova, T., Lebedenko, I., & Milani, M. et al. (2013). Nanoparticles production and inclusion in *S. aureus* incubated with polyurethane: An electron microscopy analysis. *Open Journal of Medical Imaging*, *3*(2), 69–73. doi:10.4236/ojmi.2013.32010

Didenko, L. V., Avtandilov, G. A., Shevlyagina, N. V., Smirnova, T. A., Lebedenko, I. Y., Tatti, F., & Milani, M. (2012). Biodestruction of polyurethane by *Staphylococcus aureus* (an investigation by SEM, TEM and FIB). In A. Méndez-Vilas (Ed.), *Current Microscopy Contributions to Advances in Science and Technology* (pp. 323–334). Spain: Formatex Research Center.

Didenko, L. V., Shevlyagina, N. V., Curia, R., Erega, A., & Milani, M. (2014). Electron Microscopy Furthers the Investigation of Bacteria-Nanoparticles Interactions Subcellular Dynamics. Global Journal of Medical Research: D Radiology. *Diagnostic Imaging and Instrumentation*, *14*(4), 13–23.

Döhner, K., Nagel, C., & Sodeik, B. (2005). Viral stop-and-go along microtubules: Taking a ride with dynein and kinesins. *Trends in Microbiology*, *13*(7), 320–327. doi:10.1016/j.tim.2005.05.010 PMID:15950476

Engqvist-Goldstein, A. E. Y., & Drubin, D. G. (2003). Actin Assembly and Endocytosis: From Yeast to Mammals. *Annual Review of Cell and Developmental Biology*, *19*(1), 287–332. doi:10.1146/annurev.cellbio.19.111401.093127 PMID:14570572

Foster, T. J. (2005). Immune evasion by Staphylococci. *Nature Reviews. Microbiology, 3*(12), 948–958. doi:10.1038/nrmicro1289 PMID:16322743

Fux, C. A., Costerton, J. W., Stewart, P. S., & Stoodley, P. (2005). Survival strategies of infectious biofilms. *Trends in Microbiology, 13*(1), 34–40. doi:10.1016/j.tim.2004.11.010 PMID:15639630

Ghannoum, M. A., & O'Toole, G. A. (2004). *Microbial Biofilms*. Washington, DC, USA: Amer. Society for Microbiology Press.

Gordon, R. J., & Lowy, F. D. (2008). Pathogenesis of Methicillin-Resistant *Staphylococcus aureus* Infection. *Clinical Infectious Diseases, 46*(Suppl. 5), S350–S359. doi:10.1086/533591 PMID:18462090

Granger, E., McNee, G., Allan, V., & Woodman, P. (2014). The role of the cytoskeleton and molecular motors in endosomal dynamics. *Seminars in Cell & Developmental Biology, 31*, 20–29. doi:10.1016/j.semcdb.2014.04.011 PMID:24727350

Gurung, M., Moon, D. C., Choi, C. W., Lee, J. H., Bae, Y. C., Kim, J., & Lee, J. C. (2011). *Staphylococcus aureus* Produces Membrane-Derived Vesicles That Induce Host Cell Death. *PLoS ONE, 6*(11), e27958. doi:10.1371/journal.pone.0027958 PMID:22114730

Howard, G. T. (2011). Microbial biodegradation of polyurethane. In A. Fainleib & O. Grigoryeva (Eds.), *Recent Developments in Polymer Recycling* (pp. 215–238). Kerala, India: Transworld Research Network.

Huotari, J., & Helenius, A. (2011). Endosome maturation. *The EMBO Journal, 30*(17), 3481–3500. doi:10.1038/emboj.2011.286 PMID:21878991

Iyer, S., Romanowicz, B., & Laudon, M. (2004). *Biomolecular Motors, - A DARPA commissioned overview with highlights from the special session at Nanotech 2004*. Boston, MA, USA. Retrieved fromhttp://www.nsti.org/outreach/Biomolecular_Motors/

Kamal, A., & Goldstein, L. S. B. (2000). Connecting vesicle transport to the cytoskeleton. *Current Opinion in Cell Biology, 12*(4), 503–508. doi:10.1016/S0955-0674(00)00123-X PMID:10873823

Krumeick, F. (n. d.) Properties of electrons, their interactions with matter and applications in electron microscopy. *Microcopy.ethz.ch.* Retrieved from http://www.microscopy.ethz.ch/downloads/Interactions.pdf

Lee, E., Choi, D., Kim, D., Kim, J., Park, J. O., Kim, S., & Gho, Y. S. et al. (2009). Gram-positive bacteria produce membrane vesicles: Proteomics-based characterization of *Staphylococcus aureus*-derived membrane vesicles. *Proteomics*, *9*(24), 5425–5436. doi:10.1002/pmic.200900338 PMID:19834908

Lee, J. C. (2012). *Staphylococcus aureus* Membrane Vesicles and Its Potential Role in Bacterial Pathogenesis. *Journal of Bacteriology and Virology*, *42*(3), 181–188. doi:10.4167/jbv.2012.42.3.181

Löwe, J., van den Ent, F., & Amos, L. A. (2004). Molecules of the bacterial cytoskeleton. *Annual Review of Biophysics and Biomolecular Structure*, *33*(1), 177–198. doi:10.1146/annurev.biophys.33.110502.132647 PMID:15139810

Lowy, F. D. (1998). *Staphylococcus aureus* infections. *The New England Journal of Medicine*, *339*(8), 520–532. doi:10.1056/NEJM199808203390806 PMID:9709046

Lowy, F. D. (2000). Is *Staphylococcus aureus* an intracellular pathogen? *Trends in Microbiology*, *8*(8), 341–343. doi:10.1016/S0966-842X(00)01803-5 PMID:10920387

Lynch, I., Salvati, A., & Dawson, K. A. (2009). Protein-nanoparticle interactions: What does the cell see? *Nature Nanotechnology*, *4*(9), 546–547. doi:10.1038/nnano.2009.248 PMID:19734922

May, R. C., & Machesky, L. M. (2001). Phagocytosis and the actin cytoskeleton. *Journal of Cell Science*, *114*(6), 1061–1077. PMID:11228151

Michie, K. A., & Löwe, J. (2006). Dynamic Filaments of the Bacterial Cytoskeleton. *Annual Review of Biochemistry*, *75*(1), 467–492. doi:10.1146/annurev.biochem.75.103004.142452 PMID:16756499

Milani, M., Didenko, L. V., Avtandilov, G. A., Curia, R., Erega, A., & Shevlyagina, N. V. (2016). Electron Microscopy Documents the Microorganisms' Biodestructive Action on Polyurethane and the Production, Internalization and Vesicular Trafficking of Nanoparticles. *British Journal of Applied Science & Technology*, *12*(3), 1–19. doi:10.9734/BJAST/2016/20290

Milani, M., Drobne, D., & Tatti, F. (2007). How to study biological samples by FIB/SEM? In A. Méndez-Vilas & J. Díaz (Eds.), *Modern Research and Educational Topics in Microscopy* (pp. 787–794). Spain: Formatex Research Center.

Møller-Jensen, J., & Löwe, J. (2005). Increasing complexity of the bacterial cytoskeleton. *Current Opinion in Cell Biology*, *17*(1), 75–81. doi:10.1016/j.ceb.2004.11.002 PMID:15661522

Nel, A. E., Mädler, L., Velegol, D., Xia, T., Hoek, E. M. V., Somasundaran, P., & Thompson, M. et al. (2009). Understanding biophysicochemical interactions at the nano–bio interface. *Nature Materials*, *8*(7), 543–557. doi:10.1038/nmat2442 PMID:19525947

Revell, P. A. (2006). The biological effects of nanoparticles. *Nanotechnology Perceptions*, *2*, 283–298.

Rooijakkers, S. H. M., van Kessel, K. P. M., & van Strijp, J. A. G. (2005). Staphylococcal innate immune evasion. *Trends in Microbiology*, *13*(12), 596–601. doi:10.1016/j.tim.2005.10.002 PMID:16242332

Sahneh, F. D., Scoglio, C., & Riviere, J. (2013). Dynamics of Nanoparticle-Protein Corona Complex Formation: Analytical Results from Population Balance Equations. *PLoS ONE*, *8*(5), e64690. doi:10.1371/journal.pone.0064690 PMID:23741371

Sămaj, J., Baluška, F., Boris, B., Schlicht, M., Volkmann, D., & Menzel, D. (2004). Endocytosis, Actin Cytoskeleton, and Signaling. *Plant Physiology*, *135*(3), 1150–1161. doi:10.1104/pp.104.040683 PMID:15266049

Schliwa, M., & Woehlke, G. (2003). Molecular motors. *Nature*, *422*(6933), 759–765. doi:10.1038/nature01601 PMID:12700770

Shah, A. A., Hasan, F., Hameed, A., & Ahmed, S. (2008). Biological degradation of plastics: A comprehensive review. *Biotechnology Advances*, *26*(3), 246–265. doi:10.1016/j.biotechadv.2007.12.005 PMID:18337047

Tenzer, S., Docter, D., Kuharev, J., Musyanovych, A., Fetz, V., Hecht, R., & Stauber, R. H. et al. (2013). Rapid formation of plasma protein corona critically affects nanoparticle pathophysiology. *Nature Nanotechnology*, *8*(10), 772–781. doi:10.1038/nnano.2013.181 PMID:24056901

Utsunomiya, S., & Ewing, R. C. (2003). Application of high-angle annular dark field scanning transmission electron microscopy, scanning transmission electron microscopy-energy dispersive X-ray spectrometry, and energy-filtered transmission electron microscopy to the characterization of nanoparticles in the environment. *Environmental Science & Technology*, *37*(4), 786–791. doi:10.1021/es026053t PMID:12636280

Vale, R. D. (2003). The Molecular Motor Toolbox for Intracellular Transport. *Cell*, *112*(4), 467–480. doi:10.1016/S0092-8674(03)00111-9 PMID:12600311

van den Ent, F., Amos, L. A., & Löwe, J. (2001). Prokaryotic origin of the actin cytoskeleton. *Nature*, *413*(6851), 39–44. doi:10.1038/35092500 PMID:11544518 van der Pol, E., Böing, A. N., Harrison, P., Sturk, A., & Nieuwland, R. (2012). Classification, Functions, and Clinical Relevance of Extracellular Vesicles. *Pharmacological Reviews*, *64*(3), 676–705. doi:10.1124/pr.112.005983 PMID:22722893

KEY TERMS AND DEFINITIONS

Biodestruction: Process carried out by bacteria and fungi at the expenses of biomaterials. Biodestruction can be caused by bacterial metabolic and enzymatic activities (change of pH in the environment, breaking of chemical bonds between atoms). Biodestruction provokes defects on a material's surface with the detachment of tiny debris (micro- and nano-particles).

Biofilm: Aggregate of microorganisms in which cells frequently embedded within a self-produced exopolysaccharide matrix adhere to each other and/or to a surface.

Cytoskeleton: Complex network of actin filaments, intermediate filaments and microtubules that gives support to the cell, maintaining its shape, and backs off important processes such as cell division, chromosome segregation and vesicular trafficking.

Membrane Vesicle: Cell-derived vesicle containing a biological (proteins, phospholipids, signaling molecules, virulence factors, nucleic acids and genetic information) or xenobiotic cargo secreted by bacteria in the extracellular medium. Membrane vesicles can transport their cargoes without the activation of the immune system and are involved in different processes: cell signaling, biofilm formation, sensing of nutrients, cytotoxicity and killing of other bacteria, invasion of eukaryotic host cells, cause of infection and disease, antibiotic resistance.

Nanoparticle: Object which size is in the order of tens of nanometers (less than 100 nm). 1 nm is as large as 10 atoms across, hence nanoparticles represent a bridge between bulk materials and their atomic structures. Nanoparticles have a higher surface/volume ratio in comparison with bigger particles, a high percentage of atoms is exposed to the surface and therefore nanoparticles are chemically highly reactive. Due to the numerous chemical bonds that nanoparticles can establish with approaching molecules, the properties of a nanosized material change in comparison to those of the same bulk material, showing unexpected behavior.

Protein Corona: Protein cover that surrounds nanoparticles when they are in a biological fluid. Interactions occurring between nanoparticles and biomolecules are led by colloidal forces and dynamic biophysicochemical relations that can be seen as complex combinations of elementary electromagnetic interactions. Two major features to be kept into account are the time dependence of all processes, and the cellular and intracellular presence of endogenous electromagnetic fields. Shielding nanoparticles, the protein corona highly affects the particles' charge, influencing their following interactions.

Vesicular Trafficking: Movements of membranous vesicles along the cytoskeletal structures of a cell. Vesicles' direction and their switching from filaments to tubules are regulated by motor proteins moving along the cytoskeleton.

Chapter 3
Applications of Polymeric Micro- and Nano-Particles in Dentistry

Balasankar Meera Priyadarshini
National University of Singapore, Singapore

Nileshkumar Dubey
National University of Singapore, Singapore

ABSTRACT

The use of micro- and nanoparticles is rapidly advancing and has been most commonly used in medical and biological research that offers an encouraging scope in broad range of disciplines. Manipulation of the biomaterials to their micro- and nano-scale renders their properties and behavior different from that of the same material in the mass scale and make them more reactive than large particles. The removal of tooth structure and its restoration with synthetic material to solve the problems caused by dental caries, trauma and fracture is a practice nearly as old as dentistry. Efforts are made to create micro- and nanomaterials that can revolutionize these ancestral therapies and dental procedures. The use of these materials had shown some promising applications in caries control, endodontic therapy, regenerative dentistry, periodontology and oral biofilm management. This review aims to discuss the recent advances and future potential of polymer-based micro- and nanoparticles in dentistry.

DOI: 10.4018/978-1-5225-1043-7.ch003

INTRODUCTION

The prevention of tooth decay, treatment of bone loss and periodontal disease are ongoing challenges in dentistry. According to CDC, tooth decay is the most frequent childhood disease 5 times as common as asthma. LA Times came up with the fact that 42% of children are expected to have some decay by the age of 2-12 (Mascarelli, 2011). For example, in 2015-16, a UK report shows that 9.8% rise in removal of one or more teeth in children aged 10 and under (Howell, 2016). Periodontal diseases are the most common infections due to the overgrowth of the dental plaque that may increase the risk of heart disease and it is estimated that 47.2% of adults aged 30 years and older has some form of periodontal disease (Dhadse, Gattani, & Mishra, 2010).

With the emergence and increase of such dental problems and continuing concerns on healthcare costs, many researchers have tried to develop new and effective treatment regimes. Such problems and needs have led to the resurgence in the use of micro- and nanomaterials in order to revolutionize the ancestral therapies and dental procedures. For example, liquids and pastes containing nano-apatite's have been used remineralization of early sub micrometer sized enamel lesions and the role of nanoparticle in the control of the oral biofilm has also been recognized (Figure 1) (Besinis, De Peralta, Tredwin, & Handy, 2015; Hannig & Hannig, 2010). Similarly, micro-/nanoparticles that can deliver antibiotics and bioactive compounds have been used to treat periodontal diseases (Yao et al., 2014).

Application of micro-/nanotechnology in dentistry aims at using these materials and theories to enhance prevention, diagnosis and treatment of injured tissues at the molecular and micro molecular levels. The high surface area to volume ratio of these materials has reportedly shown to improve the biomaterial-biological interactions (Yong, 2014). Their ability to penetrate inside locations that are inaccessible make them best suited for various site-specific delivery (Cristina Buzea & Kevin, 2007). Moreover, their stability during storage as well as in biological fluids makes them advantageous in many ways. The shape of these materials enhances their surface reactivity and renders high antibacterial action in comparison to other formulations (Khodashenas & Ghorbani). Although still evolving with their developments at infancy, these fields provide an array of possibilities that benefit the patient's health by eliminating or reducing pain associated with conventional procedures.

The use of polymeric micro-and nanoparticles has been the epicenter of extensive research in the recent years. Polymers are high performance materials predominantly used in controlled drug release due to their versatility and unique physical, chemical and structural properties that hold tremendous promise for improvements in many areas of scientific research. The use of polymers plays a major role in improving the mucoadhesive properties of nano and micro biomaterials because this improvises the interaction of polymeric particles with biological systems, especially tooth

Figure 1. (A) Diagram shows the presence of NPs (isolated particles or agglomerates) in saliva and the structure of dental tissues. The pellicle covers the superficial layer of enamel, and the oral biofilm develops on the pellicle surface. The characteristic hexagonal shape of the enamel crystallites is apparent and also the presence of the dentinal tubules in the underlying tissue of dentine. The NP–ion–protein complexes do not adhere directly to the tooth surfaces, but adhesion occurs either to the pellicle layer or the developing biofilm. (B) Schematic diagram of the oral environment, oral biofilm, and dental mineralized tissues showing the distribution of NPs and ions. Natural saliva normally contains a range of ions and proteins. In the presence of NPs, NP–ion–protein complexes are formed. Oral conditions promote particle agglomeration that results in particle sedimentation onto the dental surfaces. The pellicle has a globular structure and its proteinaceous layer facilitates the adherence of the early colonizing species necessary for the oral biofilm development. The oral biofilm and pellicle act as diffusion/permeation barriers to NPs preventing them from reaching the enamel–pellicle interface. Certain ions (F–, Cl–, SiO44–, Zn2+) are more abundant near the external surface of enamel, while others (Na+, Mg2+, CO32–) are found at higher concentrations near the dentino–enamel junction. The most commonly ions found in dentine are F–, Na+, Mg2+, and CO32–.

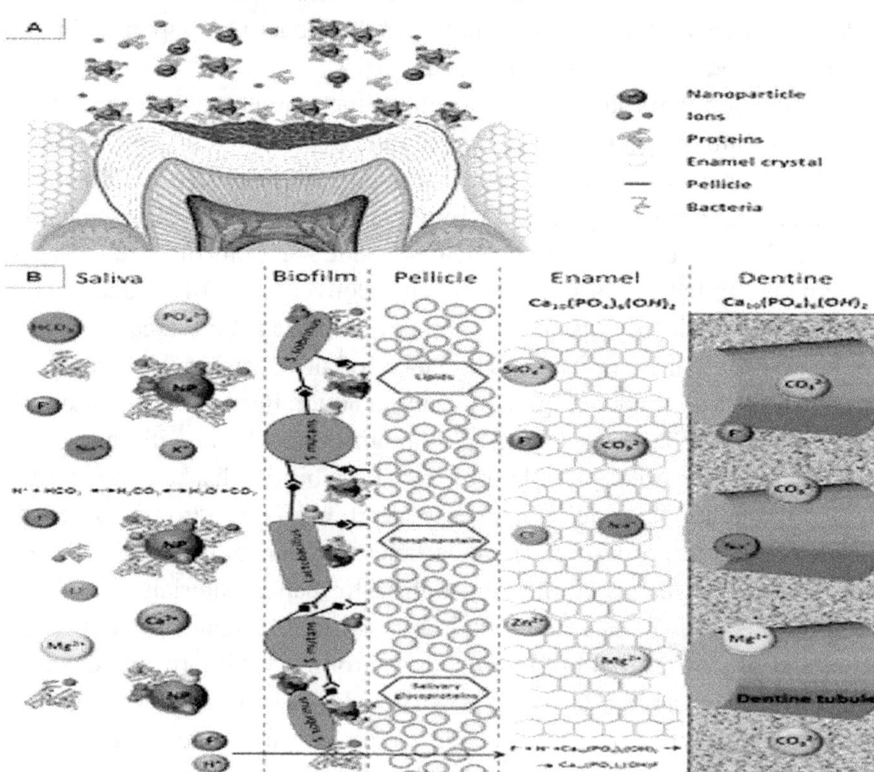

structures that are predominantly hard tissues, where it is difficult for particles to adhere. For example, both natural (e.g., chitosan and alginate) and synthetic (e.g., poly(lactic-co-glycolic acid) and poly(ε-caprolactone)) polymers have been used as carrier materials, that can be either biodegradable (e.g., poly(lactic-co-glycolic acid) or non-biodegradable (e.g., ethylcellulose) based on their ability to breakdown in a given environment.

Polymer nano and microparticles are most commonly synthesized by various methods shown in Figure 2. Other innovative procedures like dripping technique, microfluidic systems and high pressure homogenization have also been used (de Francisco, Cerquetani, & Bruschi, 2012). The dramatic impact of these materials on medicine and biology has led to applications including coatings on medical devices, wound dressing materials, antimicrobial agents, drug or gene delivery systems and bio imaging (Ambrosio & Payne, 2013; Hutter & Maysinger, 2013; Rai, Yadav, & Gade, 2009; Tsukada, Goan, & Furusawa, 2008; S. Z. Wu, Hossainy, Harish, Sanders-Millare, & Mirzaee, 2003; Zheng, Chen, Jin, Ye, & Liu, 2016). With such convincing developments in technology, micro-and nanoparticles have been anticipated to bring advances (Figure 3) and possible innovations in existing oral diagnostic and therapeutic techniques (Ozak & Ozkan, 2013). Some of the clinical aspects mentioned above (e.g., antibacterial, drug delivery) find relevance in the field of dentistry and has received significant momentum in the past few years. In this book chapter, we will mainly focus on the applications of polymer micro- and nanoparticles in dentistry.

Figure 2. Various methods to synthesis micro and nanoparticles

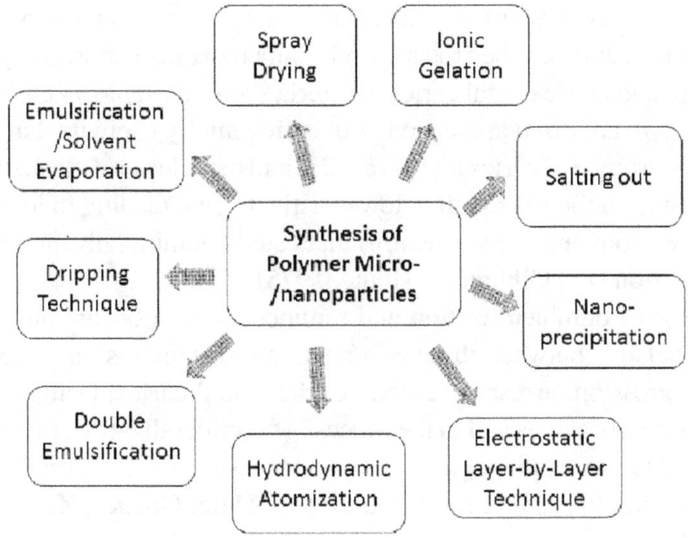

Figure 3. Different ways micro and nanoparticle can be applied to dentistry

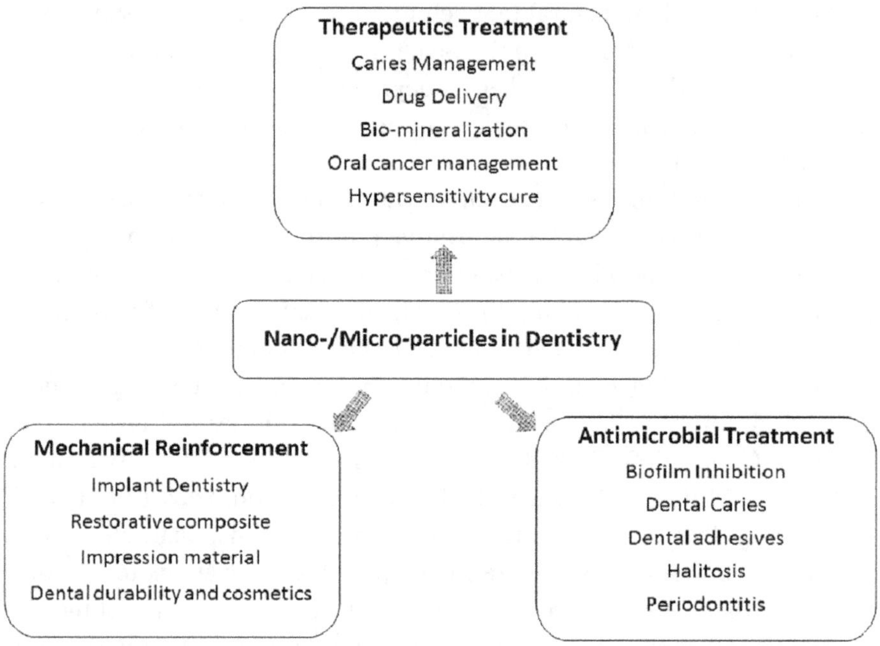

Caries Control

The microbiota present in the oral cavity is the most complex part of the human system comprising over 300 species of different bacterial, fungal and viral species (Listgarten, 1987). Most of these species associate to form biofilms that are resistant to antibiotic exposure or mechanical stress, depending on their toxicity (Avila, Ojcius, & Yilmaz, 2009). When the acid-releasing biofilms mature, they destroy the tooth structure and cause dental caries. Bacteria such as *Streptococcus mutans*, and *Lactobacillus spp* are often associated with cariogenicity (Selwitz, Ismail, & Pitts, 2007). The occurrence of caries is site-specific and the release of weak organic acids influences a drop in the pH to fall below a critical value leading to loss of calcium and phosphate (from enamel and dentin) that causes demineralization of tooth and eventual cavitation (R J Gibbons & Houte, 1975).

The process of demineralization and remineralization occurs on a daily basis and it is the balance between these two phenomena which is an important determinant of progression, arrest or reversal of the dental caries. Demineralization is reversible in its early stages and in the process of remineralization, the lost minerals can be restored to the tooth by gingival fluid or by salivary calcium and phosphate that buffers the toothbrush removal of biofilms (Melo, Guedes, Xu, & Rodrigues,

2013; Sharma, Puranik, & K.R, 2015). Various remineralization strategies have been proposed earlier focusing on restoring the mineral content of enamel and/or dentin, but the extent to which the remineralized dentin could be structurally and mechanically reconstituted remains a challenge. Although engineered nanoparticles of calcium fluoride, hydroxyapatite, bioactive glass and phosphates have frequently reported to replenish the mineral content of tooth, various studies have also utilized biodegradable polymers in order to achieve the similar effect.

The use of fluoride in inhibition of demineralization and enhancement of dental tissue remineralization has been widely acknowledged (Jorgensen, Shariati, Shields, Durr, & Proskin, 1989). Significant advancements have been made in controlling the occurrence of dental caries with the use of anti-caries fluoride such as sodium fluoride, sodium monofluorophosphate, amine fluoride and stannous fluoride (Campus, Lallai, & Carboni, 2003; Diarra, Pourroy, Boymond, & Muster, 2003; Van Rijkom, Truin, & Van't Hof, 1998; van Strijp, Buijs, & Ten Cate, 1999). The use of fluoride alone only reduces the caries progression but does not delay its development. Drawbacks such as short-term fluoride retention on the tooth surface might restrict fluoride penetration of residual plaque deposits and limits its anti-plaque activity in inaccessible areas (Watson et al., 2005). Moreover, repeated intakes of agents containing low fluoride ion concentrations might induce cytotoxicity over time. Therefore, the need for a pharmaceutical form that can improve retention and deliver sustained levels of fluoride arises (Scheifele, Studen-Pavlovich, & Markovic, 2002). The continuous release of sodium fluoride from biodegradable micron-sized polymeric carriers such as gelatin microspheres (NaF-GMS) and chitosan microparticles have shown potential anti-caries activity (Keegan et al., 2012; H. Wu et al., 2004) as indicated in Table 1. Their low molecular weight (MW) with faster rates of degradation has demonstrated efficacy against dental plaque biofilms (Chávez de Paz, Resin, Howard, Sutherland, & Wejse, 2011). Non-biodegradable polymers such as ethylcellulose have also been used to entrap various sources of fluoride such as sodium fluoride (NaF), sodium monofluorophosphate (MFP) and amino fluoride that provide a platform for inhibition of caries progression (de Francisco et al., 2012). In vivo studies on rabbit oral cavity showed that the prolonged retention at the site of dental plaque has contributed to its potential to restrain demineralization even at low concentrations (H. Wu et al., 2004). The bioadhesive property of gelatin and chitosan microparticles previously reported as a crucial factor for prolonging the retention time in mucosal tissues, might serve as an important clinical advantage (Kawaguchi, 2000; J. J. Wang et al., 2011). Farnesol-loaded nanoparticles have shown to disrupt *Streptococcus mutans* 4 fold more effectively than free Farnesol. The release of Farnesol is pH defendant as nanoparticles undergo core destabilization at acidic pH (Figure 4) (Horev et al., 2015).

Figure 4. Structure and function of nanoparticles and properties of used polymers. (A) Depiction of the chemistry and self-assembly of diblock copolymers. Cationic and pH-responsive ~20 kDa diblock copolymers with equivalent first- to second-block molecular weights and PDI of 1.1 were synthesized by two-step RAFT polymerizations, as indicated, and self-assembled into micelle-based nanoparticles in aqueous solutions via sonication. (B) Structures of control polymers utilized to isolate required physicochemical characteristics for binding to dental surfaces. (C) Proposed mode of action of pH-responsive nanoparticles for prevention and/or treatment of biofilms. RAFT is reversible addition–fragmentation chain transfer polymerization. PDI is polydispersity index; ECT is the chain transfer agent (CTA), 4-cyano-4-[(ethylsulfanylthiocarbonyl)sulfanyl]pentanoic acid; AIBN is the initiator, 2,2-azobis(isobutyronitrile); DMF is dimethylformamide; DP is degree of polymerization

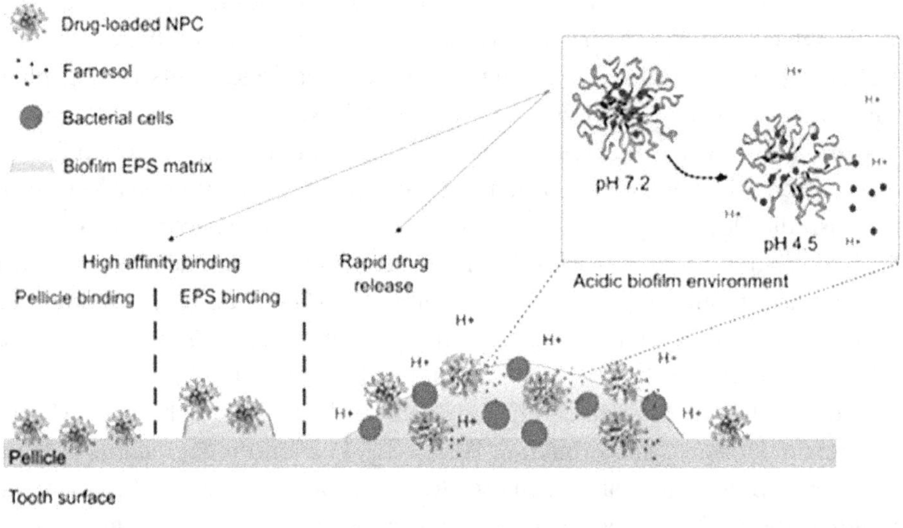

With the recognition of these agents as effective antibacterial and antibiofilm agents from the province of research workers to clinicians, newer approaches with a potential to eliminate bacteria and supersede these agents were investigated. In such context, the use of light as an effective antibacterial agent came under scrutiny (Wilson, 1993). Photodynamic therapy (PDT) is a therapeutic process that combines the effect of light and photosensitive agents called photosensitizers and renders an antibacterial effect (Longo et al., 2012). Nano and micro-sized particles functionalized with various photoactive compounds have shown to combat the

Table 1. Use of polymer microparticles in dentistry

Application	Polymer/Drug	Microparticle Characteristics	Key Findings	Citations
Cariostatic studies	Gelatin/NaF	Double-phase emulsified condensation polymerization Average dia: 11.33±5.56 µm NaF content: 8.80% Encapsulation efficiency: 76.73%	Gelatin/NaF prolongs fluoride release, delays demineralization and improves fluoride retention in dental plaque	(H. Wu et al., 2004)
Controlled delivery of fluoride	Chitosan/NaF	Spray drying Average dia: <6 µm NaF loading: 14% Encapsulation efficiency: 84.1%	Chitosan/NaF increases fluoride uptake that provides enhanced protection against dental caries	(Keegan et al., 2012)
Controlled delivery of fluoride	Ethylcellulose/ NaF, MFP, AF	Emulsification/solvent evaporation Average dia: <145.35 µm Encapsulation efficiency: 137.45±0.00	Microparticles of Ethylcellulose and gelatin deliver fluoride ions in a controlled fashion that enhanced the protection against caries.	(de Francisco et al., 2012)
	Gelatin/NaF, MFP, AF	Dripping technique Average dia: <124.22 µm Encapsulation efficiency: 79.79±0.68%		
Dental caries vaccination	Chitosan-coated PLGA/rGbpD	Water/oil/water emulsification Average dia:<2 µm Loading efficiency: 1.21%	Chitosan coating of PLGA/rGbpD enhanced protein loading and intranasal immunization demonstrated an effective vaccine delivery system	(Zhao et al., 2006)
Periodontal MIN delivery	PLGA/MIN (ARESTIN)	Average dia: 20 to 60 µm Arestin™ exhibits excellent adherence to the periodontal pocket and releases therapeutic doses of MIN for over 14 days	ARESTIN has proved to be a safe and efficient adjunct to scaling and root planing in the treatment of chronic periodontitis	(Gopinath et al., 2009)
Prolonged release of MIN	Alginate/MIN	Ionic gelation Average dia: 100µm Encapsulation efficiency: 74%	Alginate/MIN as an effective therapeutic modality for periodontitis	(Park et al., 2005)
Controlled delivery of DOX	PLGA-PCL blend/ DOX	Water/oil/water double emulsification Average dia: 90 to 200 µm Encapsulation efficiency: 9.00±0.20 to 24.7±0.53%	PLGA-PCL blend/ DOX can be effectively used in pocket therapy and treatment of refractory periodontal sites	(Mundargi et al., 2007)
Controlled delivery of Tc	Alginate-coated PLGA/ Tetracycline	Double emulsion/solvent evaporation Average dia: 102.4±3.4 µm Encapsulation efficiency: 17.9±0.5%	Enhanced delivery of Tetracycline to periodontal pockets	(Liu et al., 2005)
Treatment of Periodontal Disease	Gelatin/Propolis (PM)	Spray drying Average dia: 2.48±0.81µm Drug loading: 6.41%	The rheological properties of Gelatin/Propolis have shown to be conducive for easy penetration into periodontal pockets and thereby its retention	(Bruschi et al., 2008)

continued on following page

Table 1. Continued

Application	Polymer/Drug	Microparticle Characteristics	Key Findings	Citations
Periodontal regeneration	PLGA-PDLLA/ PDGF and Simvastatin	Co-axial electrohydrodynamic atomization Average dia: ~15μm Encapsulation efficiency: Core-81.3±4.5 and Shell-72.6±2.4	Subsequent PGDF and Simvastatin delivery accelerates osteogenesis and bone maturation	(Chang et al., 2013)
Root canal disinfection	PLGA-zein/AMX	Spray drying Average dia: 5-38 μm Drug Loading: 2.1±0.8% to 5.8±0.8% Encapsulation efficiency: 38-51%	PLGA-zein/AMX demonstrated use as an effective antibacterial vehicles and showed efficacy against *E.faecalis*	(Sousa et al., 2010)
Osteogenic differentiation	PLGA/Estrogen	Emulsion/solvent evaporation Average dia: ~1.8μm Zeta Potential: +21±3.42 mV Encapsulation efficiency: 87.381%_2.97%	Intracellular delivery of estrogen to for effective regulation of MSCs	(Hong et al., 2011)
Bone defects	PLGA/ Simvastatin	Emulsion/solvent evaporation Average dia: 21.07μm Drug loading: 91 ± 3.78%	Microspheres of PLGA/ Simvastatin can enhance bone formation in the defective areas of the rabbit calvaria	(Naito et al., 2014)
Osseointegration of dental implants	PLGA/rrIGF-1	Double emulsion/solvent evaporation Average dia: 1.25 ± 0.64μm Encapsulation efficiency: 78.3±2%	Osseointegration of dental implants in type 2 diabetes during the early post implantation healing phase and prolonged release of IGF-1 has resulted in new bone formation around the implants	(F. Wang et al., 2010)
Osseointegration of dental implants	PLGA/bFGF	Double emulsion/solvent evaporation Average dia: 1.138± 0.304μm	Early osseointegration observed in diabetic rats. Potential to stimulate bone formation in human patients.	(Zou et al., 2012)
Local delivery of dexamethasone (DEX)	PLGA/DEX	Emulsion/solvent evaporation Average dia: 0.8 to 4μm Zeta Potential: -24±2.1mV	PLGA/DEX showed osteoconductivity and proved to be an excellent platform for bone regeneration	(Son et al., 2013)
Implantation in Osteolysis	Chitosan/AS	Emulsion polymerization Average dia: 110.28±0.4μm Loading efficacy: 3.30%	Local delivery of these microspheres as implants may inhibit bone resorption during dental and orthopedic procedures	(Samdancioglu et al., 2006)
	PLGA/AS	Emulsion/solvent evaporation Average dia: 47.78± μm Loading efficacy: 7.70%		

antimicrobial resistance and enhance antimicrobial PDT (a-PDT). Since dental caries are localized, the limitations of traditional dental treatments have prompted the use of PDT as an alternative protocol (Banfi et al., 2006).

Chitosan is a natural polymer obtained from the exoskeleton of crustaceans and often used in drug delivery. Annie Shrestha et al has shown that the dual property of photoactivable chitosan nanoparticles surface functionalized with rose-bengal (CSRBnp) has resulted in the targeted elimination of *E. faecalis* biofilms and simultaneous restoration of the structural as well as mechanical integrity of the infected root dentin matrix (Annie Shrestha, Hamblin, & Kishen, 2014). The same group of researchers has demonstrated diminished antibacterial activity of chitosan nanoparticulates (CS-NP)-modified root canal sealers in comparison to that incorporated with nanoparticles of zinc oxide (ZnO-NP). The limited release of CS-NP from the sealer and the variation in their bactericidal mechanisms could have attributed to this difference (Anil Kishen, Shi, Shrestha, & Neoh, 2008). Moreover, CS-NP require longer interaction time for complete elimination of biofilms and their penetration might be restricted by the extracellular polymeric substance surrounding the biofilm mass (Upadya, Shrestha, & Kishen, 2011). The involvement of FDA-approved synthetic polymers like poly(D,L-lactide-co-glycolide) (PLGA) in dentistry have been well recognized (Maria Justina Roxana Virlan & Bogdan, 2015). Studies have shown that the encapsulation of active compounds like photosensitizers (methylene blue) and recombinant proteins in PLGA micro and nano-sized particles have improved the anti-planktonic as well as anti-biofilm activities of these delivery carriers exhibiting enhanced bacterial uptake and phototoxicity (Klepac-Ceraj et al., 2011; Zhao et al., 2006). The potential of these nanocarriers to diffuse inside biofilms and release the drug in an active form is an important criteria allowing the encapsulation phenomena to be extended for various bacteriostatic and bactericidal agents with promising prospects in vaccine delivery (Zhao et al., 2006). Polymers modified with quaternary ammonium salts have been frequently used in dentistry along with the use of composite resins (He, Söderling, Vallittu, & Lassila, 2013; Sun et al., 2011). Quaternary ammonium-loaded polyethylenimine (QAS-PEI) nanoparticles incorporated in dental composites were effective against *S. mutans* for over a month without altering the original mechanical properties of the composite (Nurit Beyth, Yudovin-Farber, Bahir, Domb, & Weiss, 2006; N. Beyth, Yudovin-Farber, Perez-Davidi, Domb, & Weiss, 2010).

Periodontics

Periodontal diseases (PD) or Periodontitis is a collective term that refers to a combination of multiple inflammatory diseases leading to slow, irreversible damage that affect the supportive structures surrounding the teeth such as gum, alveolar bone and the periodontal ligament (Armitage, 1999; Socransky & Haffajee, 1994). Different forms of periodontitis such as gingivitis, chronic or aggressive periodontitis prevail, based on the differences in clinical diagnosis (AlJehani, 2014). Bacterial adhesion and invasion of the gingival epithelium lays the foundation for pathogenesis of periodontitis (Colombo, Silva, Haffajee, & Colombo, 2006). The proximity of the subgingival biofilms to the underlying epithelial cells determines the extent to which the periodontal attachment may get affected, in order to illicit an inflammatory response (Lamont & Yilmaz, 2002). Periodontitis is often characterized by formation of periodontal pockets, resorption of alveolar bone and loss of periodontal support.

The moist, warm periodontal pockets provides a nutritious and anaerobic ambience conducive for microbial colonization and replication (Kinane, 2003; Williams, 1990). Antibiotic or antiseptic drugs loaded in biocompatible polymer micro and nano delivery systems enable the introduction of formulations directly into the root canal or inside the periodontal pocket with constant release of drugs for control of infections (Álvarez, Espinar, & Méndez, 2011). Minocycline (MIN), a tetracycline derivative, has been used in the treatment of periodontitis for a long time due to its anti-inflammatory and antibiotic activity, extended half-life, better absorption and reduced excretion rate (Ciancio, Mather, & McMullen, 1980). Considerable research has focused on using MIN loaded polymer nanoparticles for eliminating the bacteria associated to periodontitis. Minocycline-loaded PLGA microspheres (MIN-PLGA NPs), in the commercially successful form known as ARESTIN®, has demonstrated effective killing of the periodontal pathogens with their capacity to deliver MIN to periodontal tissues (Goodson et al., 2007; Kashi et al., 2012) as mentioned in Table 1. It has also been used as part of various oral health program including proper brushing, flossing and as an adjunct in scaling and root planing (Oringer et al., 2002). Their clinical efficacy and success in smokers with chronic periodontitis have also been discussed (Gopinath et al., 2009; Paquette et al., 2003). Moreover, the elimination of *Porphyromonas gingivalis* from the periodontal pocket with a potential to improve the clinical probable pocket depth, gingival and plaque index, in combination with surgery have also been proven (Hellstrom et al., 2008). Sustained release of MIN from alginate-core and chitosan-coated microcapsules have shown effective microfloral suppression for 1 week with their potential as an effective therapeutic modality (Park et al., 2005). An in vivo study conducted in beagle dogs has shown better anti-periodontitis activity of MIN-loaded PEG-PLA nanoparticles (MIN-NPs) when compared to that of commercial Periocline and MIN solution (Yao et al., 2014). Srirangarajan and workers have proved that the DOX-loaded PLGA/PCL microspheres can be in the treatment of refractory periodontal disease (Mundargi et al., 2007; Srirangarajan et al., 2011). Alginate coating of tetracycline hydrochloride-releasing PLGA microspheres have shown to enhance drug encapsulation and subsequent release following delivery in the periodontal pockets

(Liu et al., 2005). PLGA microparticles containing chlorhexidine free base (Chx), chlorhexidine di gluconate (Chx-Dg) were effective against periodontal infections and delivered chlorhexidine salts in a targeted fashion (Yue et al., 2004). Hydroxyapatite and ofloxacin-loaded PLGA microspheres were able to eliminate *S. aureus* and *E. coli* (Jamal et al., 2012).

Natural polymers like gelatin have also been used in the treatment of periodontitis because they are mucoadhesive and are used for encapsulation of drugs with high toxicity or a short half-life (Gould, Westbury, & Tillman, 1982). A liquid crystalline phase modified with propolis-incorporated gelatin nanoparticles have shown stable retention in the periodontal pocket and their eventual controlled release thereof (Bruschi et al., 2008). Moulari and workers used flavonoid-rich Harungana madagascariensis leaf extract (HLE) and showed that its encapsulation in PLGA nanoparticles has not only increased the antibacterial efficiency but also implicated its potential use as an anticaries agent and in the prevention of gingivitis (Moulari et al., 2006). Rapid release of Triclosan (2,4,4′-trichloro-hydroxydiphenylether), an antimicrobial agent with anti-plaque properties, from PLGA nanoparticles were effective against induced periodontal defects and reduced the gingival inflammation with an ability to penetrate through the junctional epithelium (Pinon-Segundo, Ganem-Quintanar, Alonso-Perez, & Quintanar-Guerrero, 2005). When the periodontal defects in rat maxillae were filled with PDLLA-PLGA microparticles containing growth (PDGF, mitogen) and differentiation factors (simvastatin), acceleration of periodontal regeneration has been observed (Chang et al., 2013). Chitosan-based microspheres and nanoparticles have been used as a local therapeutic in periodontal diseases (Dung et al., 2007) and in vivo rat studies indicate enhanced implant absorption with potential applications in wound healing (L. C. Wang et al., 2008). Thrombin, an anticoagulant, when released in a sustained fashion, acts as a local bioactive coagulation device for use during anticoagulation therapy and periodontal surgical procedures (Smeets et al., 2011).

Endodontic Therapy

Endodontic disease is often biofilm-mediated and primarily aims at the elimination of root canal microorganisms by cleaning and sterilizing the root canals with irrigant antimicrobial agents followed by filling with gutta percha (Jaberi-Ansari, Ekrami, & Nojehdehian, 2013; Jhajharia, Parolia, Shetty, & Mehta, 2015). The presence of resistant microbes such as *Enterococcus faecalis, Staphylococcus aureus, Pseudomonas aeruginosa, Bacillus subtilis, Streptococcus spp., and Candida spp.* in the root canal is of particular concern (Molander, Reit, & Dahlén, 1999; Pinheiro et al., 2003; Portenier, Waltimo, & Haapasalo, 2003; Saleh, Ruyter, Haapasalo, & Ørstavik, 2004; Stuart, Schwartz, Beeson, & Owatz, 2006). The bacterial load present in the complex anatomy of root canal poses an important challenge for penetration of antibiotic medicaments, eventually causing endodontic failure (Graves, 2008; Lin, Skribner, & Gaengler, 1992). The difficulty in killing these persistent bacteria has led to the development of improvised formulations capable of prolonged release of antibiotics within the canal. Various studies have shown to benefit from using chitosan-based nanoparticles in the prevention and neutralization of root canal bacterial recolonization and biofilm formation (Anil Kishen et al., 2008; A. Kishen & Shrestha, 2013; Annie Shrestha, Fong, Khoo, & Kishen, 2009) as shown in Table 2. Their penetration inside dentinal tubules have been significantly enhanced by high intensity focused ultrasound and aided in the biofilm removal (Annie Shrestha et al., 2009). It has also shown to enhance the mechanical properties, improve the chemical stability of the connective tissues and/or the surface/interfacial integrity between filling materials and connective tissue when used as a carrier for photo-activatable antibacterial/antibiofilm compounds (A. Kishen & Shrestha, 2013). When used in combination with cationic ZnO nanoparticles, chitosan nanoparticles exhibit enhanced antibacterial action (Anil Kishen et al., 2008). PLGA nanoparticles loaded with endodontic antimicrobials have also shown potential activity against *E. faecalis* biofilms in the experimentally infected root canal (Pagonis et al., 2010; Sousa, Luzardo-Alvarez, Perez-Estevez, Seoane-Prado, & Blanco-Mendez, 2010). Polyethylene glycol coated maghemite nanoparticles have been used for dentinal tubule occlusion as a long-lasting remedy for dental hypersensitivity (Dabbagh et al., 2014).

Dentin Tissue Stabilization

The scope of micro and nanomaterials to impart the effect on biofilm elimination, hinder demineralization process and stimulate the process of tooth structure remineralization in order to restore the stability of previously affected or infected dentin matrix is promising. Nanoparticles made of natural bioactive polymers such as chitosan has shown the potential to intensify the mechanical and structural properties of dentin collagen. These nanoparticles are easy to functionalize due to the presence of highly reactive amino and hydroxyl groups on the surface. Cross-linking dentin with various agents has shown a considerable increase in the tensile properties, hardness, elastic modulus and toughness of demineralized dentin (Bedran-Russo, Pashley, Agee, Drummond, & Miescke, 2007; Bedran-Russo, Pereira, Duarte, Drummond, & Yamauchi, 2006; Bedran-Russo, Yoo, Ema, & Pashley, 2009; U. Daood, Iqbal, Nitisusanta, & Fawzy, 2012; Umer Daood, Swee Heng, Neo Chiew Lian, & Fawzy, 2015; Fawzy, Nitisusanta, Iqbal, Daood, Beng, et al., 2012; Fawzy, Nitisusanta, Iqbal, Daood, & Neo, 2012; A. Shrestha, Friedman, & Kishen, 2011; Annie Shrestha et al., 2014). Chitosan nanoparticles and their derivatives have shown

Table 2. Use of polymer nanoparticles in dentistry

Application	Polymer/ Drug	Nanoparticle Characteristics	Key Findings	Citations
Antimicrobial and antibiofilm activity	Chitosan	Ionic gelation Average dia: 20 to 1000 nm	LMW chitosan nanoparticles show uniform delivery of chitosan, affect the cell membrane integrity of entire *S. mutans* biofilm and therefore exhibits antibiofilm and antimicrobial activity	(Chávez de Paz et al., 2011)
Antibacterial/ biofilm activity	Chitosan/ Rose Bengal (CSRBnp)	Ionic gelation Average dia: 60 ± 20 nm Zeta potential: +30 ± 0.8 mV	CSRBnp eliminates biofilm formation and restore mechanical integrity of infected dentin tissue	(Annie Shrestha et al., 2014)
Antibacterial/ biofilm activity	Chitosan (CS-NP)	Electrostatic layer-by-layer technique Average dia: ~70 nm Zeta potential: + 49 mV	Nanoparticles-treated dentin showed significant reduction in Enterococcus faecalis exhibiting a potential for root canal disinfection	(Anil Kishen et al., 2008)
		Electrostatic layer-by-layer technique Average dia: <100 nm		(Annie Shrestha et al., 2009)
Anti-plaque activity	PLGA/MB	Solvent displacement Average dia: <220nm Zeta Potential: -38.3 and +43.5 mV for anionic and cationic nanoparticles respectively	Cationic PLGA/ MB nanoparticles in combination with PDT demonstrate effective elimination of biofilms	(Klepac-Ceraj et al., 2011)
Antibacterial nanoparticles incorporated in dental composites	Quatnery ammonium PEI	Crosslinking, alkylation and quaternarization Average dia: <100nm	Composite resin restorations when incorporated with antibacterial nanoparticles have potential in eliminating biofilm formation and secondary caries.	(Nurit Beyth et al., 2006)
Periodontal Minocycline delivery	PLGA/MIN	Emulsification, nanoprecipitation Average dia: < 500 nm Drug Loading: 0.26 ± 0.06% to 1.92 ± 0.19% Encapsulation efficiency: 0.81 ± 0.18% to 29.95 ± 2.47%	PLGA/MIN synthesized by ion pairing method showed highest drug loading and higher antibacterial activity against *Aggregatibacter actinomycetemcomitans*	(Kashi et al., 2012)
Local delivery of MIN	MPEG-PLA/ MIN	Emulsion/solvent evaporation Average dia: 98±12 nm Drug Loading: 1.92%±0.19% Encapsulation efficiency: 29.95%±2.47%	MPEG-PLA/MIN increased periodontal retention of MIN and showed anti-periodontitis effects	(Yao et al., 2014)

continued on following page

Table 2. Continued

Application	Polymer/ Drug	Nanoparticle Characteristics	Key Findings	Citations
Antibacterial activity	PLGA/HSE	Interfacial polymer deposition/ solvent diffusion Average dia: 255.09±31.1nm Encapsulation efficiency: 75% Zeta Potential: -31.1±0.4mV	PLGA/HSE demonstrated bactericidal activity against cariogenic bacteria and could be an effective anticaries agent	(Moulari et al., 2006)
Periodontal treatment	PLGA/ Triclosan	Emulsification diffusion Average dia: <500nm Drug Loading: 0.84 ± 0.06% to 23.75 ± 0.06% Encapsulation efficiency: 63.88% to 85.12%	Preliminary in vivo study has shown that PLGA/Triclosan can reduce gingival inflammation but further experiments have to be conducted to confirm these results for periodontal treatment	(Pinon-Segundo et al., 2005)
Antimicrobial	PLGA/MB	Solvent displacement Average dia: ~150-200 nm Zeta Potential: -31.87 mV Bacterial cell wall toxicity: 34% to 58.5%	Synergistic action of light and PLGA/MB reduced the colony-forming units in planktonic phase and root canals to 2 and 1 log10, respectively	(Pagonis et al., 2010)
Treatment of dental hypersensitivity	PEG/ Maghemite	"Graft-to" method Average dia: <25nm Dentin treatment of nanoparticles for 120min showed high occlusion of tubules	PEG/Maghemite can be used as potential biomedical carriers for dentin remineralization	(Dabbagh et al., 2014)
Dentin Adhesion	PolymP-nActive	Commercial PolymP-nActive from NanoMyP (Granada, Spain) Average dia: 350nm	Nanocarriers potentially inhibits MMP collagen degradation and enhances mineral growth	(Osorio et al., 2014)
Dentin Adhesion	Polymeric nanogels	Average dia: 10- to 100-nm	Nanogel-reinforcement of adhesives significantly improved the wet mechanical properties	(Morães et al., 2012)
Enhancing osteogenesis	BPPD/ simvastatin	Average dia: 231.5 nm, Zeta potential: -6.7±2.8mV Encapsulation efficiency: 42.3 ± 5.5%	BPPD/Simvastatin can be used as potential hydrophobic drug carriers to promote osteogenesis	(C.-Z. Wang et al., 2014)

to interact with endogenous MMPs and/or bacterial collagenase with a potential to neutralize them, inhibit collagen degradation and enhance the biodegradation resistance of dentin (M. M. Kim & Kim, 2006; Persadmehr et al., 2014). The inclusion of zinc-loaded polymeric nanocarriers into a dental adhesive system has shown to reduce the MMPs activity without affecting bond strength (Osorio, Osorio, Medina-Castillo, & Toledano, 2014). The use of poly (methylmethacrylate)-grafted nanoclay (PMMA-g-nanoclay) as a successful reinforcing filler has been investigated and shown to result in higher shear bond strength (Atai et al., 2009; Solhi, Atai, Nodehi, & Imani, 2012; Solhi, Atai, Nodehi, Imani, et al., 2012). Polymeric nanogels when reinforced in the experimental adhesives improve the short-term micro-tensile bond strength (Morães et al., 2012). Collagen fibrils present in dentin are the principal tensile stress–bearers and contribute a major part to the mechanical properties. Intermolecular cross-linking of collagen fibrils is one of the most commonly put forth mechanisms for developing resistance to proteolytic cleavage for improving the mechanical stability of dentin (Nakabayashi, 1982; N. Nakabayashi, Kojima, & Masuhara, 1982). The inclusion of water-soluble chitosan in the cross-linking procedure helped enhance the mechanical properties (A. Shrestha et al., 2011), use of Rose bengal-functionalized chitosan nanoparticles has achieved photodynamic cross-linking in a single step (Annie Shrestha et al., 2014). Bioactive polymeric nanoparticles disinfect the dentin with simultaneous tissue stabilization.

Regenerative Dentistry

Regenerative dentistry focus on tissue regeneration to compensate for lost tissues rather than replacement which is more beneficial and involves biology-based treatments (Galler & D'Souza, 2010; Rosa, Bona, Cavalcanti, & Nör, 2012). It is a well-known fact that small bone defects can spontaneously regenerate to their default anatomic configuration, however, regeneration of defects that exceed a certain size and degree of its severity, requires external augmentation using bone substitutes to accomplish this task (Schwartz-Arad & Chaushu, 1997; Zeren, 2006). For enhancement of bone regeneration, many research groups have attempted to incorporate bioactive agents and drugs into the bone substitutes (Boden, Martin Jr, Morone, Ugbo, & Moskovitz, 1999; Ginebra, Traykova, & Planell, 2006). In order to control the release profiles, studies have shown the encapsulation of these compounds in biodegradable polymers, attempting to prolong their pharmacological activity. PLGA microparticles are utilized frequently in this area of research as shown in Table 1. PDLLA-PLGA microparticles containing growth and differentiation factors induce osteogenesis, bone maturation and periodontal regeneration in rat maxillae (Chang et al., 2013). Estrogen-containing PLGA microparticles upregulates osteogenic differentiation of human bone marrow mesenchymal stromal cells (MSCs) (Hong et al., 2011). PLGA particles loaded with simvastatin (Naito et al., 2014; C.-Z. Wang et al., 2014), growth factors (F. Wang et al., 2010; Yonamine et al., 2010; Zou et al., 2012) and dexamethasone (Piao et al., 2014; Son et al., 2013) have significantly improved bone formation. Orchestrating the release of growth factors have shown improved osteointegration to titanium implants (F. Wang et al., 2010; Zou et al., 2012). Another study has shown that local delivery of biphosphonate PLGA microspheres as implants that can be an adjunct in the treatment of bone loss in the area of dental prostheses (Samdancioglu, Calis, Sumnu, & Atilla Hincal, 2006). Fulvastatin-releasing injectable PLGA microspheres prevents bone resorption during dental procedures and increases the mechanical stability of bone (Masuzaki et al., 2010) whereas insulin-containing PLGA microparticles improved the biomechanical retention of implants (Han et al., 2012). Arestin® and MIN-loaded PLGA have been used as an adjunct to mechanical methods for treatment of human peri-implantitis (Persson, Salvi, Heitz-Mayfield, & Lang, 2006; Stefan Renvert, Lessem, Dahlén, Lindahl, & Svensson, 2006; S. Renvert, Lessem, Dahlen, Renvert, & Lindahl, 2008). Heparin-conjugated PLGA nanoparticles loaded with bone morphogenetic protein-2 (BMP-2) has shown to regenerate bone in vivo (S. E. Kim et al., 2008). Collagen-conjugated polycaprolactone microparticles have shown controlled delivery of DOX as well as a potential to induce cell proliferation and tissue regeneration in the periodontal pockets (Aishwarya, Mahalakshmi, & Sehgal, 2008). Since regenerative dentistry is at its infancy, these procedures have great potential and applications in the future.

Prosthodontics

Prosthodontics is the branch of oral medicine that mainly deals with dental defects, treatments following tooth loss, such as crowns, and dentures, including the use of artificial prostheses for periodontal disease, maxillofacial tissue defects and mandibular joint disease (Mehra, Vahidi, & Berg, 2013; Saavedra et al., 2007). Dentures restore the dental functions and facial appearance of the wearer. Therefore, their composition and performance determines the long-term stability of these prostheses (Budtz-Jörgensen, 1996). Although polymethyl methacrylate (PMMA) has been used in the fabrication of denture teeth for various advantages such as lower susceptibility to fracture, chemical bond to denture base and reduced clicking (Adams, Jooste, Thomas, & Harris, 1996; Beall, 1943; Kawano, Ohguri, Ichikawa, Mizuno, & Hasegawa, 2002; Reis et al., 2008), its wear resistance when compared to ceramic teeth has been questioned (Ogle, David, & Ortman, 1985; Whitman, McKinney, Hinman, Hesby, & Pelleu, 1987). Cooper et al. showed that a PMMA matrix modified with carbon nanotubes (CNTs) or carbon nanofibrils significantly improved the strength of the composites (Cooper, Ravich, Lips, Mayer, & Wagner, 2002). Such enhancement from using small amounts of single-wall nanotubes may be a considerable economic advantage (W. Wang et al., 2015). Since dentures come in direct contact with oral mucosa, their biocompatibility is very crucial for proper functioning. Incorporation of nanoparticles of TiO_2, Fe_2O_3 and silver in PMMA materials have shown to reduce bimolecular adherence in order to eliminate the plaque formation on PMMA based resins that eventually lead to oral cavity infections (L. S. Acosta-Torres, Mendieta, Nunez-Anita, Cajero-Juarez, & Castano, 2012; Laura S. Acosta-Torres et al., 2011; Monteiro et al., 2011). It has also shown to simultaneously improve the color of PMMA based materials (Fujishima, Rao, & Tryk, 2000; Hoffmann, Martin, Choi, & Bahnemann, 1995). Therefore, addition of metal oxides has shown to improve performance of PMMA formulations (Laura S. Acosta-Torres et al., 2011). Silver nanoparticles conjugated with polymers like PMMA and Polyamide-6 form nanocomposites that perform as effective antibacterial materials preventing the growth of infections involving the oral mucosal tissues in denture wearers (Damm, Münstedt, & Rösch, 2007; Magalh et al., 2012).

CONCLUSION

The use of adjunctive micro- and nanoparticle-based system has increased the effect of conventional dental therapies with clear clinical benefits. The preliminary preclinical studies in the various areas are rather exciting. As micro-and nanoparticle can be easily functionalized, they can be combined with biomolecules or loaded with antibiotics and used as additives in several dental materials to improve their bioactivity, physical, chemical and mechanical properties The polymeric micro- and nano-particle maintain effective concentrations of therapeutic agents to reach the site of action namely the periodontal pocket, dental cavity, or surrounding bone. These encapsulation technologies may open up new directions in the field of dentistry.

Despite the excitement created by the promising capabilities of polymeric materials it exhibits some challenges which still need to be addressed before translating into clinical reality. Firstly, toxicity of the polymeric particle induced by release of degradation products and residual surfactant used during the synthesis process needs to be addressed. The efficacy of loaded drugs and antibiotics must be systematically evaluated in animal model studies and clinical trials. In addition to biosafety, there is also the need of gaining profound understanding of the mechanisms involved in polymeric material-mediated biological behavior at both cellular and tissue levels. Although these challenges remain open, the exciting findings presented here and beyond show that polymeric particle hold great potential for dental applications.

REFERENCES

Acosta-Torres, L. S., López-Marín, L. M., Núñez-Anita, R. E., Hernández-Padrón, G., & Castaño, V. M. (2011). Biocompatible Metal-Oxide Nanoparticles: Nanotechnology Improvement of Conventional Prosthetic Acrylic Resins. *Journal of Nanomaterials*, *2011*, 8. doi:10.1155/2011/941561

Acosta-Torres, L. S., Mendieta, I., Nunez-Anita, R. E., Cajero-Juarez, M., & Castano, V. M. (2012). Cytocompatible antifungal acrylic resin containing silver nanoparticles for dentures. *International Journal of Nanomedicine*, *7*, 4777–4786. PMID:22969297

Adams, L. P., Jooste, C. H., Thomas, C. J., & Harris, A. M. (1996). Biostereometric quantification of clinical denture tooth wear. *Journal of Oral Rehabilitation*, *23*(10), 667–674. doi:10.1046/j.1365-2842.1996.d01-178.x PMID:8933383

Aishwarya, S., Mahalakshmi, S., & Sehgal, P. K. (2008). Collagen-coated polycaprolactone microparticles as a controlled drug delivery system. *Journal of Microencapsulation*, *25*(5), 298–306. doi:10.1080/02652040801972004 PMID:18465301

AlJehani, Y. A. (2014). Risk factors of periodontal disease: Review of the literature. *International Journal of Dentistry*, *2014*, 182513. doi:10.1155/2014/182513 PMID:24963294

Álvarez, A. L., Espinar, F. O., & Méndez, J. B. (2011). The Application of Microencapsulation Techniques in the Treatment of Endodontic and Periodontal Disease. *Pharmaceutics*, *3*(3), 538–571. doi:10.3390/pharmaceutics3030538 PMID:24310596

Ambrosio, A., & Payne, J. (2013). Method of promoting new tissue growth and/or wound healing using bioresorbable dressing comprising microspheres or microparticles: Google Patents.

Armitage, G. C. (1999). Development of a classification system for periodontal diseases and conditions. *Annals of Periodontology*, *4*(1), 1–6. doi:10.1902/annals.1999.4.1.1 PMID:10863370

Atai, M., Solhi, L., Nodehi, A., Mirabedini, S. M., Kasraei, S., Akbari, K., & Babanzadeh, S. (2009). PMMA-grafted nanoclay as novel filler for dental adhesives. *Dental Materials*, *25*(3), 339–347. doi:10.1016/j.dental.2008.08.005 PMID:18829096

Avila, M., Ojcius, D. M., & Yilmaz, Ö. (2009). The Oral Microbiota: Living with a Permanent Guest. *DNA and Cell Biology*, *28*(8), 405–411. doi:10.1089/dna.2009.0874 PMID:19485767

Banfi, S., Caruso, E., Buccafurni, L., Battini, V., Zazzaron, S., Barbieri, P., & Orlandi, V. (2006). Antibacterial activity of tetraaryl-porphyrin photosensitizers: An in vitro study on Gram negative and Gram positive bacteria. *Journal of Photochemistry and Photobiology. B, Biology*, *85*(1), 28–38. doi:10.1016/j.jphotobiol.2006.04.003 PMID:16737820

Beall, J. R. (1943). Wear of Acrylic Resin Teeth (Progress Report). *The Journal of the American Dental Association*, *30*(3), 252–256. doi:10.14219/jada.archive.1943.0043

Bedran-Russo, A. K., Pashley, D. H., Agee, K., Drummond, J. L., & Miescke, K. J. (2007). Changes in stiffness of demineralized dentin following application of collagen crosslinkers. *Journal of Biomedical Materials Research. Part B, Applied Biomaterials*, *86*(2), 330–334. PMID:18161815

Bedran-Russo, A. K., Pereira, P. N., Duarte, W. R., Drummond, J. L., & Yamauchi, M. (2006). Application of crosslinkers to dentin collagen enhances the ultimate tensile strength. *Journal of Biomedical Materials Research. Part B, Applied Biomaterials*, *80*(1), 268–272. PMID:16767722

Bedran-Russo, A. K., Yoo, K. J., Ema, K. C., & Pashley, D. H. (2009). Mechanical properties of tannic-acid-treated dentin matrix. *Journal of Dental Research*, *88*(9), 807–811. doi:10.1177/0022034509342556 PMID:19767576

Besinis, A., De Peralta, T., Tredwin, C. J., & Handy, R. D. (2015). Review of Nanomaterials in Dentistry: Interactions with the Oral Microenvironment, Clinical Applications, Hazards, and Benefits. *ACS Nano*, *9*(3), 2255–2289. doi:10.1021/nn505015e PMID:25625290

Beyth, N., Yudovin-Farber, I., Bahir, R., Domb, A. J., & Weiss, E. I. (2006). Antibacterial activity of dental composites containing quaternary ammonium polyethylenimine nanoparticles against Streptococcus mutans. *Biomaterials*, *27*(21), 3995–4002. doi:10.1016/j.biomaterials.2006.03.003 PMID:16564083

Beyth, N., Yudovin-Farber, I., Perez-Davidi, M., Domb, A. J., & Weiss, E. I. (2010). Polyethyleneimine nanoparticles incorporated into resin composite cause cell death and trigger biofilm stress in vivo. *Proceedings of the National Academy of Sciences of the United States of America*, *107*(51), 22038–22043. doi:10.1073/pnas.1010341107 PMID:21131569

Boden, S. D., Martin, G. J. Jr, Morone, M. A., Ugbo, J. L., & Moskovitz, P. A. (1999). Posterolateral lumbar intertransverse process spine arthrodesis with recombinant human bone morphogenetic protein 2/hydroxyapatite–tricalcium phosphate after laminectomy in the nonhuman primate. *Spine*, *24*(12), 1179–1185. doi:10.1097/00007632-199906150-00002 PMID:10382242

Bruschi, M. L., de Freitas, O., Lara, E. H., Panzeri, H., Gremiao, M. P., & Jones, D. S. (2008). Precursor system of liquid crystalline phase containing propolis microparticles for the treatment of periodontal disease: Development and characterization. *Drug Development and Industrial Pharmacy*, *34*(3), 267–278. doi:10.1080/03639040701655911 PMID:18363142

Budtz-Jörgensen, E. (1996). Restoration of the partially edentulous mouth — a comparison of overdentures, removable partial dentures, fixed partial dentures and implant treatment. *Journal of Dentistry*, *24*(4), 237–244. doi:10.1016/0300-5712(95)00075-5 PMID:8783527

Campus, G., Lallai, M. R., & Carboni, R. (2003). Fluoride concentration in saliva after use of oral hygiene products. *Caries Research*, *37*(1), 66–70. doi:10.1159/000068220 PMID:12566642

Chang, P. C., Dovban, A. S., Lim, L. P., Chong, L. Y., Kuo, M. Y., & Wang, C. H. (2013). Dual delivery of PDGF and simvastatin to accelerate periodontal regeneration in vivo. *Biomaterials*, *34*(38), 9990–9997. doi:10.1016/j.biomaterials.2013.09.030 PMID:24079892

Chávez de Paz, L. E., Resin, A., Howard, K. A., Sutherland, D. S., & Wejse, P. L. (2011). Antimicrobial Effect of Chitosan Nanoparticles on Streptococcus mutans Biofilms. *Applied and Environmental Microbiology*, *77*(11), 3892–3895. doi:10.1128/AEM.02941-10 PMID:21498764

Ciancio, S. G., Mather, M. L., & McMullen, J. A. (1980). An Evaluation of Minocycline in Patients With Periodontal Disease*. *Journal of Periodontology*, *51*(9), 530–534. doi:10.1902/jop.1980.51.9.530 PMID:6999148

Colombo, A. V., Silva, C. M., Haffajee, A., & Colombo, A. P. (2006). Identification of oral bacteria associated with crevicular epithelial cells from chronic periodontitis lesions. *Journal of Medical Microbiology*, *55*(Pt 5), 609–615. doi:10.1099/jmm.0.46417-0 PMID:16585650

Cooper, C. A., Ravich, D., Lips, D., Mayer, J., & Wagner, H. D. (2002). Distribution and alignment of carbon nanotubes and nanofibrils in a polymer matrix. *Composites Science and Technology*, *62*(7–8), 1105–1112. doi:10.1016/S0266-3538(02)00056-8

Cristina Buzea, I. I. P. B., & Kevin, R. (2007). Nanomaterials and nanoparticles: Sources and toxicity. *Biointerphases*, *2*, MR17–MR172. PMID:20419892

Virlan, M. J. R., Miricescu, D., Totan, A., Greabu, M., Tanase, C., Sabliov, C. M., ... & Calenic, B. (2015). Current Uses of Poly(lactic-co-glycolic acid) in the Dental Field: A Comprehensive. *Review Journal of Chemistry*, 2015.

Dabbagh, A., Kasim, N. H. A., Bakri, M. M., Wakily, H., Ramasindarum, C., & Abdullah, B. J. J. (2014). Polyethylene-glycol coated maghemite nanoparticles for treatment of dental hypersensitivity. *Materials Letters*, *121*, 89–92. doi:10.1016/j.matlet.2014.01.120

Damm, C., Münstedt, H., & Rösch, A. (2007). Long-term antimicrobial polyamide 6/silver-nanocomposites. *Journal of Materials Science*, *42*(15), 6067–6073. doi:10.1007/s10853-006-1158-5

Daood, U., Iqbal, K., Nitisusanta, L. I., & Fawzy, A. S. (2012). Effect of chitosan/riboflavin modification on resin/dentin interface: Spectroscopic and microscopic investigations. *Journal of Biomedical Materials Research. Part A*, *101*(7), 1846–1856. PMID:23184366

Daood, U., Swee Heng, C., Neo Chiew Lian, J., & Fawzy, A. S. (2015). In vitro analysis of riboflavin-modified, experimental, two-step etch-and-rinse dentin adhesive: Fourier transform infrared spectroscopy and micro-Raman studies. *Journal of Oral Science*, *7*(2), 110–124. doi:10.1038/ijos.2014.49 PMID:25257880

de Francisco, L. M., Cerquetani, J. A., & Bruschi, M. L. (2012). Development and characterization of gelatin and ethylcellulose microparticles designed as platforms to delivery fluoride. *Drug Development and Industrial Pharmacy*, *39*(11), 1644–1650. doi:10.3109/03639045.2012.728610 PMID:23034061

Dhadse, P., Gattani, D., & Mishra, R. (2010). The link between periodontal disease and cardiovascular disease: How far we have come in last two decades? *Journal of Indian Society of Periodontology*, *14*(3), 148–154. doi:10.4103/0972-124X.75908 PMID:21760667

Diarra, M., Pourroy, G., Boymond, C., & Muster, D. (2003). Fluoride controlled release tablets for intrabuccal use. *Biomaterials*, *24*(7), 1293–1300. doi:10.1016/S0142-9612(02)00444-1 PMID:12527271

Dung, T. H., Lee, S. R., Han, S. D., Kim, S. J., Ju, Y. M., Kim, M. S., & Yoo, H. (2007). Chitosan-TPP nanoparticle as a release system of antisense oligonucleotide in the oral environment. *Journal of Nanoscience and Nanotechnology*, *7*(11), 3695–3699. doi:10.1166/jnn.2007.041 PMID:18047039

Fawzy, A. S., Nitisusanta, L. I., Iqbal, K., Daood, U., Beng, L. T., & Neo, J. (2012). Chitosan/Riboflavin-modified demineralized dentin as a potential substrate for bonding. *Journal of the Mechanical Behavior of Biomedical Materials*, *17*, 278–289. doi:10.1016/j.jmbbm.2012.09.008 PMID:23127636

Fawzy, A. S., Nitisusanta, L. I., Iqbal, K., Daood, U., & Neo, J. (2012). Riboflavin as a dentin crosslinking agent: Ultraviolet A versus blue light. *Dental Materials*, *28*(12), 1284–1291. doi:10.1016/j.dental.2012.09.009 PMID:23031483

Fujishima, A., Rao, T. N., & Tryk, D. A. (2000). Titanium dioxide photocatalysis. *Journal of Photochemistry and Photobiology C, Photochemistry Reviews*, *1*(1), 1–21. doi:10.1016/S1389-5567(00)00002-2

Galler, K. M., & D'Souza, R. N. (2010). Tissue engineering approaches for regenerative dentistry. *Regenerative Medicine*, *6*(1), 111–124. doi:10.2217/rme.10.86 PMID:21175291

Gibbons, R. J., & Houte, J. V. (1975). Dental Caries. *Annual Review of Medicine*, *26*(1), 121–136. doi:10.1146/annurev.me.26.020175.001005 PMID:1096752

Ginebra, M.-P., Traykova, T., & Planell, J. (2006). Calcium phosphate cements as bone drug delivery systems: A review. *Journal of Controlled Release*, *113*(2), 102–110. doi:10.1016/j.jconrel.2006.04.007 PMID:16740332

Goodson, J. M., Gunsolley, J. C., Grossi, S. G., Bland, P. S., Otomo-Corgel, J., Doherty, F., & Comiskey, J. (2007). Minocycline HCl microspheres reduce red-complex bacteria in periodontal disease therapy. *Journal of Periodontology*, *78*(8), 1568–1579. doi:10.1902/jop.2007.060488 PMID:17668977

Gopinath, V., Ramakrishnan, T., Emmadi, P., Ambalavanan, N., Mammen, B., & Vijayalakshmi, . (2009). Effect of a controlled release device containing minocycline microspheres on the treatment of chronic periodontitis: A comparative study. *Journal of Indian Society of Periodontology*, *13*(2), 79–84. doi:10.4103/0972-124X.55844 PMID:20407655

Gould, T., Westbury, L., & Tillman, J. (1982). Dentin matrix gelatin (DMG) as a possible "universal" grafting material in periodontics. *Journal of Periodontology*, *53*(1), 22–25. doi:10.1902/jop.1982.53.1.22 PMID:6948946

Graves, D. (2008). Cytokines that promote periodontal tissue destruction. *Journal of Periodontology*, *79*(8), 1585–1591. doi:10.1902/jop.2008.080183 PMID:18673014

Han, Y., Zeng, Q., Lingling, E., Wang, D., He, H., & Liu, H. (2012). Sustained Topical delivery of insulin from fibrin gel loaded with poly (lactic-co-glycolic acid) microspheres improves the biomechanical retention of titanium implants in type 1 diabetic rats. *Journal of Oral and Maxillofacial Surgery*, *70*(10), 2299–2308. doi:10.1016/j.joms.2012.05.028 PMID:22871310

Hannig, M., & Hannig, C. (2010). Nanomaterials in preventive dentistry. *Nature Nanotechnology*, *5*(8), 565–569. doi:10.1038/nnano.2010.83 PMID:20581832

He, J., Söderling, E., Vallittu, P. K., & Lassila, L. V. J. (2013). Preparation and Evaluation of Dental Resin with Antibacterial and Radio-Opaque Functions. *International Journal of Molecular Sciences*, *14*(3), 5445–5460. doi:10.3390/ijms14035445 PMID:23470923

Hellstrom, M. K., McClain, P. K., Schallhorn, R. G., Bellis, L., Hanlon, A. L., & Ramberg, P. (2008). Local minocycline as an adjunct to surgical therapy in moderate to severe, chronic periodontitis. *Journal of Clinical Periodontology*, *35*(6), 525–531. doi:10.1111/j.1600-051X.2008.01219.x PMID:18384391

Hoffmann, M. R., Martin, S. T., Choi, W., & Bahnemann, D. W. (1995). Environmental Applications of Semiconductor Photocatalysis. *Chemical Reviews*, *95*(1), 69–96. doi:10.1021/cr00033a004

Hong, L., Krishnamachari, Y., Seabold, D., Joshi, V., Schneider, G., & Salem, A. K. (2011). Intracellular Release of 17-β Estradiol from Cationic Polyamidoamine Dendrimer Surface-Modified Poly (Lactic-co-Glycolic Acid) Microparticles Improves Osteogenic Differentiation of Human Mesenchymal Stromal Cells. *Tissue Engineering. Part C, Methods*, *17*(3), 319–325. doi:10.1089/ten.tec.2010.0388 PMID:20883116

Horev, B., Klein, M. I., Hwang, G., Li, Y., Kim, D., Koo, H., & Benoit, D. S. W. (2015). pH-activated nanoparticles for controlled topical delivery of farnesol to disrupt oral biofilm virulence. *ACS Nano*, *9*(3), 2390–2404. doi:10.1021/nn507170s PMID:25661192

Howell, D. (2016). Tooth decay still on the rise among English children. from http://www.bbc.com/news/health-35672775

Hutter, E., & Maysinger, D. (2013). Gold-nanoparticle-based biosensors for detection of enzyme activity. *Trends in Pharmacological Sciences*, *34*(9), 497–507. doi:10.1016/j.tips.2013.07.002 PMID:23911158

Jaberi-Ansari, Z., Ekrami, M., & Nojehdehian, H. (2013). Controlled Release of Drugs for Management of Pulpitis. *Avicenna Journal of Dental Research*, *6*(1).

Jamal, T., Rahman, M. A., Mirza, M. A., Panda, A. K., Talegaonkar, S., & Iqbal, Z. (2012). Formulation, antimicrobial and toxicity evaluation of bioceramic based ofloxacin loaded biodegradable microspheres for periodontal infection. *Current Drug Delivery*, *9*(5), 515–526. PMID:22758655

Jhajharia, K., Parolia, A., Shetty, K. V., & Mehta, L. K. (2015). Biofilm in endodontics: A review. *Journal of International Society of Preventive & Community Dentistry*, *5*(1), 1–12. doi:10.4103/2231-0762.151956 PMID:25767760

Jorgensen, J., Shariati, M., Shields, C. P., Durr, D. P., & Proskin, H. M. (1989). Fluoride uptake into demineralized primary enamel from fluoride-impregnated dental floss in vitro. *Pediatric Dentistry*, *11*(1), 17–20. PMID:2626334

Kashi, T. S., Eskandarion, S., Esfandyari-Manesh, M., Marashi, S. M., Samadi, N., & Fatemi, S. M. et al.. (2012). Improved drug loading and antibacterial activity of minocycline-loaded PLGA nanoparticles prepared by solid/oil/water ion pairing method. *International Journal of Nanomedicine*, *7*, 221–234. PMID:22275837

Kawaguchi, H. (2000). Functional polymer microspheres. *Progress in Polymer Science*, *25*(8), 1171–1210. doi:10.1016/S0079-6700(00)00024-1

Kawano, F., Ohguri, T., Ichikawa, T., Mizuno, I., & Hasegawa, A. (2002). Shock absorbability and hardness of commercially available denture teeth. *The International Journal of Prosthodontics*, *15*(3), 243–247. PMID:12066486

Keegan, G. M., Smart, J. D., Ingram, M. J., Barnes, L. M., Burnett, G. R., & Rees, G. D. (2012). Chitosan microparticles for the controlled delivery of fluoride. *Journal of Dentistry*, *40*(3), 229–240. doi:10.1016/j.jdent.2011.12.012 PMID:22212236

Khodashenas, B., & Ghorbani, H. R. Synthesis of silver nanoparticles with different shapes. *Arabian Journal of Chemistry*.

Kim, M. M., & Kim, S. K. (2006). Chitooligosaccharides inhibit activation and expression of matrix metalloproteinase-2 in human dermal fibroblasts. *FEBS Letters*, *580*(11), 2661–2666. doi:10.1016/j.febslet.2006.04.015 PMID:16647062

Kim, S. E., Jeon, O., Lee, J. B., Bae, M. S., Chun, H. J., Moon, S. H., & Kwon, I. K. (2008). Enhancement of ectopic bone formation by bone morphogenetic protein-2 delivery using heparin-conjugated PLGA nanoparticles with transplantation of bone marrow-derived mesenchymal stem cells. *Journal of Biomedical Science*, *15*(6), 771–777. doi:10.1007/s11373-008-9277-4 PMID:18773307

Kinane, D. L. J. (2003). Clinical Periodontology and Implant Dentistry (4 ed.). Copenhagen: Blackwell Munksgaard.

Kishen, A., Shi, Z., Shrestha, A., & Neoh, K. G. (2008). An investigation on the antibacterial and antibiofilm efficacy of cationic nanoparticulates for root canal disinfection. *Journal of Endodontics*, *34*(12), 1515–1520. doi:10.1016/j.joen.2008.08.035 PMID:19026885

Kishen, A., & Shrestha, A. (2013). Multi-functional micro and nanoparticles for use in root canal therapies: Google Patents.

Klepac-Ceraj, V., Patel, N., Song, X., Holewa, C., Patel, C., Kent, R., & Soukos, N. S. et al. (2011). Photodynamic effects of methylene blue-loaded polymeric nanoparticles on dental plaque bacteria. *Lasers in Surgery and Medicine*, *43*(7), 600–606. doi:10.1002/lsm.21069 PMID:22057487

Lamont, R. J., & Yilmaz, O. (2002). In or out: The invasiveness of oral bacteria. *Periodontology 2000*, *30*(1), 61–69. doi:10.1034/j.1600-0757.2002.03006.x PMID:12236896

Lin, L. M., Skribner, J. E., & Gaengler, P. (1992). Factors associated with endodontic treatment failures. *Journal of Endodontics*, *18*(12), 625–627. doi:10.1016/S0099-2399(06)81335-X PMID:1298804

Listgarten, M. A. (1987). Nature of periodontal diseases: Pathogenic mechanisms. *Journal of Periodontal Research*, *22*(3), 172–178. doi:10.1111/j.1600-0765.1987.tb01560.x PMID:2955092

Liu, D. Z., Chen, W. P., Lee, C. P., Wu, S. L., Wang, Y. C., & Chung, T. W. (2005). Effects of alginate coated on PLGA microspheres for delivery tetracycline hydrochloride to periodontal pockets. *Journal of Microencapsulation*, *21*(6), 643–652. doi:10.1080/02652040400000512 PMID:15762321

Longo, J. P. F., Leal, S. C., Simioni, A. R., Almeida-Santos, M. F. M., Tedesco, A. C., & Azevedo, R. B. (2012). Photodynamic therapy disinfection of carious tissue mediated by aluminum-chloride-phthalocyanine entrapped in cationic liposomes: An in vitro and clinical study. *Lasers in Medical Science*, *27*(3), 575–584. doi:10.1007/s10103-011-0962-6 PMID:21809069

Magalh, xe, es, A. P. R., Santos, L. B., Lopes, L. G., Estrela, C. R. d. A., et al. (2012). Nanosilver Application in Dental Cements. *ISRN Nanotechnology, 2012*, 6.

Mascarelli, A. (2011). Tooth decay is prevalent among poor children. Retrieved from http://articles.latimes.com/2011/jul/01/health/hk-he-dental-underserved-20110701

Masuzaki, T., Ayukawa, Y., Moriyama, Y., Jinno, Y., Atsuta, I., Ogino, Y., & Koyano, K. (2010). The effect of a single remote injection of statin-impregnated poly (lactic-co-glycolic acid) microspheres on osteogenesis around titanium implants in rat tibia. *Biomaterials*, *31*(12), 3327–3334. doi:10.1016/j.biomaterials.2010.01.016 PMID:20129662

Mehra, M., Vahidi, F., & Berg, R. W. (2013). A complete denture impression technique survey of postdoctoral prosthodontic programs in the United States. *Journal of Prosthodontics*, *23*(4), 320–327. doi:10.1111/jopr.12099 PMID:24118134

Melo, M. A. S., Guedes, S. F. F., Xu, H. H. K., & Rodrigues, L. K. A. (2013). Nanotechnology-based restorative materials for dental caries management. *Trends in Biotechnology*, *31*(8), 459–467. doi:10.1016/j.tibtech.2013.05.010 PMID:23810638

Molander, A., Reit, C., & Dahlén, G. (1999). The antimicrobial effect of calcium hydroxide in root canals pretreated with 5% iodine potassium iodide. *Dental Traumatology*, *15*(5), 205–209. doi:10.1111/j.1600-9657.1999.tb00775.x PMID:10825827

Monteiro, D. R., Gorup, L. F., Takamiya, A. S., de Camargo, E. R., Filho, A. C., & Barbosa, D. B. (2011). Silver distribution and release from an antimicrobial denture base resin containing silver colloidal nanoparticles. *Journal of Prosthodontics*, *21*(1), 7–15. doi:10.1111/j.1532-849X.2011.00772.x PMID:22050139

Morães, R. R., Garcia, J. W., Wilson, N. D., Lewis, S. H., Barros, M. D., Yang, B., & Stansbury, J. W. et al. (2012). Improved Dental Adhesive Formulations Based on Reactive Nanogel Additives. *Journal of Dental Research*, *91*(2), 179–184. doi:10.1177/0022034511426573 PMID:22019910

Moulari, B., Lboutounne, H., Chaumont, J.-P., Guillaume, Y., Millet, J., & Pellequer, Y. (2006). Potentiation of the bactericidal activity of Harungana madagascariensis Lam. ex Poir.(Hypericaceae) leaf extract against oral bacteria using poly (D, L-lactide-co-glycolide) nanoparticles: In vitro study. *Acta Odontologica Scandinavica*, *64*(3), 153–158. doi:10.1080/00016350500483152 PMID:16809192

Mundargi, R. C., Srirangarajan, S., Agnihotri, S. A., Patil, S. A., Ravindra, S., Setty, S. B., & Aminabhavi, T. M. (2007). Development and evaluation of novel biodegradable microspheres based on poly(d,l-lactide-co-glycolide) and poly(ε-caprolactone) for controlled delivery of doxycycline in the treatment of human periodontal pocket: In vitro and in vivo studies. *Journal of Controlled Release*, *119*(1), 59–68. doi:10.1016/j.jconrel.2007.01.008 PMID:17331611

Naito, Y., Terukina, T., Galli, S., Kozai, Y., Vandeweghe, S., Tagami, T., & Jimbo, R. et al. (2014). The effect of simvastatin-loaded polymeric microspheres in a critical size bone defect in the rabbit calvaria. *International Journal of Pharmaceutics*, *461*(1–2), 157–162. doi:10.1016/j.ijpharm.2013.11.046 PMID:24296045

Nakabayashi, N., Kojima, K., & Masuhara, E. (1982). The promotion of adhesion by the infiltration of monomers into tooth substrates. *Journal of Biomedical Materials Research*, *16*(3), 265–273. doi:10.1002/jbm.820160307 PMID:7085687

Nakabayashi. (1982). Bonding mechanism of resins and the tooth. *Kokubyo Gakkai Zasshi., 49*(410).

Ogle, R. E., David, L. J., & Ortman, H. R. (1985). Clinical wear study of a new tooth material: Part II. *The Journal of Prosthetic Dentistry*, *54*(1), 67–75. doi:10.1016/S0022-3913(85)80073-1 PMID:3894642

Oringer, R. J., Al-Shammari, K. F., Aldredge, W. A., Iacono, V. J., Eber, R. M., Wang, H. L., & Giannobile, W. V. et al. (2002). Effect of locally delivered minocycline microspheres on markers of bone resorption. *Journal of Periodontology*, *73*(8), 835–842. doi:10.1902/jop.2002.73.8.835 PMID:12211491

Osorio, R., Osorio, E., Medina-Castillo, A. L., & Toledano, M. (2014). Polymer Nanocarriers for Dentin Adhesion. *Journal of Dental Research*, *93*(12), 1258–1263. doi:10.1177/0022034514551608 PMID:25227634

Ozak, S. T., & Ozkan, P. (2013). Nanotechnology and dentistry. *European Journal of Dentistry*, *7*(1), 145–151. PMID:23408486

Pagonis, T. C., Chen, J., Fontana, C. R., Devalapally, H., Ruggiero, K., Song, X., & Yamazaki, H. et al. (2010). Nanoparticle-based endodontic antimicrobial photodynamic therapy. *Journal of Endodontics*, *36*(2), 322–328. doi:10.1016/j.joen.2009.10.011 PMID:20113801

Paquette, D., Oringer, R., Lessem, J., Offenbacher, S., Genco, R., Persson, G. R., & Williams, R. C. et al. (2003). Locally delivered minocycline microspheres for the treatment of periodontitis in smokers. *Journal of Clinical Periodontology*, *30*(9), 787–794. doi:10.1034/j.1600-051X.2003.00375.x PMID:12956654

Park, Y. J., Lee, J. Y., Yeom, H. R., Kim, K. H., Lee, S. C., Shim, I. K., & Lee, S. J. et al. (2005). Injectable polysaccharide microcapsules for prolonged release of minocycline for the treatment of periodontitis. *Biotechnology Letters*, *27*(22), 1761–1766. doi:10.1007/s10529-005-3550-7 PMID:16314967

Persadmehr, A., Torneck, C. D., Cvitkovitch, D. G., Pinto, V., Talior, I., Kazembe, M., & Kishen, A. et al. (2014). Bioactive chitosan nanoparticles and photodynamic therapy inhibit collagen degradation in vitro. *Journal of Endodontics*, *40*(5), 703–709. doi:10.1016/j.joen.2013.11.004 PMID:24767568

Persson, G. R., Salvi, G. E., Heitz-Mayfield, L. J., & Lang, N. P. (2006). Antimicrobial therapy using a local drug delivery system (Arestin®) in the treatment of peri-implantitis. I: Microbiological outcomes. *Clinical Oral Implants Research*, *17*(4), 386–393. doi:10.1111/j.1600-0501.2006.01269.x PMID:16907769

Piao, Z.-G., Kim, J.-S., Son, J.-S., Lee, S.-Y., Fang, X.-H., Oh, J.-S., & Kim, S.-G. et al. (2014). Osteogenic evaluation of collagen membrane containing drug-loaded polymeric microparticles in a rat calvarial defect model. *Tissue Engineering. Part A*, *20*(23-24), 3322–3331. doi:10.1089/ten.tea.2013.0717 PMID:24967649

Pinheiro, E., Gomes, B., Ferraz, C., Sousa, E., Teixeira, F., & Souza-Filho, F. (2003). Microorganisms from canals of root-filled teeth with periapical lesions. *International Endodontic Journal*, *36*(1), 1–11. doi:10.1046/j.1365-2591.2003.00603.x PMID:12656508

Pinon-Segundo, E., Ganem-Quintanar, A., Alonso-Perez, V., & Quintanar-Guerrero, D. (2005). Preparation and characterization of triclosan nanoparticles for periodontal treatment. *International Journal of Pharmaceutics*, *294*(1-2), 217–232. doi:10.1016/j.ijpharm.2004.11.010 PMID:15814246

Portenier, I., Waltimo, T. M., & Haapasalo, M. (2003). Enterococcus faecalis–the root canal survivor and 'star' in post-treatment disease. *Endodontic topics, 6*(1), 135-159.

Rai, M., Yadav, A., & Gade, A. (2009). Silver nanoparticles as a new generation of antimicrobials. *Biotechnology Advances*, *27*(1), 76–83. doi:10.1016/j.biotechadv.2008.09.002 PMID:18854209

Reis, K. R., Bonfante, G., Pegoraro, L. F., Conti, P. C. R., Oliveira, P. C. G., & Kaizer, O. B. (2008). In vitro wear resistance of three types of polymethyl methacrylate denture teeth. *Journal of Applied Oral Science*, *16*(3), 176–180. doi:10.1590/S1678-77572008000300003 PMID:19089214

Renvert, S., Lessem, J., Dahlén, G., Lindahl, C., & Svensson, M. (2006). Topical minocycline microspheres versus topical chlorhexidine gel as an adjunct to mechanical debridement of incipient peri-implant infections: A randomized clinical trial. *Journal of Clinical Periodontology*, *33*(5), 362–369. doi:10.1111/j.1600-051X.2006.00919.x PMID:16634959

Renvert, S., Lessem, J., Dahlen, G., Renvert, H., & Lindahl, C. (2008). Mechanical and repeated antimicrobial therapy using a local drug delivery system in the treatment of peri-implantitis: A randomized clinical trial. *Journal of Periodontology*, *79*(5), 836–844. doi:10.1902/jop.2008.070347 PMID:18454662

Rosa, V., Bona, A. D., Cavalcanti, B. N., & Nör, J. E. (2012). Tissue engineering: from research to dental clinics. *Dental materials: official publication of the Academy of Dental Materials, 28*(4), 341-348.

Saavedra, G., Valandro, L. F., Leite, F. P., Amaral, R., Ozcan, M., & Bottino, M. A. et al.. (2007). Bond strength of acrylic teeth to denture base resin after various surface conditioning methods before and after thermocycling. *The International Journal of Prosthodontics*, *20*(2), 199–201. PMID:17455445

Saleh, I., Ruyter, I., Haapasalo, M., & Ørstavik, D. (2004). Survival of Enterococcus faecalis in infected dentinal tubules after root canal filling with different root canal sealers in vitro. *International Endodontic Journal*, *37*(3), 193–198. doi:10.1111/j.0143-2885.2004.00785.x PMID:15009409

Samdancioglu, S., Calis, S., Sumnu, M., & Atilla Hincal, A. (2006). Formulation and in vitro evaluation of bisphosphonate loaded microspheres for implantation in osteolysis. *Drug Development and Industrial Pharmacy*, *32*(4), 473–481. doi:10.1080/03639040500528871 PMID:16638686

Scheifele, E., Studen-Pavlovich, D., & Markovic, N. (2002). Practitioner's guide to fluoride.[xi.]. *Dental Clinics of North America*, *46*(4), 831–846. doi:10.1016/S0011-8532(02)00019-8 PMID:12436834

Schwartz-Arad, D., & Chaushu, G. (1997). Placement of Implants Into Fresh Extraction Sites: 4 to 7 Years Retrospective Evaluation of 95 Immediate Implants. *Journal of Periodontology*, *68*(11), 1110–1116. doi:10.1902/jop.1997.68.11.1110 PMID:9407405

Selwitz, R. H., Ismail, A. I., & Pitts, N. B. (2007). Dental caries. *Lancet*, *369*(9555), 51–59. doi:10.1016/S0140-6736(07)60031-2 PMID:17208642

Sharma, G., Puranik, M. P., & K.R, S. (2015). Approaches to Arresting Dental Caries: An Update. *Journal of Clinical and Diagnostic Research: JCDR*, *9*(5), ZE08-ZE11.

Shrestha, A., Fong, S.-W., Khoo, B.-C., & Kishen, A. (2009). Delivery of Antibacterial Nanoparticles into Dentinal Tubules Using High-intensity Focused Ultrasound. *Journal of Endodontics*, *35*(7), 1028–1033. doi:10.1016/j.joen.2009.04.015 PMID:19567328

Shrestha, A., Friedman, S., & Kishen, A. (2011). Photodynamically Crosslinked and Chitosan-incorporated Dentin Collagen. *Journal of Dental Research*, *90*(11), 1346–1351. doi:10.1177/0022034511421928 PMID:21911787

Shrestha, A., Hamblin, M. R., & Kishen, A. (2014). Photoactivated rose bengal functionalized chitosan nanoparticles produce antibacterial/biofilm activity and stabilize dentin-collagen. *Nanomedicine; Nanotechnology, Biology, and Medicine*, *10*(3), 491–501. doi:10.1016/j.nano.2013.10.010 PMID:24200522

Smeets, R., Gerhards, F., Stein, J., Pereira Paz, R. M., Vogt, S., Pautke, C., & Kolk, A. et al. (2011). A novel hemostatic delivery device for thrombin: Biodegradable poly(D,L-lactide-co-glycolide) 50:50 microspheres. *Journal of Biomedical Materials Research. Part A, 96A*(1), 177–185. doi:10.1002/jbm.a.32970 PMID:21105166

Socransky, S. S., & Haffajee, A. D. (1994). Evidence of bacterial etiology: A historical perspective. *Periodontology 2000, 5*(1), 7–25. doi:10.1111/j.1600-0757.1994.tb00016.x PMID:9673160

Solhi, L., Atai, M., Nodehi, A., & Imani, M. (2012). A novel dentin bonding system containing poly(methacrylic acid) grafted nanoclay: Synthesis, characterization and properties. *Dental Materials, 28*(10), 1041–1050. doi:10.1016/j.dental.2012.06.004 PMID:22795491

Solhi, L., Atai, M., Nodehi, A., Imani, M., Ghaemi, A., & Khosravi, K. (2012). Poly (acrylic acid) grafted montmorillonite as novel fillers for dental adhesives: Synthesis, characterization and properties of the adhesive. *Dental Materials, 28*(4), 369–377. doi:10.1016/j.dental.2011.11.010 PMID:22169675

Son, J. S., Choi, Y.-A., Park, E.-K., Kwon, T.-Y., Kim, K.-H., & Lee, K.-B. (2013). Drug delivery from hydroxyapatite-coated titanium surfaces using biodegradable particle carriers. *Journal of Biomedical Materials Research. Part B, Applied Biomaterials, 101B*(2), 247–257. doi:10.1002/jbm.b.32834 PMID:23143817

Sousa, F. F., Luzardo-Alvarez, A., Perez-Estevez, A., Seoane-Prado, R., & Blanco-Mendez, J. (2010). Development of a novel AMX-loaded PLGA/zein microsphere for root canal disinfection. *Biomedical Materials (Bristol, England), 5*(5), 055008. doi:10.1088/1748-6041/5/5/055008 PMID:20844319

Srirangarajan, S., Mundargi, R. C., Ravindra, S., Setty, S. B., Aminabhavi, T., & Thakur, S. (2011). Randomized, controlled, single-masked, clinical study to compare and evaluate the efficacy of microspheres and gel in periodontal pocket therapy. *Journal of Periodontology, 82*(1), 114–121. doi:10.1902/jop.2010.100324 PMID:20681818

Stuart, C. H., Schwartz, S. A., Beeson, T. J., & Owatz, C. B. (2006). Enterococcus faecalis: Its role in root canal treatment failure and current concepts in retreatment. *Journal of Endodontics, 32*(2), 93–98. doi:10.1016/j.joen.2005.10.049 PMID:16427453

Sun, X. Z., Wang, N., Cao, D., Hu, Z. Y., Mao, P., Zhang, H. B., & Zhou, X. H. (2011). The antimicrobial activities of a series of bis-quaternary ammonium compounds. *Chinese Chemical Letters, 22*(8), 887–890. doi:10.1016/j.cclet.2011.01.030

Tsukada, K., Goan, K., & Furusawa, N. (2008). Fluorescent semiconductor microparticle, method for production of the microparticle, fluorescent labeling agent for biological substance using the microparticle, and bioimaging method using the microparticle: Google Patents.

Upadya, M., Shrestha, A., & Kishen, A. (2011). Role of Efflux Pump Inhibitors on the Antibiofilm Efficacy of Calcium Hydroxide, Chitosan Nanoparticles, and Light-activated Disinfection. *Journal of Endodontics*, *37*(10), 1422–1426. doi:10.1016/j.joen.2011.06.017 PMID:21924195

Van Rijkom, H., Truin, G., & Van't Hof, M. (1998). A meta-analysis of clinical studies on the caries-inhibiting effect of fluoride gel treatment. *Caries Research*, *32*(2), 83–92. doi:10.1159/000016436 PMID:9544855

van Strijp, A., Buijs, M., & Ten Cate, J. (1999). In situ fluoride retention in enamel and dentine after the use of an amine fluoride dentifrice and amine fluoride/sodium fluoride mouthrinse. *Caries Research*, *33*(1), 61–65. doi:10.1159/000016496 PMID:9831781

Wang, C.-Z., Fu, Y.-C., Jian, S.-C., Wang, Y.-H., Liu, P.-L., Ho, M.-L., & Wang, C.-K. (2014). Synthesis and characterization of cationic polymeric nanoparticles as simvastatin carriers for enhancing the osteogenesis of bone marrow mesenchymal stem cells. *Journal of Colloid and Interface Science*, *432*, 190–199. doi:10.1016/j.jcis.2014.06.037 PMID:25086394

Wang, F., Song, Y. L., Li, C. X., Li, D. H., Zhang, H. P., Ma, A. J., & Zhou, W. et al. (2010). Sustained release of insulin-like growth factor-1 from poly(lactide-co-glycolide) microspheres improves osseointegration of dental implants in type 2 diabetic rats. *European Journal of Pharmacology*, *640*(1-3), 226–232. doi:10.1016/j.ejphar.2010.04.024 PMID:20438725

Wang, J. J., Zeng, Z. W., Xiao, R. Z., Xie, T., Zhou, G. L., & Zhan, X. R. et al.. (2011). Recent advances of chitosan nanoparticles as drug carriers. *International Journal of Nanomedicine*, *6*, 765–774. PMID:21589644

Wang, L. C., Wu, H., Chen, X. G., De Li, L., Ji, Q. X., & Liu, C. S. et al.. (2008). Biological evaluation of a novel chitosan-PVA-based local delivery system for treatment of periodontitis. *Journal of Biomedical Materials Research. Part A*, *91*(4), 1065–1076. PMID:19107793

Wang, W., Liao, S., Zhu, Y., Liu, M., Zhao, Q., & Fu, Y. (2015). Recent applications of nanomaterials in prosthodontics. *Journal of Nanomaterials*, *2015*, 3. doi:10.3390/nano6010003

Watson, P. S., Pontefract, H. A., Devine, D. A., Shore, R. C., Nattress, B. R., Kirkham, J., & Robinson, C. (2005). Penetration of fluoride into natural plaque biofilms. *Journal of Dental Research*, *84*(5), 451–455. doi:10.1177/154405910508400510 PMID:15840782

Whitman, D., McKinney, J., Hinman, R., Hesby, R., & Pelleu, G. Jr. (1987). In vitro wear rates of three types of commercial denture tooth materials. *The Journal of Prosthetic Dentistry*, *57*(2), 243–246. doi:10.1016/0022-3913(87)90154-5 PMID:3470516

Williams, R. C. (1990). Periodontal Disease. *The New England Journal of Medicine*, *322*(6), 373–382. doi:10.1056/NEJM199002083220606 PMID:2405268

Wilson, M. (1993). Photolysis of oral bacteria and its potential use in the treatment of caries and periodontal disease. *The Journal of Applied Bacteriology*, *75*(4), 299–306. doi:10.1111/j.1365-2672.1993.tb02780.x PMID:8226389

Wu, H., Zhang, Z., Zhao, H., Wu, D., Wu, B., & Cong, R. (2004). Preparation of sodium fluoride-loaded gelatin microspheres, characterization and cariostatic studies. *Journal of Microencapsulation*, *21*(8), 889–903. doi:10.1080/02652040400015445 PMID:15799544

Wu, S. Z., Hossainy, S. F., Harish, S., Sanders-Millare, D., & Mirzaee, D. (2003). Microparticle coated medical device: Google Patents.

Yao, W., Xu, P., Pang, Z., Zhao, J., Chai, Z., & Li, X. et al.. (2014). Local delivery of minocycline-loaded PEG-PLA nanoparticles for the enhanced treatment of periodontitis in dogs. *International Journal of Nanomedicine*, *9*, 3963–3970. PMID:25170266

Yonamine, Y., Matsuyama, T., Sonomura, T., Takeuchi, H., Furuichi, Y., Uemura, M., & Noguchi, K. et al. (2010). Effectable application of vascular endothelial growth factor to critical sized rat calvaria defects. *Oral Surgery, Oral Medicine, Oral Pathology, Oral Radiology, and Endodontology*, *109*(2), 225–231. doi:10.1016/j.tripleo.2009.09.010 PMID:20034820

Yong, C. W. (2014). Study of interactions between polymer nanoparticles and cell membranes at atomistic levels. *Philosophical Transactions of the Royal Society of London B: Biological Sciences, 370*(1661). Doi:10.1098/rstb.2014.0036

Yue, I. C., Poff, J., Cortes, M. E., Sinisterra, R. D., Faris, C. B., Hildgen, P., & Shastri, V. P. et al. (2004). A novel polymeric chlorhexidine delivery device for the treatment of periodontal disease. *Biomaterials*, *25*(17), 3743–3750. doi:10.1016/j.biomaterials.2003.09.113 PMID:15020150

Zeren, K. J. (2006). Minimally invasive extraction and immediate implant placement: The preservation of esthetics. *The International Journal of Periodontics & Restorative Dentistry*, *26*(2), 171–181. PMID:16642906

Zhao, H., Wu, B., Wu, H., Su, L., Pang, J., Yang, T., & Liu, Y. (2006). Protective immunity in rats by intranasal immunization with Streptococcus mutans glucan-binding protein D encapsulated into chitosan-coated poly(lactic-co-glycolic acid) microspheres. *Biotechnology Letters*, *28*(16), 1299–1304. doi:10.1007/s10529-006-9086-7 PMID:16804764

Zheng, Y., Chen, Y., Jin, L. W., Ye, H. Y., & Liu, G. (2016). Cytotoxicity and Genotoxicity in Human Embryonic Kidney Cells Exposed to Surface Modify Chitosan Nanoparticles Loaded with Curcumin. *AAPS PharmSciTech*. PMID:26718819

Zou, G.-K., Song, Y.-L., Zhou, W., Yu, M., Liang, L.-H., Sun, D.-C., et al. (2012). Effects of local delivery of bFGF from PLGA microspheres on osseointegration around implants in diabetic rats. *Oral surgery, oral medicine, oral pathology and oral radiology, 114*(3), 284-289.

Chapter 4
Sensing the Presence and Amount of Microbes Using Double Walled Carbon Nanotubes

Anand Y Joshi
G.H. Patel College of Engineering and Technology, India

Ajay M Patel
G.H. Patel College of Engineering & Technology, India

ABSTRACT

The principle of mass detection using nano biosensors is based on the fact that the resonant frequency is very much sensitive to the mass of the bio-molecule, as with mass changes stiffness varies. The change of the attached mass on the CNT causes a shift to the resonant frequency. The key issue of mass detection is in quantifying the shift in the resonant frequency due to the mass of the attached molecule.This study, explores the vibration responses of the cantilever single and double-walled carbon nanotube with various attached microbes on the tip with an aim of developing a sensor. The biological objects studied include Alanine with Amino terminal residue, Deoxyadeonosine with free residue, Coronaviridae, Bartonella bacilliformis etc.. This sensor will be utilized to facilitate the identification of bacteria or virus that may be attached to CNT.

DOI: 10.4018/978-1-5225-1043-7.ch004

INTRODUCTION

Detection of the mass of bio-molecules has become an increasing growing field in the biological and biomedical sciences. It is recognized as one of the key technologies for predictive and preventive medicine as discussed by Hood et al. (2004). The common strategy for prediction of disease is based on sensing the corresponding biomolecules as indicated by Iijima et al. (1991, 1993). It is reasonable to believe that biosensors with real-time sensing capability and ease of use can change the future of disease detection and health monitoring. Since the discovery of carbon nanotubes (CNTs) they have emerged as a new class of nanomaterials as indicated by Dresselhaus, and Haddon (2001, 2002). Due to their unique physical properties suggested by Ajayan et al, (1999a, 1999b) Lin et al (2004). Gu et al (2005) they have been used in a variety of applications in fields such as scanning probes Wang et al. (2005)Kumar et al. (2008)Yun, et al. (2007) nano electronics, storage devices Rivas et al. (2007)Baughman, et al. (1999)Fennimore, et al. (2003) and nano electromechanical systems.

Recently there has been a rapidly growing interest of CNTs in biological applications Tsang et al. (1995)Davis, et al. (1998)Wong, et al. (1998)Mattson et al. (2000) in the field of medicine Lu, et al. (2009) and sensing mechanism specifically as biosensors. The development of nano-biosensors lee and Yoon et al (2007) based on CNTs has been driven by the fact that biological objects such as proteins, enzymes, bacteria can be immobilized either in the hollow cavity or on the surface of carbon nanotubes.

Two common methods are used for discovering the mechanical behaviour of nanostructures, namely continuum mechanics and atomistic-based modeling approaches. MD (molecular dynamics) is an accurate method capable of simulating the full mechanical DWCNT performance. Sohlberg et al. (1998) have investigated the free vibrations of CNTs. Blevins used the Euler–Bernoulli beam theory to derive corresponding eigenvalues of the CNT vibration Blevins et al. (1979). An elasticity solution of thin-walled cylinder for vibration analysis has been proposed by Mahan et al (2002). It is noted that the atomistic approaches are more reliable than the continuum-based methods as a consequence of implementing interactions at the atomic scale Sanchez-Portal et al. (1999) Dereli et al. (2003). Li and Chou (2003) developed an equivalent structural beam to mimic interatomic forces of the covalently bonded carbon atoms. Later, Tserpes and Papanikos (2005) presented the atomistic finite element (FE) model of single-walled carbon nanotubes (SWCNT) based on the Li and Chou equivalent structural beam concept. The suitability of the single walled carbon nanotubes as a mass detector device for many sensing applications has been explored using Rayleigh Ritz method Joshi et al (2008). It is reported that SWCNT based sensors exhibit super-harmonic and sub-harmonic response with different level of mass Joshi et al (2010a). Periodic and different nonlinear behavior of SWCNT is exposed as mass is attached at different positions along the length Joshi et al. (2010c). It is also reported that with the increase in mass attached to CNT, the peak excitation appears in the vibration signature in chaotic nature with reduced vibration amplitude Joshi et al. (2010d). Recently the effect of defects like atomic vacancies Joshi et al. (2011) and pinhole defects Joshi et al. (2010b) on the dynamic characteristics of carbon nanotubes is investigated using continuum mechanics and the molecular structural mechanics approach. Lassagne et al. (2008) utilized a 1nm diameter nanotube as a mechanical resonator for mass sensing of evaporated chromium atoms and measured a mass responsivity of 11Hz/ yoctogram. Jensen et al. (2008) have demonstrated a room-temperature, carbon-nanotube-based nanomechanical resonator with atomic mass resolution. It is essentially a mass spectrometer with a mass sensitivity of 1.3×10^{-25} kg $Hz^{-1/2}$ or, equivalently, 0.40 gold atoms $Hz^{-1/2}$. Hierold et al (2008) have presented the concepts to be followed for carbon nanotube sensors for mechanical and chemical quantities. A nanomechanical resonator incorporating intrinsically geometric nonlinearity and operated in a highly nonlinear regime has been modeled and developed by Cho et al (2010).

The mass sensing capabilities of multi-walled CNTs have been investigated by Georgantzinos et al. (2010). The main issue to be handled is whether multi-walled CNTs are more or less sensitive to an added mass in terms of frequency shift than single-walled CNTs with the same geometry, i.e. same outer diameter and length. Based on the analysis and the data available it is seen that a significant difference in frequency shift exists for the different masses attached to the CNTs. There are cases which show that the frequency shift of multiwalled CNTs is much higher than the corresponding shift present in a multi-walled CNT. These observations suggest that double walled CNTs are more suitable for sensing of biological objects than multi-walled CNTs.

The different biological objects in the form of microbes whose presence and intensity requires to be found have been discussed as under:

It has been observed that viruses like Coronaviridae contain a wide range of major pathogens that affect mammals and birds causing a variety of disease conditions including: respiratory disease (human, chicken and rats), gastro-enteritis (cattle, horses, pigs, dogs, turkey and mice), encephalomyelitis (pigs and mice), hepatitis (mice), myocarditis (rabbits), peritonitis (cats) and nephritis (chicken). In humans, coronaviruses are part of the spectrum of viruses that cause the common cold Haitham et al. (2011).

Also, Bartonella bacilliformis is the etiologic agent of Carrion´s disease or Oroya fever (acute phase of infection) and Verruga peruana or Peruvian wart (chronic phase of infection). The acute phase of the disease is a life threatening disease characterized by massive invasion of bartonella to human red blood cells and consequently an acute hemolysis and fever Maguina et al. Bartonella, the agent of trench fever and a cause of endocarditis and bacillary angiomatosis in humans, has the highest reported in vitro hemin requirement for any bacterium Jame et al. (2000). Hence, it is very important to safeguard against these viruses by sensing the presence and amount of the same. As CNTs possess a very high sensitivity in terms of mass sensing characteristics, double walled carbon nanotubes are used in this chapter for sensing the presence of these biological objects.

Patel and Joshi (2015a, 2015b, 2014a, 2014b, and 2015c) have investigated influence of resonant frequency of double walled carbon nanotubes with atomic vacancy defect and pinhole defect. This investigation is helpful in applications involving ultra-high frequency nano resonators which contain one or other type of manufacturing defects in it. Yoon et al. (2003) studied the resonant frequency and associated vibrational modes of an individual multi-walled carbon nanotube (MW-CNTs) embedded in elastic medium. Xu et al. (2008) worked on the free vibrations of DWCNT modeled as elastic beams due to different boundary conditions between inner and outer tubes. Li and Chou (2003a) developed an equivalent structural beam to mimic interatomic forces of the covalently bonded carbon atoms. Patel and Joshi (2013, 2014c) analysed vibrational characteristics of double walled carbon Nano-tube (DWCNT) modelled using spring elements and lumped masses. The inner and outer walls of carbon Nanotube were modeled as two individual elastic beams connecting each other by van der Waals forces. Patel and Joshi (2014d) have been investigated resonant frequency of double walled carbon nanotubes with deviations along it is axis and different boundary conditions namely cantilever and bridged. The sensitivity of the apparently deviated double walled carbon nanotubes, different masses attached to the end of outer tube tip on DWCNT and center outer tube tip of the bridged DWCNT and different lengths has been explored and presented. The effect of waviness along the axis has been evaluated by subjecting the nanotube to different boundary conditions namely bridged, cantilever and simply supported and the vibration responses of straight and wavy DWCNTs are investigated by Patel and Joshi (2014e).

MODELLING AND SIMULATION TECHNIQUES

Due to availability of highly computational facility over the past few years and improvements in theoretical method, algorithm and finite element techniques, the capability of simulation has greatly expended. Three techniques are described here and out of this three techniques molecular structural mechanics approach is better compared to describe other techniques. This approach is more accurate and efficient which authenticates the authors to use it for the purpose of analysis.

Structural Mechanics (SM) Approach

An additional atomistic model is developed under the Structural Mechanics (SM) approach Tserpes and Papanikos (2005). In this model, the bonds between every two nearest-neighboring atoms act like load-bearing beams and each atom acts as the joint of the related beams. This method is used to predict the fundamental frequencies of SWCNT and MWCNT mechanical resonators Li and Chou (2003a, 2003b). SM is used to study the frequency shift in cantilevered and bridged SWCNT resonators induced by a mass change due to an attached particle. SM is also used to estimate the Q-factor of CNT resonators as a function of temperature Jiang et al. (2004). The findings of SM simulations are in good agreement with the existing theoretical and experimental results. However, due to their atomistic detail, there are reports of simulations of CNTs only, which are few nano-meters long.

Continuum Mechanics

MD computations are still limited to simulations on the order of 10^6 atoms for only a few nanoseconds Qian et al. (2002). The simulation of larger systems or longer times can only be achieved by other methods, typically, continuum methods. In continuum methods, mechanical simulations are completely discarded the variations over lengths on the order of the atomic scale as well as the atomic structure of the CNTs. Several advanced methods are developed toward continuum simulations and were applied to CNTs. The basic continuum shell model is also shown to describe the mechanics of CNT accurately, when the proper parameters are chosen Yakobson et al. (1996). Finite elements (FE) analysis is implemented to produce results which are very similar to experimental observations, such as the rippling of nanotubes at large deformations Wang and Wang (2004).

Molecular Dynamics

In Molecular Dynamics (MD) simulations, each of the CNT atoms is assigned with a potential and its equation of motion is evaluated at every time step the molecular mechanics force field was used by Allinger et al. (1989). As MD retains the atomic details of the molecule, it enables to simulate the effect of the chirality of CNTs and also the interactions between CNTs with other materials such as proteins, hydrogen Darkrim et al. (2002) and other atoms at the atomic level. MD also facilitates simulations concerning nano- scale effects, such as radial deformation caused by vander waals forces Hertel et al. (1998), CNT buckling Ijima et al., (1996) nanotube self-healing Ding et al. (2007) closed ring structures Hod et al. (2003), Another goal of MD simulations is to obtain the macroscopic properties of CNTs in order to create continuum models. This was done by "virtual tests" of axial compression Yakobson et al. (1996) or lateral, axial Sears and Batra (2004) and torsional deformations Yu et al. (2004). The continuum parameters are particularly desired because if the tube behaves in a continuum mechanical manner, simple analytical relations can be used to describe its mechanics. MD was also used to study small NEMS structures, such as switches. Due to the atomic nature of MD, large scale simulations are computationally demanding, thus, realistic NEMS are currently not simulated with MD

Atomic Structure of Double Walled Carbon Nanotube

A Double walled carbon nanotube consists of two layers of graphite rolled to form a tube. In other words, a DWCNT consists of two coaxial layers, where the interlayer distance is close to the distance between graphene layers in graphite, approximately 0.34 nm.

As shown in Figure 1, CNTs are categorized as tubes formed by rolling a sheet of graphene about the $\vec{T}$ vector. A vector perpendicular to the $\vec{T}$ is the chiral vector denoted by $\vec{C_h}$. The chiral vector and the corresponding chiral angle define the type of CNT i.e. zigzag, armchair, chiral. $\vec{C_h}$ can be expressed with respect to two base vector $\vec{a_1}$ and $\vec{a_1}$ as under

Figure 1. Hexagonal lattice of graphene sheet including base vectors

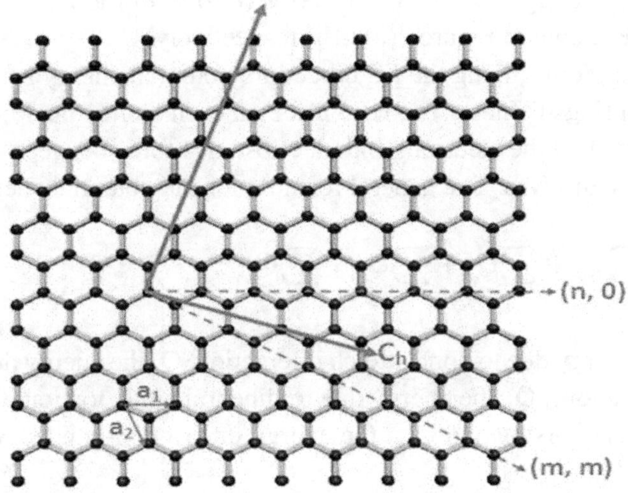

The chiral vector C_h of the nanotube is shown in Figure 1, defined by

$$\vec{C}_h = n\vec{a}_1 + m\vec{a}_2 \tag{1}$$

where $\vec{a}_1$ and $\vec{a}_2$ are unit vectors in the two-dimensional hexagonal lattice, and n and m are integers. If the indices of translation are such that m=0 and n=m then the corresponding CNTs are categorized as zigzag and armchair, respectively.

The outer diameter D_o of DWCNT can be calculated from the diameter of the corresponding outer (n,m) SWCNT as given by the following equation:

$$D_o = \frac{1}{\pi} a_{c-c} \sqrt{3(n^2 + nm + m^2)} \tag{2}$$

where a_{c-c} is the distance between two neighboring carbon atoms equals to 0.1421nm.

ATOMIC FINITE ELEMENT MODELING OF DWCNT

A FE model made of beam elements and atom mass is employed and the vibrational characteristics of cantilever and bridged DWCNT with different type biosensors are explored. The elastic properties of the beam elements are calculated to consider the interatomic covalent forces between carbon atoms. The atom masses which represent the mass of the carbon atom are placed at the end of the beams, which represent covalent bond between the carbon atoms. The schematic diagram of armchair DW-CNT (8,8)@(5,5) & zigzag DWCNT (12,0)@(8,0) with the beam elements and point masses are shown in Figure 2(a)&(b) respectively.

In this chapter, for utilizing the FE procedure potential energy is used to evaluate linear nanospring stiffness. The total force on each atomic nuclei is the sum of the force generated by the electrons and electrostatics force between the positively charged nuclei themselves. The general formula for the potential energy is

$$\pi = \sum Q_r + \sum Q_\theta + \sum Q_\phi + \sum Q_\omega + \sum Q_{vdw} \tag{3}$$

where Q_r is the energy due to bond stretch interaction, Q_θ the energy due to bending (bond angle variation), Q_Φ the energy due to dihedral angle torsion, Q_ω the energy due to out-of-plane torsion and Q_{vdw} the energy due to non-bonded Vander Waals interaction.

Figure 2. (a) and (b) Atomic finite element model of (8, 8)@(5,5) zigzag and (12,0)@ (8,0) armchair DWCNT with beam elements and point masses

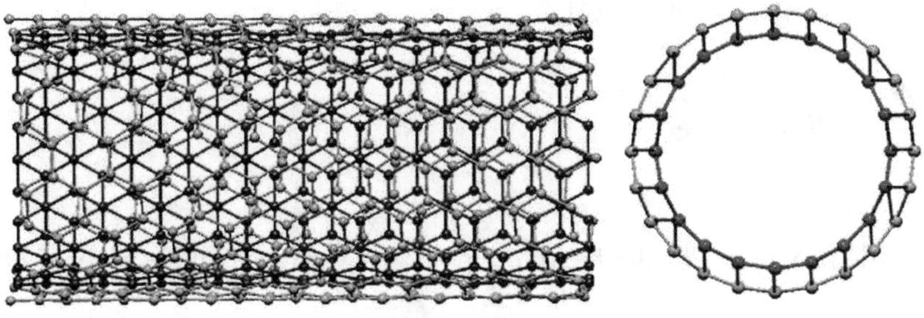

(a) (8, 8)@(5,5) armchair DWCNT

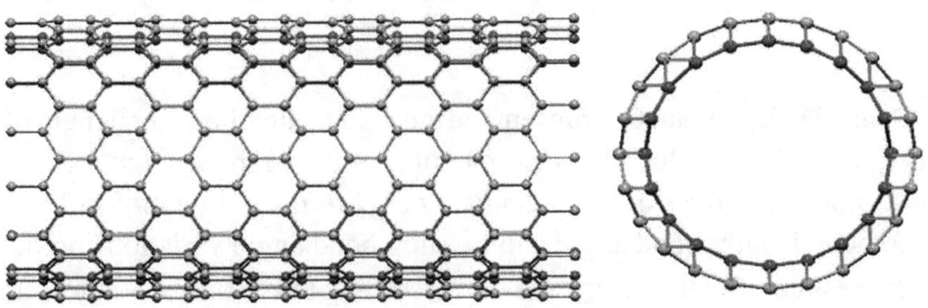

(a) (12,0)@(8,0) zigzag DWCNT

$$Q_r = \frac{1}{2}K_r(r - r_0)^2 = \frac{1}{2}K_r(\Delta r)^2 \tag{4}$$

$$Q_\theta = \frac{1}{2}K_\theta(\theta - \theta_0)^2 = \frac{1}{2}K_\theta(\Delta\theta)^2 \tag{5}$$

$$Q_\tau = Q_\phi + Q_\omega = \frac{1}{2}K_\tau(\Delta\phi)^2 \tag{6}$$

where K_r, K_θ, and $K\tau$ are the bond stretching, bond bending and torsional resistance force constants, respectively, while Δr, Δθ and Δϕ represent bond stretching increment, bond angle variation and angle variation of bond twisting, respectively. In contrast to the other bonded interactions, the van der Waals interactions and the electrostatic forces may be neglected due their minor effects in terms of predicting the dynamics response of CNTs.

As the potential energy in the two approaches is independent, energy equivalence of the stored energy of the two approaches, i.e. molecular mechanics and structural mechanics Li, C. and Chou (2003)

$$\frac{E\left(A_1 + A_2\right)}{L_{1,2}} = K_r, \quad \frac{E\left(I_1 + I_2\right)}{L_{1,2}} = K_\theta, \quad \frac{GJ}{L_{1,2}} = K_\tau. \tag{7}$$

The elastic properties of the beam element are given as Yakobson et al (1996)

$$D_1 = \sqrt{\frac{k_\theta}{k_r}}, D_2 = \sqrt{\frac{k_\theta}{k_r}}, \quad E = \frac{k_r^2 L_{1,2}}{4\pi k_\theta}, \quad G = \frac{k_r^{\,2} k_\phi L_{1,2}}{8\pi k_\theta^{\,2}}, \tag{8}$$

where, D_1 and D_2, L_{12}, E and G represent the inner and outer diameter, length of inner and outer, Young's modulus and shear modulus of the beam element for DWCNT.

The second derivatives of the potential energy terms in Equations (4)-(6) with respect to bond length, bond angle and twisting bond angle variations produce the spring stiffness coefficients k_r, k_θ, and K_τ according to Castigliano's theorem. The angle bending interaction is simulated with an axial nanospring, using the simplification described in Giannopoulos et al. (2008). The stiffness K_s of special spring is defined by following equation:

$$K_s = \left(\frac{1}{a_{c-c}\cos(60)^0}\right)^2 K_\theta \tag{9}$$

The angle may be varying for each C-C-C microstructure in a CNT according to its type and radius, due to cylindrical shape. Figure 3 Shows a DWCNT of the outer tube radius R_2, inner tube radius R_1 with spring element used to represent the week Van der Waals (vdW) force and atomic mass at node, using the Leonard-Jones model. The van der Waals force field between the interfacial layers is represented by a spring element COMBIN14.

The spring stiffness coefficient of Equations (4)-(6) are given in Table 1 Cornell et al. (1995).

In order to simulate the inertia effects, a point mass equal to a whole or half of the carbon atomic nucleus mass (M= 1.9943 x 10^{-26} gm) is added to a node in a particular elements. For simulating the structure, a structural point mass is added on all nodes as element.

Figure 3. DWCNT of the outer tube radius R_2, inner tube radius R_1 with spring element COMBIN14 and atomic mass at node
Source: Patel and joshi et al. (2015a).

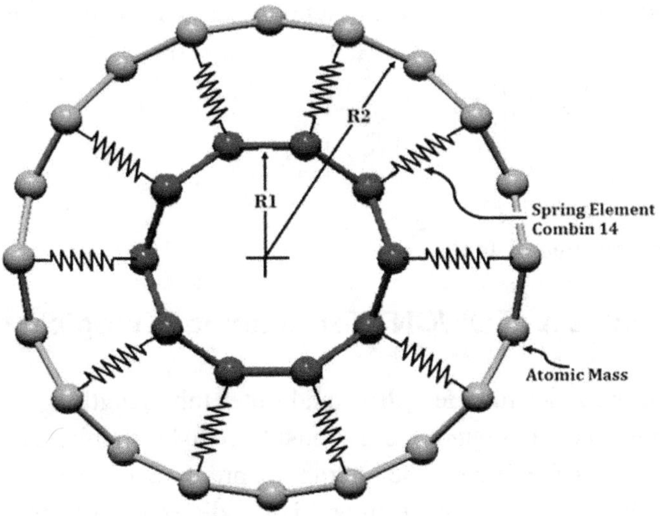

Table 1. Properties of spring stiffness coefficient

k_R	6.52 x 10^{-7} N nm^{-1}
k_θ	8.76 x 10^{-10} N nm rad^{-2}
k_ϕ	2.78 x 10^{-10} N nm rad^{-2}
t	0.34 nm

The interlayer interaction is described by the van der Waals potential. The Lenard-Jones 6-12 potential is utilized to express the interaction of carbon atoms located on the different walls

$$Q(R) = 4\varepsilon\left[\left(\frac{\rho}{R}\right)^{12} - \left(\frac{\rho}{R}\right)^{6}\right], \tag{10}$$

where R is the interatomic distance and $\varepsilon = 3.8655\times10^{-13}$ N nm and $\sigma = 0.34$ nm, respectively. Let R_1 be the initial distance between two carbon atoms located in different walls. We consider nanospring with initial length R_1 in order to describe the van der walls interaction. Consistence with the stiffness definition of the suggested nanospring, we describe K_{vdw} as the stiffness coefficient, acquired from the second derivative of the van der Waal potential energy for distance R_1 as follows:

$$K_{vdw} = \frac{dQ(R_1)}{dR_1} = 24\frac{\varepsilon}{\sigma}\left[2\left(\frac{\sigma}{R_1}\right)^{13} - \left(\frac{\sigma}{R_1}\right)^{7}\right] \tag{11}$$

for R_1 is the interatomic distance.

Vibration Analysis of DWCNT for Sensing Biological Objects

DWCNTs with the inner tube length L_1 and outer tube length L_2, the interaction between the inner and outer nanotubes is considered to be coupled together through the van der Waals (vdW) forces. The interaction pressure at any point between any two adjacent tubes is a linear function and is the difference of their deflection at that point. By Euler-Bernoulli beam theory, the differential equations controlling the motion for DWCNTs are described by two-coupled differential equations:

$$\rho A_1 \frac{\partial^2 u_1(x,t)}{\partial t^2} + EI\frac{\partial^4 u_1(x,t)}{\partial x^4} = c(u_2(x,t) - u_1(x,t)) \tag{12}$$

$$\rho A_2 \frac{\partial^2 u_2(x,t)}{\partial t^2} + EI_2\frac{\partial^4 u_2(x,t)}{\partial x^4} = -c(u_2(x,t) - u_1(x,t)) \tag{13}$$

where the subscripts 1 and 2 define the quantities associated with the inner and outer nanotubes, respectively. $u_i(x,t)$ (i=1,2) are the transverse deflection of the inner and outer nanotubes, respectively. C stands for the Van der Waals interaction coefficient between nanotubes. Which can be obtained from the interlayer energy potential and given as He et al (2005)

$$c = \frac{\Pi\varepsilon R_1 R_2 \sigma^6}{a^4}\left[\frac{1001\sigma^6}{3}H^{13} - \frac{1120}{9}H^7\right] \tag{14}$$

where

$$H^m = (R_1 + R_2)^{-m} \int_0^{\Pi/2} \frac{d\theta}{(1 - K\cos^2\theta)^{m/2}}, \text{ (m=7, 13)} \quad (15)$$

$$K = \frac{4R_1R_2}{(R_1 + R_2)^2} \quad (16)$$

where R_1 and R_2 are the inner radius and outer radius of the DWCNT.

$$c = \frac{320 \times 2R_1}{0.16a^2} \quad (17)$$

where c=6.943×10^{11} erg/cm^3= 69.43GPa, a =0.142nm is the carbon- carbon bond length and σ and ε are the vdW radius and the well depth of the Lennard- Jones potential, respectively. The vdW parameters in the Lennard- Jones potential are ε =2.967 meV and σ =0.34 nm as reported by saito et al. (2001)

For a DWCNT of inner tube and outer tube length are equal $L_{1,2}$=L with attached mass, based on the fundamental resonance frequency can be expressed as

$$f_n = \frac{1}{2\pi}\sqrt{\frac{k_{eq}}{m_{eq}}} \quad (18)$$

where, k_{eq} and m_{eq} are equivalent stiffness and mass of DWNCT with attached mass in the first mode of vibration, respectively.

Cantilever DWCNT with a Mass at the Tip

The resonant frequency of vibration for DWCNT was calculated by the well-known vibration equation:

$$f_n = \frac{\xi_n^{\,2}}{L^2}\sqrt{\frac{E(I_1 + I_2)}{\rho(A_1 + A_2)}}, \quad (19)$$

where E the Young's modulus, $I_{1,2}$ the second moment of the cross-sectional area $A_{1,2}$, and ρ is the density of DWCNTs. The values ξ_n (n=1, 2,…) of the cantilevered beam can be found by solving the equation,

$$\cos\xi \cosh\xi + 1 = 0, \xi_1 = 1.875, \xi_2 = 4.694 \text{ and } \xi_3 = 7.854$$

for the first, second and third modes respectively. $\xi_n = (n + 0.5)\pi$ for the larger mode n.

Figure 4 shown that cantilevered DWCNT resonator of length L with an attached biomass Alanine with amino terminal residue with a of 1.213071E^{-22} grams mass at the end of tip.

The vibration mode can be expressed as cantilever DWCNT

$$U_n(x) = \left(\cosh\xi_n \frac{x}{L} - \cos\xi_n \frac{x}{L}\right) - \left(\frac{\sinh\xi_n - \sin\xi_n}{\cosh\xi_n + \cos\xi_n}\right) \times \left(\sinh\xi_n \frac{x}{L} - \sin\xi_n \frac{x}{L}\right) \tag{20}$$

The mode shape in Equation (20) is normalized such that

$$\frac{1}{L}\int_0^L U_n^2(x)dx = 1 \tag{21}$$

The equivalence between the actual mass loaded DWCNT and the virtual single-DOF (SDOF) oscillator can be established be equating the potential and kinetic energies of both the systems. One needs to consider a deflection shape for the DW-CNT. Since the cantilevered DWCNT is vibrating in the first mode, it is natural to consider the deflection proportional to the first mode of vibration given by (20) with ξ=1.8751. This value is obtained Meirovitch et al. (1997) by solving the transcendental equation $\cos\xi \cosh\xi + 1 = 0$. Therefore, the assumed deflection is

Figure 4. Cantilevered DWCNT resonator of length L with an attached biomass Alanine with amino terminal residue at the end of tip

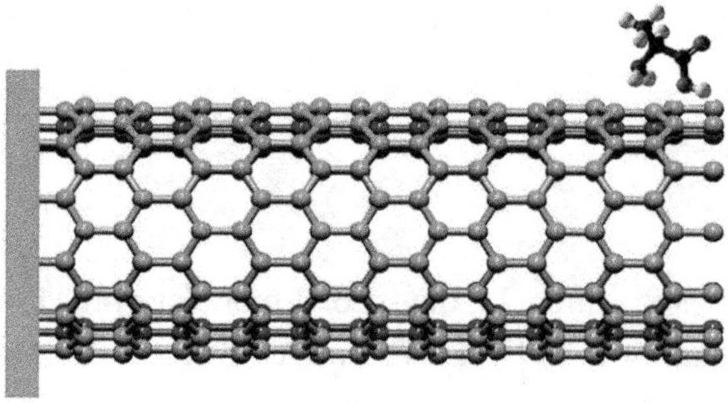

$$U(x) = aU_{1,2}(x), \tag{22}$$

where a is constant amplitude.

The potential energy of the outer walled DWCNT can be obtained as

$$P_1 = \frac{1}{2}EI_1\int_0^L \left[\frac{\partial^2 U_1(x)}{\partial x^2}\right]^2 dx = \frac{1}{2}EI_1\alpha^2\int_0^L \left[\frac{\partial^2 U_1(x)}{\partial x^2}\right]^2 dx \tag{23}$$

The potential energy of the inner walled DWCNT can be obtained as

$$P_2 = \frac{1}{2}EI_2\int_0^L \left[\frac{\partial^2 U_2(x)}{\partial x^2}\right]^2 dx = \frac{1}{2}EI_2\alpha^2\int_0^L \left[\frac{\partial^2 U_2(x)}{\partial x^2}\right]^2 dx \tag{24}$$

Differentiating Un(x) in Equation (20) twice with respect to x and using ξ =1.8751, this integral can be obtains as total potential energy of the cantilever DWCNT, $P_c = P_1 + P_2$:

$$P_c = 12.3623a^2\frac{E}{L^3}\left(I_1 + I_2\right) \tag{25}$$

Since the potential energy of the equivalent SDOF oscillator is $(1/2)k_{eq}\,a^2$, from the previous equation, we have

$$k_{eq} = 12.3623\frac{E}{L^3}\left(I_1 + I_2\right) \tag{26}$$

The value of the added mass is M in outer walled of cantilever DWCNT. Assuming harmonic motion, i.e. u(x,t) =U(x)exp(iωt), where ω is the frequency, the kinetic energy of the inner walled DWCNT can be obtained as

$$K_1 = \frac{\omega^2}{2}\int_0^L \rho A_1 U_1^2(x)dx = \frac{\omega^2 a^2}{2}\rho A_1 L \tag{27}$$

The kinetic energy of the outer walled DWCNT can be obtained as

$$K_2 = \frac{\omega^2}{2}\int_0^L \rho A_2 U_2^2(x)dx + \frac{\omega^2}{2} M U_2^2(L) = \frac{\omega^2 a^2}{2}(\rho A_2 L + 4M) \tag{28}$$

The total kinetic energy of the DWCNT can be obtained as

$$K_c = K_1 + K_2 = \frac{\omega^2 a^2}{2}\left[\rho\left(A_1 + A_2\right)L + 4M\right] \tag{29}$$

Since the kinetic energy of the equivalent SDOF oscillator is (1/2) $m_{eq}\omega^2 a^2$, we have

$$m_{eq} = \left[\rho\left(A_1 + A_2\right)L + 4M\right] \tag{30}$$

Substituting values in Equation (18), the resultant frequency can be obtained of the DWCNT with attached mass can be obtained by

$$f_n = \frac{1}{2\pi}\sqrt{\frac{k_{eq}}{m_{eq}}} = \frac{1}{2\pi}\sqrt{\frac{12.3623\frac{E}{L^3}\left(I_1 + I_2\right)}{\rho\left(A_1 + A_2\right)L + 4M}} \tag{31}$$

$$f_n = \frac{1}{2\pi}\frac{\alpha_k \beta}{\sqrt{1 + \alpha_m \Delta M}} \tag{32}$$

where, $\alpha_k = \sqrt{12.3623} = 3.5160$, $\beta = \sqrt{\frac{EI}{\rho(A_1 + A_2)L^4}}$, $\Delta M = \frac{M}{\rho(A_1 + A_2)L}$, and $\alpha_m = 4$

ΔM is known as relative mass change.

The resonant frequency for a cantilevered DWCNT without any added mass is obtained by substituting $\Delta M = 0$ in Equation (32) as

$$f_{on} = \frac{1}{2\pi}\alpha_k \beta \tag{33}$$

Bridged DWCNT with a Biological Object at the Midpoint

In case of a bridged DWCNT the maximum deflection occurs under the added mass at center of the length =L/2. The bridged beam can be found by explaining the equation,

$$\cos\xi\cosh\xi - 1 = 0, \xi_1 = 4.7300\,.$$

The vibration mode can be expressed as

$$U_n(x) = \left(\cosh\xi_n\frac{x}{L} - \cos\xi_n\frac{x}{L}\right) - \left(\frac{\cosh\xi_n - \cos\xi_n}{\sinh\xi_n + \sin\xi_n}\right)\times\left(\sinh\xi_n\frac{x}{L} - \sin\xi_n\frac{x}{L}\right) \tag{34}$$

The mode shape is normalized according to Equation (34) same like as cantilever. From the total potential energy of the DWCNT, the equivalent stiffness can be obtained as

$$P_b = P_1 + P_2 = 500.5638a^2\frac{E}{L^3}\left(I_1 + I_2\right) \tag{35}$$

$$k_{eq} = 500.5638\frac{E}{L^3}\left(I_1 + I_2\right) \tag{36}$$

The total kinetic energy of the bridged DWCNT can be obtained as

$$K_b = K_1 + K_2 = \frac{\omega^2 a^2}{2}\left[\rho\left(A_1 + A_2\right)L + 2.52220M\right] \tag{37}$$

$$m_{eq} = \left[\rho\left(A_1 + A_2\right)L + 2.52220M\right] \tag{38}$$

The resultant frequency can be obtained of the DWCNT with attached mass can be obtained by substituting value in Equation (18)

Figure 5. Bridged DWCNT resonator of length L with an attached biomass Alanine with amino terminal residue at the centre

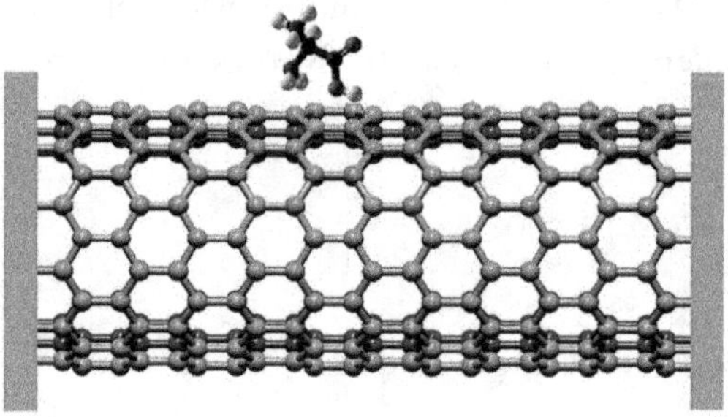

$$f_n = \frac{1}{2\pi}\sqrt{\frac{k_{eq}}{m_{eq}}} = \frac{1}{2\pi}\sqrt{\frac{500.5638\frac{E}{L^3}\left(I_1 + I_2\right)}{\rho\left(A_1 + A_2\right)L + 2.52220M}} = \frac{1}{2\pi}\frac{\alpha_k \beta}{\sqrt{1 + \alpha_m \Delta M}} \tag{39}$$

where,

$\alpha_k = \sqrt{500.5638} = 22.3732$, $\beta = \sqrt{\frac{EI}{\rho(A_1 + A_2)L^4}}$, $\Delta M = \frac{M}{\rho(A_1 + A_2)L}$, and $\alpha_m = 2.52220$.

Figure 5 shown that bridged DWCNT resonator of length L with an attached biomass Alanine with amino terminal residue with a of $1.213071E^{-22}$ grams mass at the center.

Mass Detection and Sensitivity Calculation of DWCNT

The expressions of the natural frequencies of the mass loaded DWCNT are now used to obtain the mass based on the frequency shift Chowdhury et al. (2009, 2011). Combining (32) and (33), one obtains the relationship between the resonant frequencies as

$$f_n = \frac{f_{on}}{\sqrt{1 + \alpha_m \Delta M}} \tag{40}$$

$$\Delta f = f_{on} - f_n = f_{on} - \frac{f_{on}}{\sqrt{1 + \alpha_m \Delta M}} \tag{41}$$

Rearranging gives the expression

$$\Delta M = \frac{1}{\alpha_m (1 - (\Delta f / f_{on}))^2} - \frac{1}{\alpha_m} \tag{42}$$

This equation completely relates the change in mass with the frequency shift. The actual value of the added mass can be obtained from (42) as

$$M = \frac{\rho(A_1 + A_2)L}{\alpha_m} \frac{(\alpha_k^2 \beta^2)}{(\alpha_k \beta - 2\pi \Delta f)^2} - \frac{\rho(A_1 + A_2)L}{\alpha_m} \tag{43}$$

This is the common equation which completely relates the added mass and the frequency shift. The constants appearing in this equation can be summarized as

$$S = \frac{\partial M}{\partial \Delta f} = \frac{\rho(A_1 + A_2)L}{\alpha_m} \frac{4\pi(\alpha_k^2 \beta^2)}{(\alpha_k \beta - 2\pi \Delta f)^3} \tag{44}$$

It is observed that the mass-detection sensitivity of a DWCNT-based biomass nano sensor is a nonlinear function of the frequency shift. For further understanding, expanding the sensitivity in (44) in Taylor series in Δf, we have the minimum sensitivity, for the case when Δf is very small, can be obtained by taking lim Δf→ 0

$$S_{\min} = \frac{\rho(A_1 + A_2)L}{\alpha_m} \frac{4\pi}{(\alpha_k \beta)} \tag{45}$$

Applying this equation for the cantilevered and bridged DWCNT, the minimum sensitivities can be obtained as

$$S_{cantilever} = \frac{4\pi}{14.064} \frac{\rho(A_1 + A_2)L}{\beta} \; and \; S_{bridged} = \frac{4\pi}{56.430} \frac{\rho(A_1 + A_2)L}{\beta} \tag{46}$$

From this, it can be resolved that the cantilever DWCNT sensor is about four times more sensitive compared to the bridged DWCNT Chowdhury et al. (2011).

ATOMIC FINITE ELEMENT MODELING OF SWCNT

A finite element model consisting of beam elements and concentrated masses is employed and the vibrational characteristics of Fixed- Free SWCNT with different chiralities behaving as bio sensors are explored. The elastic properties of the beam elements are calculated to consider the interatomic covalent forces between carbon atoms. The concentrated masses which represent the mass of the carbon atom are placed at the end of the beams, which represent covalent bond between the carbon atoms. The schematic diagram of (6, 4), (7, 3) and (8, 1) carbon nanotubes with the beam elements and point masses are shown in Figure 6 (a), (b) and (c) respectively.

CNTs with different chiralities have been employed and its response to the addition of bio molecules at the tip has been considered

Mass can be attached to the single walled carbon nanotube using either of the following methods:

Figure 6. atomic finite element model of (6, 4), (7, 3) and (8, 1) SWCNT with beam elements and point masses

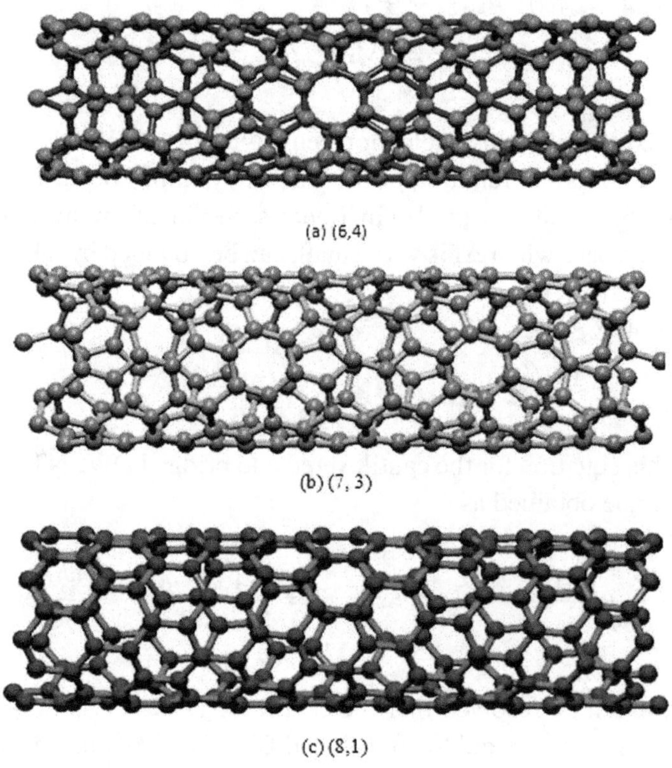

Figure 7. Carbonaceous substances induced by EBID

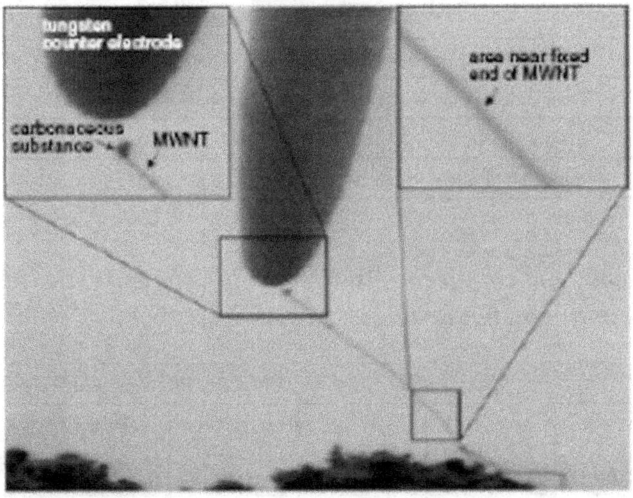

- Ion Beam induced deposition (IBID) technique.
- Electron beam induced deposition(EBID) technique.

Electron beam-induced deposition (EBID) is a process of decomposing gaseous molecules by electron beam leading to deposition of non-volatile fragments onto a nearby substrate. The electron beam is usually provided by a scanning electron microscope that results in high spatial accuracy (below one nanometer) and possibility to produce free-standing, three-dimensional structures.

Figure 7 shows Carbonaceous substances induced by EBID during electron beam scanning deposited near free end of MWNT, as indicated by arrow in upper left inset Wu et al(2008). Due to the attached mass there is a corresponding reduction in the resonant frequency which gives an indication of the attached mass.

RESULTS AND DISCUSSION FOR DWCNT

In this chapter, the vibration of the cantilever double-walled carbon nanotube (DWCNT) with attached biological object on the tip has been investigated in the view of developing the sensor. The sensor is to be developed based on DWCNT so as to be able to identify the viruses, bacteria and other biological objects that may be attached to outer wall of the nanotube. In order to perform the vibration analysis of the system, information about the mass of various biological object is required which is indicated in Table 2.

Table 2. Mass of various biological object

Sr. No.	Name	Mass Mi (g)
1	Alanine with Amino terminal residue	$1.21307\times10^{-22} \leq Mi \leq 9.70456\times10^{-22}$
2	Deoxyadeonosine with free residue	$4.16965\times10^{-22} \leq Mi \leq 2.918755\times10^{-21}$
3	Coronaviridae	$1.334\times10^{-18} \leq Mi \leq 6.579\times10^{-19}$
4	Bartonella bacilliformis	$5.400\times10^{-14} \leq Mi \leq 6.479\times10^{-13}$

Table 3. Frequency for cantilever DWCNT with different value of 08 molecules Alanine with Amino terminal residue

Mass (Zg)	Alanine with Amino Terminal Residue Frequency (Hz)					
	DWCNT (8,8) & (6,6)	DWCNT (10,10) & (5,5)	DWCNT (7,7) & (4,4)	DWCNT (14,0) & (10,0)	DWCNT (18,0) & (9,0)	DWCNT (12,0) & (7,0)
0.121307	2.48E+08	1.90E+08	1.98E+08	2.41E+08	2.00E+08	1.64E+08
0.242614	2.08E+08	1.84E+08	1.95E+08	2.11E+08	1.80E+08	1.55E+08
0.363921	1.80E+08	1.62E+08	1.91E+08	1.87E+08	1.63E+08	1.42E+08
0.485228	1.61E+08	1.46E+08	1.70E+08	1.69E+08	1.49E+08	1.28E+08
0.606535	1.46E+08	1.33E+08	1.54E+08	1.55E+08	1.37E+08	1.16E+08
0.727842	1.35E+08	1.24E+08	1.42E+08	1.44E+08	1.28E+08	1.09E+08
0.849149	1.26E+08	1.16E+08	1.33E+08	1.35E+08	1.21E+08	1.00E+08
0.970456	1.19E+08	1.09E+08	1.25E+08	1.27E+08	1.14E+08	2.90E+07

Table 3 shows frequency for various values of the attached molecules. Table 4 and 5 shows the percentage error in frequency shift for various values of the attached molecules. The percentage error is calculated by taken the difference between results obtained using the Atomic finite element simulation and analytical approach.

Table 6 signifies the resonant frequency and frequency shift of cantilever DW-CNT due to Coronaviridae virus. Table 7 shows resonant frequency for cantilever DWCNT using biological mass Bartonella Bacilliformis.

As indicated in Figure 8(a) (b) (c) and (d), four biological objects with mass in Zeptogram scales have been used to simulate the vibration responses of zigzag and armchair DWCNT based biosensors:

1. **Alanine with Amino Terminal Residue:** Alanine is an α-amino acid with the chemical formula CH_3 CH (NH_2) COOH. It is one of the 22 proteinogenic amino acids, which are considered as the building blocks of proteins. It is also

Table 4. Percentage error in the frequency shift for cantilever DWCNT with different value of 08 molecules Alanine with Amino terminal residue

Mass (Zg)	Frequency Shift (Hz)			Percentage Error (%)		
	DWCNT (8,8 @ (6,6)	DWCNT (1 0,10)@ (5,5)	DWCNT (7,7)@ (4,4)	DWCNT (8,8) @ (6,6)	DWCNT (10,10) @ (5,5)	DWCNT (7,7) @ (4,4)
0.121307	4.39E+07	2.00E+07	1.00E+07	2.489	5.632	6.236
0.242614	8.43E+07	2.56E+07	1.30E+07	4.690	3.205	2.589
0.363921	1.12E+08	3.40E+07	2.71E+07	3.425	5.045	5.26
0.485228	1.31E+08	4.62E+07	3.85E+07	6.680	8.520	7.258
0.606535	1.46E+08	5.97E+07	5.01E+07	7.465	6.253	10.256
0.727842	1.57E+08	6.86E+07	6.01E+07	2.458	6.025	7.256
0.849149	1.66E+08	7.86E+07	7.07E+07	1.025	7.250	8.269
0.970456	1.74E+08	8.73E+07	8.06E+07	2.980	2.123	1.025
Mass (Zg)	Frequency Shift (Hz)			Percentage Error (%)		
	DWCNT (14,0) @ (10,0)	DWCNT (18,0) (9,0)	DWCNT (12,0) @ (7,0)	DWCNT (14,0) @ (10,0)	DWCNT (18,0) @ (9,0)	DWCNT (12,0) @ (7,0)
0.121307	3.00E+07	2.13E+07	1.10E+07	8.691	6.358	5.360
0.242614	3.96E+07	2.98E+07	1.90E+07	5.798	4.236	2.369
0.363921	5.43E+07	4.71E+07	3.21E+07	4.236	2.456	8.236
0.485228	7.25E+07	6.50E+07	4.59E+07	5.236	3.211	7.025
0.606535	8.64E+07	8.22E+07	5.83E+07	3.236	2.365	7.025
0.727842	9.74E+07	9.13E+07	6.52E+07	1.986	4.045	4.236
0.849149	1.06E+08	9.89E+07	7.43E+07	2.368	10.23	3.023
0.970456	1.14E+08	1.05E+08	9.53E+07	2.358	8.230	2.780

well known as 2-aminopropanoic acid. The α carbon atom of alanine is bound with a methyl group ($-CH_3$), making it one of the simplest α -amino acids with respect to molecular structure and also resulting in alanine's being classified as an aliphatic amino acid. The methyl group of alanine is non-reactive and is thus almost never directly involved in protein function. It occurs in bacterial cell walls and in some peptide antibiotics. Alanine can be manufactured in the body from branched chain amino acids such as valine, leucine and isoleucine and it plays a key role in glucose–alanine cycle between tissues and liver. In muscle and other tissues that degrade amino acids for fuel, amino groups are collected by transamination.

Table 5. Percentage error in the frequency shift for cantilever DWCNT with different value of 07 molecules Deoxyadeonosine with free residue

Mass (Zg)	Frequency Shift (Hz)			Percentage Error %		
	DWCNT (8,8) @ (6,6)	DWCNT (10,10) @ (5,5)	DWCNT (7,7) @ (4,4)	DWCNT (8,8) @ (6,6)	DWCNT (10,10) @ (5,5)	DWCN (7,7) @ (4,4)
0.416965	1.21E+08	3.76E+07	4.74E+07	7.203	2.369	2.583
0.83393	1.65E+08	7.57E+07	8.46E+07	10.253	3.458	3.458
1.250895	1.87E+08	9.51E+07	1.01E+08	6.890	5.269	6.248
1.667860	2.00E+08	1.07E+08	1.12E+08	3.356	4.587	6.458
2.084825	2.09E+08	1.16E+08	1.22E+08	4.589	8.257	1.916
2.501790	2.16E+08	1.24E+08	1.29E+08	7.586	1.872	2.564
2.918755	2.25E+08	1.36E+08	1.41E+08	5.256	2.125	9.869
Mass (Zg)	**Frequency Shift (Hz)**			**Percentage Error (%)**		
	DWCNT (14,0) @ (10,0)	**DWCNT (18,0) @ (9,0)**	**DWCNT (12,0) @ (7,0)**	**DWCNT (14,0) @ (10,0)**	**DWCNT (18,0) @ (9,0)**	**DWCNT (12,0) @ (7,0)**
0.416965	6.29E+07	5.36E+07	1.00E+07	2.582	4.025	7.985
0.83393	1.05E+08	9.81E+07	3.52E+07	5.356	6.656	4.528
1.250895	1.28E+08	1.17E+08	5.41E+07	7.458	3.258	7.426
1.667860	1.41E+08	1.29E+08	6.91E+07	6.256	2.321	2.568
2.084825	1.51E+08	1.38E+08	7.96E+07	8.661	3.103	2.798
2.501790	1.59E+08	1.45E+08	8.74E+07	2.314	9.569	3.357
2.918755	1.65E+08	1.50E+08	9.36E+07	1.011	2.869	2.951

2. **Deoxyadeonosine with Free Residue:** Deoxyadenosine refers to a deoxyribonucleoside (a nucleoside component of DNA; composed of adenosine and deoxyribose) with the chemical formula $C_{10}H_{13}N_5O_3$. It is derived from the nucleoside adenosine, which differs by the replacement of a hydroxyl group (–OH) by hydrogen (–H) in the ribose part of the molecule. This modification prevents the formation of phosphodiester linkages which are needed for the completion of nucleic acid chains. A deoxyribonucleoside is a type of nucleoside including deoxyribose as a component.
3. **Coronaviridae:** Coronaviridae is a family of enveloped, positive-stranded RNA viruses. The viral genome is 26–32 kb in length. Virions are spherical, 120–160 nm across (Coronavirinae), bacilliform, and 170–200 by 75–88 nm (Bafinivirus) or found as a mixture of both, with bacilliform particles characteristically bent into crescents (Torovirus). The particles are typically

Table 6. Resonant frequency and frequency shift of cantilever DWCNT due to Coronaviridae biological object

Mass (gm)	Frequency (Hz)			Frequency Shift (Hz)		
	DWCNT (8,8) @ (6,6)	DWCNT (10,10) @ (5,5)	DWCNT (7,7) @ (4,4)	DWCNT (8,8) @ (6,6)	DWCNT (10,10) @ (5,5)	DWCNT (7,7) @ (4,4)
1.33E-19	3.32E+07	3.08E+07	3.49E+07	2.58782E+08	1.61347E+08	1.73577E+08
1.20767E-18	1.11E+07	1.02E+07	1.16E+07	2.80968E+08	1.81883E+08	1.96850E+08
2.28193E-18	8.05E+06	7.46E+06	8.44E+06	2.83981E+08	1.84675E+08	2.00010E+08
3.3562E-18	6.64E+06	6.15E+06	6.96E+06	2.85393E+08	1.85982E+08	2.01491E+08
4.43047E-18	5.78E+06	5.35E+06	6.06E+06	2.86253E+08	1.86779E+08	2.02393E+08
5.50473E-18	5.18E+06	4.80E+06	5.43E+06	2.86847E+08	1.87329E+08	2.03016E+08
6.579E-18	4.74E+06	4.39E+06	4.97E+06	2.87289E+08	1.87739E+08	2.03479E+08
Mass (gm)	**Frequency (Hz)**			**Frequency Shift (Hz)**		
	DWCNT (14,0) @ (10,0)	**DWCNT (18,0) @ (9,0)**	**DWCNT (12,0) @ (7,0)**	**DWCNT (14,0) @ (10,0)**	**DWCNT (18,0) @ (9,0)**	**DWCNT (12,0) @ (7,0)**
1.33E-19	3.64E+07	3.33E+07	3.84E+07	2.05E+08	1.86E+08	1.36E+08
1.20767E-18	1.21E+07	1.11E+07	1.28E+07	2.29E+08	2.09E+08	1.62E+08
2.28193E-18	8.82E+06	8.08E+06	9.29E+06	2.32E+08	2.12E+08	1.65E+08
3.3562E-18	7.28E+06	6.67E+06	1.75E+07	2.34E+08	2.13E+08	1.57E+08
4.43047E-18	6.33E+06	5.80E+06	7.66E+06	2.35E+08	2.14E+08	1.67E+08
5.50473E-18	5.68E+06	5.21E+06	5.99E+06	2.35E+08	2.14E+08	1.68E+08
6.579E-18	5.20E+06	4.76E+06	5.47E+06	2.36E+08	2.15E+08	1.69E+08

Table 7. Resonant frequency for cantilever DWCNT using biological mass Bartonella Bacilliformis

Mass (gm)	Frequency (Hz)_Cantilever Condition					
	DWCNT (8,8) @ (6,6)	DWCNT (10,10) @ (5,5)	DWCNT (7,7) @ (4,4)	DWCNT (14,0) @ (10,0)	DWCNT (18,0) @ (9,0)	DWCNT (12,0) @ (7,0)
5.400000E-14	5.23E+04	4.85E+04	5.49E+04	5.74E+04	5.26E+04	6.04E+04
1.529833E-13	3.11E+04	2.88E+04	3.26E+04	3.41E+04	3.12E+04	3.59E+04
2.519667E-13	2.42E+04	2.24E+04	2.54E+04	2.66E+04	2.43E+04	2.80E+04
3.509500E-13	2.05E+04	1.90E+04	2.15E+04	2.25E+04	2.06E+04	2.37E+04
4.499333E-13	1.81E+04	1.68E+04	1.90E+04	1.99E+04	1.82E+04	2.09E+04
5.489167E-13	1.64E+04	1.52E+04	1.72E+04	1.80E+04	1.65E+04	1.90E+04
6.479000E-13	1.51E+04	1.40E+04	1.58E+04	1.66E+04	1.52E+04	1.74E+04

Figure 8. (a) Alanine with amino terminal residue with a mass of 1.213071E^{-22} grams. DWCNT is loaded with upto 08 molecules of the same type. (b) deoxyadeonosine with free residue with a mass of 4.16965E^{-22} grams. DWCNT is loaded with upto 07 molecules of the same type. (c) Coronavirus virion structure (d) Bartonella baciliformis (Proteobacteria)
NIAID, NIH Source.

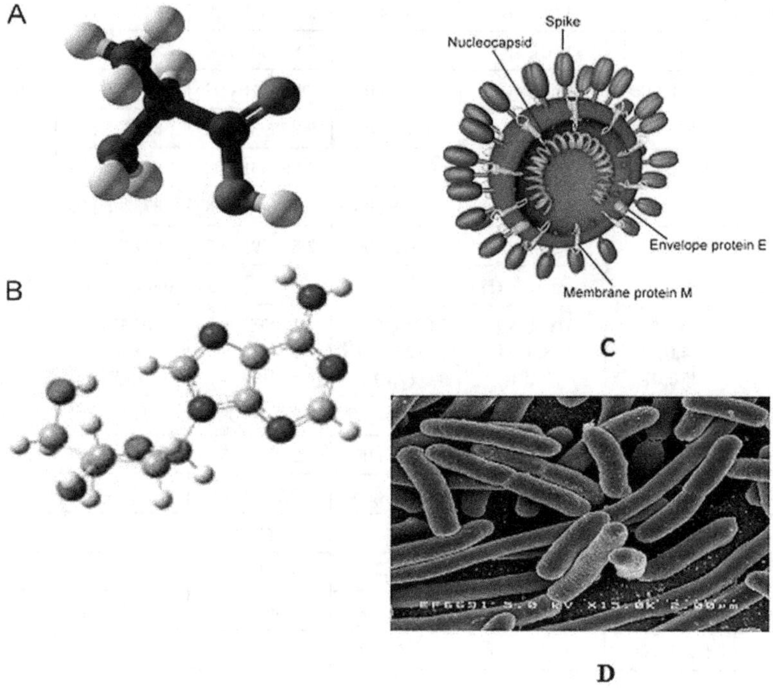

decorated with large (~20 nm), club- or petal-shaped surface projections (the "peplomers" or "spikes"), which in electron micrographs of spherical particles create an image reminiscent of the solar corona.

4. **Bartonella Bacilliformis:** Bartonella bacilliformis is a proteobacterium, Gram negative aerobic, pleomorphic, flagellated, motile, coccobacillary, 2–3 µm large and 0.2–0.5 µm wide and facultative intracellular bacterium. It is endemic in some areas of Peru, occurring outbreaks of the disease in new epidemic areas. The bacterium is transmitted by sandflies of genus Lutzomyia.

Six different type of DWCNTs, i.e three different type of armchair (8,8) @ (6,6), (10,10)@(5,5)and(7,7)@(4,4) and zigzag (14,0)@(10,0), (18,0)@(9,0) and (12,0) @ (7,0) length 25nm are loaded with 08 molecules of alanine with amino terminal residue, 07 molecules of deoxyadeonosine with free residue, 07 molecules of Coronaviridae and 07 molecules of Bartonella Bacilliformis considered for analysis.

Table 8. Mass sensitivity of cantilevered DWCNT

DWCNT Type	Outer Diameter (nm)	Inner Diameter (nm)	Sensitivity (zg/GHz)
(8,8)@(6,6)	10.78	8.092	0.1424
(10,10)@(5,5)	13.42	6.74	0.1221
(7,7)@(4,4)	9.43	5.39	0.1011
(14,0)@(10,0)	10.81	7.78	0.1359
(18,0)@(9,0	14.012	7.0063	0.1251
(12,0)@(7,0)	9.341	5.44	0.0989

Figure 9. Frequency shift of cantilever DWCNT due to 08 molecules of alanine with amino terminal vs. mass (Zg)

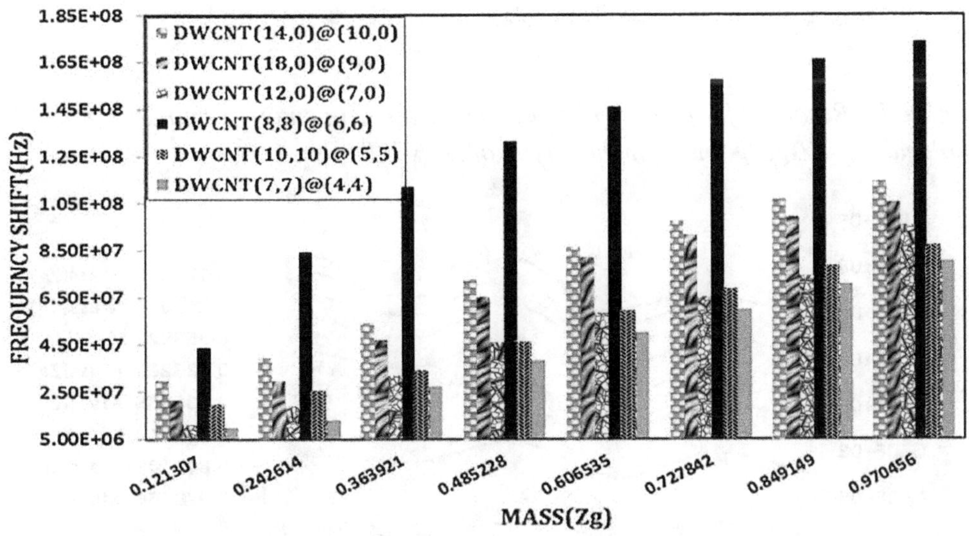

The sensitivities of six cantilevered DWCNT are calculated and represented in Table 8. It is clearly observed that zigzag DWCNTs possess better mass sensitivities as compared to armchair DWCNTs. These sensitivities are calculated using Equation (46).

Figure 9 shows the shifts in the resonant frequency due to the attached masses (08 molecules of alanine with amino terminal) for the cantilever, zigzag and armchair DWCNT. It can be visualized form that a higher frequency shift observed for armchair DWCNT (8,8)@(6,6) with the increase in the attached molecules of the biological object. Further a gradual reduction in the frequency shift is observed for other types of DWCNTs.

Figure 10. Resonant frequency shifts in the due to the attached masses (07 molecules of deoxyadeonosine) for the cantilever DWCNT

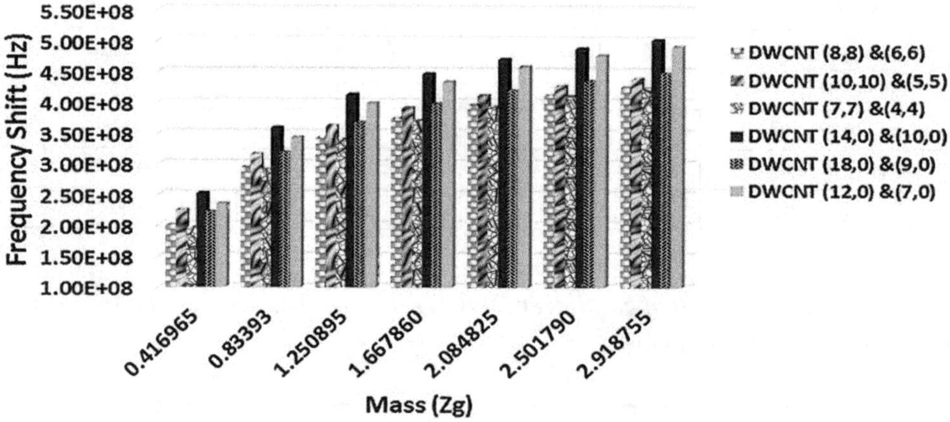

Figure 11. Resonant frequency shift due to Alanine with amino terminal residue 08 molecules vs. Zigzag and Armchair type of DWCNT

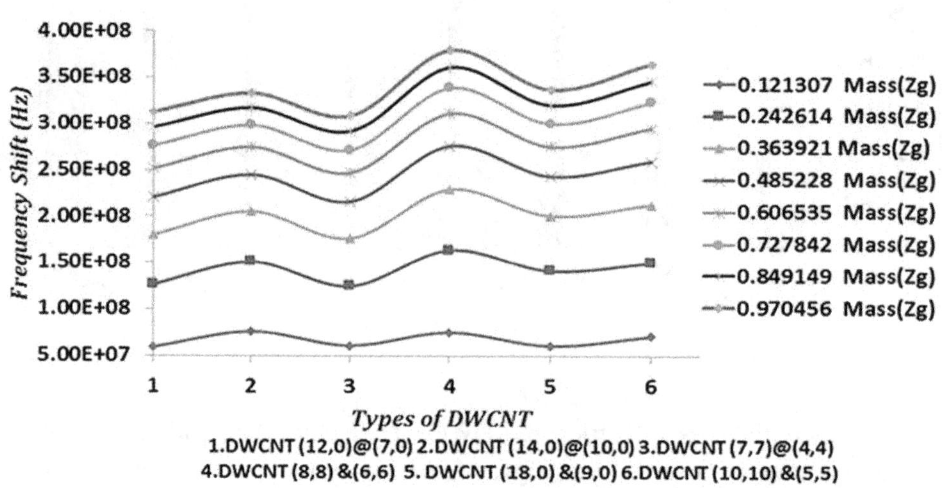

Figure 10 shows the shifts in the resonant frequency due to the attached masses (07 molecules of deoxyadeonosine) for the cantilever (zigzag and armchair) DWCNT. Similarly, Figure 11 shows frequency shift due to 08 molecules Alanine with amino terminal residue vs. Zigzag and Armchair type of DWCNT. In both the cases a higher frequency shift is observed in cantilever DWCNT (8,8)@(6,6).

Figure 12. Frequency shift of Cantilever DWCNT Zigzag and Armchair type of DWCNT vs. Bartonella Bacilliformis 07 molecules

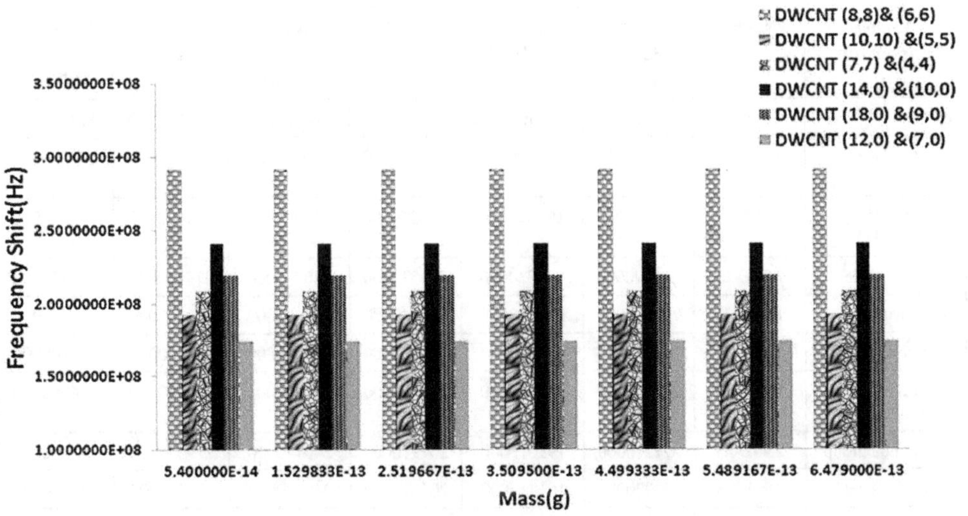

Figure 12 shows frequency shift due to 07 molecules of Coronaviridae Bartonella Bacilliformis respectively. The general tread is similar in both the cases with the maximum shift for armchair (8,8)@(6,6) CNTs, whereas the maximum shift is obtained for (14,0)@(10,0) in case of zigzag DWCNT. The reason for it can be attributed to the change in the orientation of the bonds in CNT which further leads to a change in stiffness variation and hence a change in the frequency shift. Also the sensitivity values given in Table 8 clearly suggest that zigzag CNTs are better candidates for sensing the viruses and biological objects as its sensitivity is 0.1424 Zg/GHz.

Table 9 summarizes the various bacterium/virus which have been taken for analysis along with their mass Elishakoff et al. (2011). It also shows resonant frequency of Armchair and Zigzag DWCNT with attached Bacterium/Viruses at the tip of outer walled DWCNT.

Table 10 shows resonant frequency of Chiral DWCNT with attached Bacterium/ Viruses at the tip of outer walled DWCNT.

As indicated in Figure 13 (a) Parvovirus B19 in blood. (b) Caliciviridae (c) Togaviridae (d) Polyomaviridae and (e) Strain of Staphylococcus aureus bacteria taken from a vancomycin intermediate resistant culture five Virus/ Bacterium with mass in Zeptogram scales have been used to simulate the vibration responses of zigzag, armchair and chiral DWCNT based biosensors:

Table 9. Resonant frequency of Armchair and Zigzag DWCNT with attached Bacterium/Viruses at the tip of outer walled DWCNT

Virus/ Bacterium	Mass (Kg)	DWCNT (8,8) & (6,6)	DWCNT (10,10) & (5,5)	DWCNT (7,7) & (4,4)	DWCNT (14,0) & (10,0)	DWCNT (18,0) & (9,0)	DWCNT (12,0) & (7,0)
ESAT-6	9.96E-24	291790000	192130000	208430000	240960000	219550000	174320000
Parvoviridae largest	4.28E-24	291769850	192109850	208409850	240939850	219529850	174299850
Parvoviridae smallest	1.29E-23	291710000	192050000	208350000	240880000	219470000	174240000
Picornaviridae	1.37E-23	291690000	192030000	208330000	240860000	219450000	174220000
Nodaviridae	1.71E-23	291600000	191940000	208240000	240770000	219360000	174130000
Tymovirus group	1.81E-23	291570000	191910000	208210000	240740000	219330000	174100000
Diantovirus group	1.94E-23	291530000	191870000	208170000	240700000	219290000	174060000
Nudaurelia group	2.90E-23	291270000	191610000	207910000	240440000	219030000	173800000
Caliciviridae	3.94E-23	290970000	191310000	207610000	240140000	218730000	173500000
Togaviridae largest	4.19E-23	290900000	191240000	207540000	240070000	218660000	173430000
Togaviridae smallest	2.25E-22	285370000	185710000	202010000	234540000	213130000	167900000
Staphilococcus	5.60E-23	290500000	190840000	207140000	239670000	218260000	173030000
Polyoma	6.30E-23	290300000	190640000	206940000	239470000	218060000	172830000
Orthomyxoviridae	6.23E-22	286220000	186560000	202860000	235390000	213980000	168750000

1. **ESAT-6:** The 6 kDa early secretory antigenic target produced by Mycobacterium tuberculosis, is a secretory protein and potent T cell antigen. It is used in tuberculosis diagnosis by the whole blood interferon γ test Quanti FERON-TB Gold, in conjunction with CFP-10 and TB7.7. It has been shown to directly bind to the TLR2 receptor, inhibiting downstream signal transduction Pathak et al. (2007).
2. **Parvoviruses:** Linear, non-segmented single-stranded DNA viruses, with an average genome size of 5000 nucleotides. Parvoviruses are among the smallest viruses and are 18–28 nm in diameter Leppard et al. (2007).
3. **Nodaviridae:** A family of viruses. Vertebrates and invertebrates serve as natural hosts Viral Zone. (2015). There are currently nine species in this family, divided among 2 genera. Diseases associated with this family include: viral encephalopathy and retinopathy in fishes Virus Taxonomy (2014).
4. **Caliciviridae:** A family of viruses, members of Class IV of the Baltimore scheme. They are positive-sense, single stranded RNA which is non-segmented. There are currently seven species in this family, divided among 5 genera Viral

Table 10. Resonant frequency of Chiral DWCNT with attached Bacterium/Viruses at the tip of outer walled DWCNT

Virus/ Bacterium	Mass (Kg)	DWCNT (13,2) & (9,2)	DWCNT (10,4) & (6,4)	DWCNT (11,3) & (7,3)	DWCNT (12,1) & (8,1)
ESAT-6	9.96E-24	158060000	202850000	114700000	107240000
Parvoviridae largest	4.28E-24	158039850	202829850	114679850	107219850
Parvoviridae smallest	1.29E-23	157980000	202770000	114620000	107160000
Picornaviridae	1.37E-23	157960000	202750000	114600000	107140000
Nodaviridae	1.71E-23	157870000	202660000	114510000	107050000
Tymovirus group	1.81E-23	157840000	202630000	114480000	107020000
Diantovirus group	1.94E-23	157800000	202590000	114440000	106980000
Nudaurelia group	2.90E-23	157540000	202330000	114180000	106720000
Caliciviridae	3.94E-23	157240000	202030000	113880000	106420000
Togaviridae largest	4.19E-23	157170000	201960000	113810000	106350000
Togaviridae smallest	2.25E-22	151640000	196430000	108280000	100820000
Staphilococcus	5.60E-23	156770000	201560000	113410000	105950000
Polyoma	6.30E-23	156570000	201360000	113210000	105750000
Orthomyxoviridae	6.23E-22	152490000	197280000	109130000	101670000

Zone, (2015). Diseases associated with this family include: feline calicivirus: respiratory disease; rabbit hemorrhagic disease virus: often-fatal hemorrhages; Norwalk group of viruses: gastroenteritis Virus Taxonomy. (2014).

5. **Togaviridae:** A family of viruses. Humans, mammals, birds, and mosquitoes serve as natural hosts. There are currently 32 species in this family, divided among 2 genera. Diseases associated with this family include: Alphaviruses: arthritis, encephalitis; Rubiviruses: rubella Viral Zone, (2015)
6. **Polyomaviridae:** A family of viruses. Mammals and birds serve as natural hosts. There are currently 13 species in this family, contained within a single genus, Polyomavirus, whose type species is Simian virus 40 (Viral Zone, 2015). Not all members of the family infect humans, but among those that do, several are associated with human disease: JC virus with progressive multifocal leukoencephalopathy, BK virus with mild respiratory infection (Virus Taxonomy: 2014).

Figure 13. (a) Parvovirus B19 in blood, (b) Caliciviridae, (c) Togaviridae, (d) Polyomaviridae, and (e) Strain of Staphylococcus aureus bacteria taken from a vancomycin intermediate resistant culture five virus/bacterium with mass in Zeptogram scales

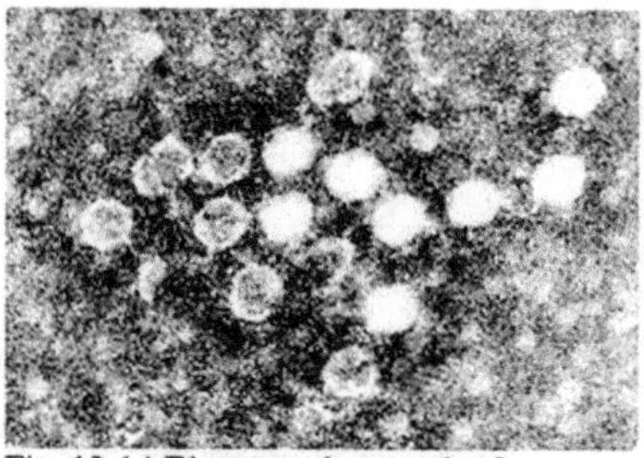

Fig. 13 (a) Electron micrograph of negatively stained Parvovirus B19 in blood. Each virus particle is about 25 to 30nM in diameter. [GrahamColm]

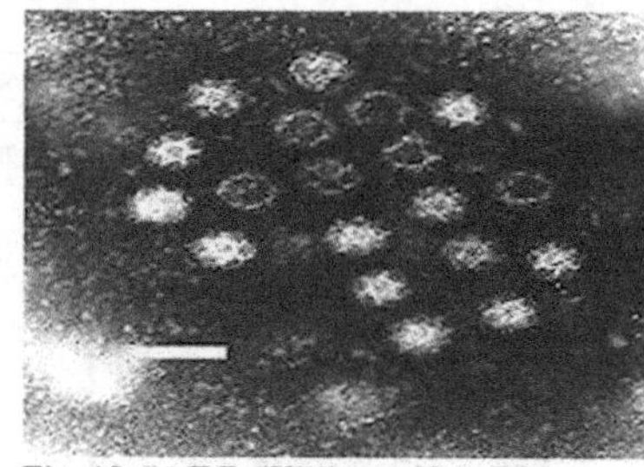

Fig. 13 (b) F.P. Williams, U.S. EPA - http://www.epa.gov/nerlcwww/calici.htm

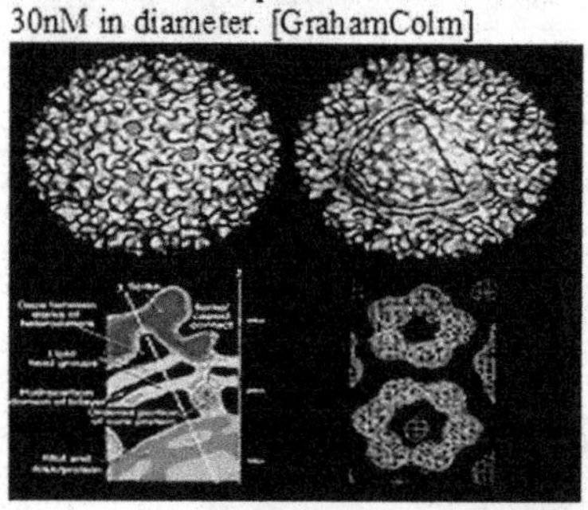

Fig13 (c) Togaviridae

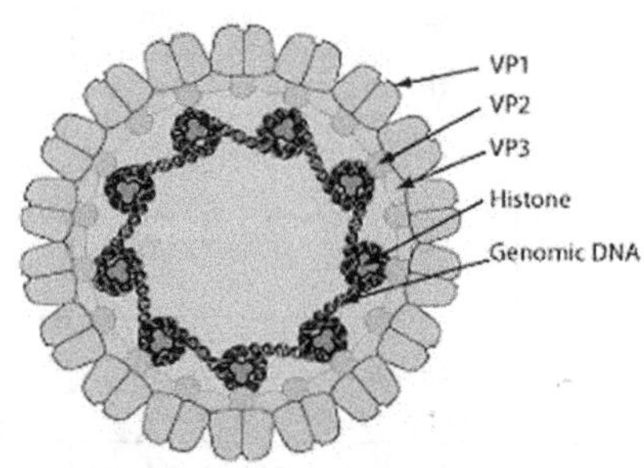

Fig 13 (d) Polyomaviridae www.virology.net

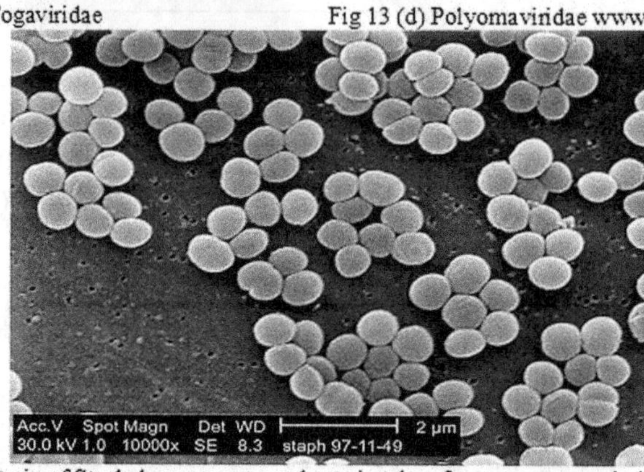

Fig 13 (e) Strain of Staphylococcus aureus bacteria taken from a vancomycin intermediate resistant culture [Janice Carr]

7. **Staphylococcus:** Genus includes at least 40 species. Of these, nine have two subspecies, one has three subspecies and one has four subspecies Harris et al. (2002) organisms. Found worldwide, they are a small component of soil microbial flora Madigan et al. (2005).

CONCLUSION

The use of double walled carbon nanotube and single walled carbon nanotube based biomass sensors for sensing Zeptogram scale mass has been investigated in this study. The shift in the resonant frequency due to the change in the attached mass has been used as a parameter for exploring DWCNT as a using four bio nano mass sensor and fourteen Virus/ Bacterium. As cantilever configurations are four times more sensitive than bridged configurations, only cantilever boundary conditions have been used in the present study.

A higher frequency shift is observed for armchair DWCNT (8,8)@(6,6) type of nanotube with the increase in the attached molecules of four the biological objects.

A greater amount of shift is observed for the molecule of alanine with amino terminal residue which has a smaller mass as compared to the deoxyadenosine with free residue, Coronaviridae and Bartonella Bacilliformis.

Based on the sensitivity and the frequency shift it can be concluded that cantilever zigzag DWCNTs are better candidates for detecting the biological objects

Vibrations of a cantilever DWCNT with attached various masses of virus at the tip have been studied in this chapter. Several bacteria or viruses masses have been chosen (as shown in Table 10) and feasibility of DWCNT as a Bacterium/ virus detector has been shown. Carbon nanotubes are the ideal material for this purpose and their small size makes them sensitive enough to resolve single atoms even at room temperature. While scientists already have the ability to measure the mass of individual atoms through a complex technique known as mass spectrometry Jensen et al. (2008) this new carbon nanotube based mass sensor offers some distinct advantages like it does not require the ionization of neutral atoms or molecules that can destroy samples such as proteins while sensing the mass. Finally, CNT and DWCNT is small enough so that, it could be incorporated onto a chip in future.

REFERENCES

Ajayan, P. (1999). Nanotubes from Carbon. *Chemical Reviews*, *99*(7), 1787–1800. doi:10.1021/cr970102g PMID:11849010

Ajayan, P. M., Charlier, J. C., & Rinzler, A. G. (1999). Carbon nanotubes: From macromolecules to nanotechnology. *Proceedings of the National Academy of Sciences of the United States of America*, *96*(25), 14199–14200. doi:10.1073/pnas.96.25.14199 PMID:10588679

(1803-1804). Allen O áHill, H. (1995). Immobilization of small proteins in carbon nanotubes: High-resolution transmission electron microscopy study and catalytic activity. *Journal of the Chemical Society. Chemical Communications*, 17.

Allinger, N. L., Yuh, Y. H., & Lii, J. H. (1989). Molecular mechanics. The MM3 force field for hydrocarbons. 1. *Journal of the American Chemical Society*, *111*(23), 8551–8566. doi:10.1021/ja00205a001

Baughman, R. H., Cui, C., Zakhidov, A. A., Iqbal, Z., Barisci, J. N., Spinks, G. M., & Jaschinski, O. et al. (1999). Carbon nanotube actuators. *Science*, *284*(5418), 1340–1344. doi:10.1126/science.284.5418.1340 PMID:10334985

Blevins, R. D., & Plunkett, R. (1980). Formulas for natural frequency and mode shape. *Journal of Applied Mechanics*, *47*(2), 461. doi:10.1115/1.3153712

Brenner, W., Shenderova, O. A., Harrison, J. A., Stuart, S. J., Ni, B., & Sinnott, S. B. (2002). Second generation reactive empirical bonds order (rebo) potential energy expression for hydrocarbons. *Journal of Physics*, *14*, 783–802.

Carroll, J. A., Coleman, S. A., Smitherman, L. S., & Minnick, M. F. (2000). Hemin-binding surface protein fromBartonella quintana. *Infection and Immunity*, *68*(12), 6750–6757. doi:10.1128/IAI.68.12.6750-6757.2000 PMID:11083791

Cho, H., Yu, M.-F., Vakakis, A. F., Bergman, L. A., & McFarland, D. M. (2010). Tunable Broadband Nonlinear Nanomechanical Resonator. *Nano Letters*, *10*(5), 1793–1798. doi:10.1021/nl100480y PMID:20384349

Chowdhury, R., & Adhikari, S. (2011). Boron nitride nanotubes as zeptogram-scale bionano sensors: Theoretical investigations. *IEEE Transactions on Nanotechnology*, *10*(4), 659–667. doi:10.1109/TNANO.2010.2060492

Chowdhury, R., Adhikari, S., & Mitchell, J. (2009). Vibrating carbon nanotube based bio-sensors. *Physica E, Low-Dimensional Systems and Nanostructures*, *42*(2), 104–109. doi:10.1016/j.physe.2009.09.007

Cornell, W. D., Cieplak, P., Bayly, C. I., Gould, I. R., Merz, K. M., Ferguson, D. M., & Kollman, P. A. (1996). A second generation force field for the simulation of proteins, nucleic acids, and organic molecules j. am. chem. soc. 1995, 117, 5179-5197. *Journal of the American Chemical Society*, *118*(9), 2309–2309. doi:10.1021/ja955032e

Darkrim, F. L., Malbrunot, P., & Tartaglia, G. P. (2002). Review of hydrogen storage by adsorption in carbon nanotubes. *International Journal of Hydrogen Energy*, *27*(2), 193–202. doi:10.1016/S0360-3199(01)00103-3

Davis, J. J., Green, M. L., Hill, H. A. O., Leung, Y. C., Sadler, P. J., Sloan, J., & Tsang, S. C. (1998). The immobilisation of proteins in carbon nanotubes. *Inorganica Chimica Acta*, *272*(1), 261–266. doi:10.1016/S0020-1693(97)05926-4

Dereli, G., & Ozdogan, C. (2003). Structural stability and energetics of single walled carbon nanotubes under uniaxial strain. *Physical Review B: Condensed Matter and Materials Physics*, *67*(3), 035416. doi:10.1103/PhysRevB.67.035416

Ding, F., Jiao, K., Wu, M., & Yakobson, B. I. (2007). Pseudoclimb and dislocation dynamics in superplastic nanotubes. *Physical Review Letters*, *98*(7), 075503. doi:10.1103/PhysRevLett.98.075503 PMID:17359035

Elishakoff, I., Versaci, C., Maugeri, N., & Muscolino, G. (2011). Clamped-free single-walled carbon nanotube-based mass sensor treated as Bernoulli–Euler beam. *Journal of nanotechnology in engineering and medicine, 2*.

Fennimore, A. M., Yuzvinsky, T. D., Han, W. Q., Fuhrer, M. S., Cumings, J., & Zettl, A. (2003). Rotational actuators based on carbon nanotubes. *Nature*, *424*(6947), 408–410. doi:10.1038/nature01823 PMID:12879064

Gelin, B. R. (1994). *Molecular modeling of polymer structures and properties*. Cincinnati, OH: Hanser/Gardner Publishers.

Georgantzinos, S. K., & Anifantis, N. K. (2010). Carbon nanotube-based resonant nanomechanical sensors: A computational investigation of their behavior. *Physica E, Low-Dimensional Systems and Nanostructures*, *42*(5), 1795–1801. doi:10.1016/j.physe.2010.02.002

Giannopoulos, G. I., Kakavas, P. A., & Anifantis, N. K. (2008). Evaluation of the effective mechanical properties of single walled carbon nanotubes using a spring based finite element approach. *Computational Materials Science*, *41*(4), 561–569. doi:10.1016/j.commatsci.2007.05.016

Gu, L., Elkin, T., Jiang, X., Li, H., Lin, Y., Qu, L., & Sun, Y.-P. et al. (2005). Single-walled carbon nanotubes displaying multivalent ligands for capturing pathogens. *Chemical Communications*, (7): 874. doi:10.1039/b415015e PMID:15700066

Haddon, R. C. (2002). Carbon nanotubes. *Accounts of Chemical Research*, *35*(12), 997–997. doi:10.1021/ar020259h PMID:12484786

Haitham, M. A., & Almajhdi, F. N. (2011). Development of a SYBR Green I based real-time RT-PCR assay for detection and quantification of bovine coronavirus. *Molecular and Cellular Probes*, *25*(2-3), 101–107. doi:10.1016/j.mcp.2011.03.001 PMID:21419844

Harris, L. G., Foster, S. J., & Richards, R. G. (2002). An introduction to Staphylococcus aureus, and techniques for identifying and quantifying S. aureus adhesins in relation to adhesion to biomaterials: review. *Eur. Cell. Mater.*, 4, 39–60.

He, X. Q., Kitipornchai, S., & Liew, K. M. (2005). Buckling analysis of multi-walled carbon nanotubes: A continuum model accounting for van der Waals interaction. *Journal of the Mechanics and Physics of Solids*, *53*(2), 303–326. doi:10.1016/j.jmps.2004.08.003

Hertel, T., Walkup, R. E., & Avouris, P. (1998). Deformation of carbon nanotubes by surface van der waals forces. *Physical Review B: Condensed Matter and Materials Physics*, *58*(20), 13870–13873. doi:10.1103/PhysRevB.58.13870

Hierold, C., Helbling, T., Roman, C., Durrer, L., Jungen, A., & Stampfer, C. (2008). CNT based sensors. In Advances in Science and Technology (Vol. 54, pp. 343-349). Trans Tech Publications. doi:10.4028/www.scientific.net/AST.54.343

Hod, O., Rabani, E., & Baer, R. (2003). Carbon nanotube closed-ring structures. *Physical Review B: Condensed Matter and Materials Physics*, *67*(19), 195408. doi:10.1103/PhysRevB.67.195408

Hood, L., Heath, J. R., Phelps, M. E., & Lin, B. (2004). Systems biology and new technologies enable predictive and preventative medicine. *Science*, *306*(5696), 640–643. doi:10.1126/science.1104635 PMID:15499008

Iijima, S. (1991). Helical microtubules of graphitic carbon. *Nature, 354,* 6348, 56.

Iijima, S., Brabec, C., Maiti, A., & Bernholc, J. (1996). Structural flexibility of carbon nanotubes. *The Journal of Chemical Physics*, *104*(5), 2089–2092. doi:10.1063/1.470966

Iijima, S., & Ichihashi, T. (1993). Single-shell carbon nanotubes of 1-nm diameter. *Nature*, *363*(6430), 603–605. doi:10.1038/363603a0

Jensen, K., Kim, K., & Zettl, A. (2008). An atomic-resolution nanomechanical mass sensor. *Nature Nanotechnology*, *3*(9), 533–537. doi:10.1038/nnano.2008.200 PMID:18772913

Joshi, A. Y., Harsha, S. P., & Sharma Satish, C. (2010c). Vibration Response Analysis of Doubly Clamped Single Walled Wavy Carbon Nanotube Based Nano Mechanical Sensors. *ASME Journal of Nanotechnology in Engineering & Medicine*, *1*, 031004-1–5.

Joshi, A. Y., Sharma, S. C., & Harsha, S. P. (2011). Effect of chirality and atomic vacancies on dynamics of nanoresonators based on SWCNT. *Sensor Review*, *31*(1), 47–57. doi:10.1108/02602281111099080

Joshi Anand, Y., Harsha, S. P., & Sharma Satish, C. (2010a). Vibration signature analysis of single walled carbon nanotube based nano mechanical sensors. *Physica E, Low-Dimensional Systems and Nanostructures*, *42*(8), 2115–2123. doi:10.1016/j.physe.2010.03.033

Joshi Anand, Y., Sharma Satish, C., & Harsha, S. P. (2008). Vibration Analysis of Pre-Stressed Single Walled CNT Based Mass Sensor. *International Journal of Electrospun Nanofibers and Applications*, *2*, 161–170.

Joshi Anand, Y., Sharma Satish, C., & Harsha, S. P. (2010b). The Effect of Pinhole Defect on Vibration Characteristics of Single Walled Carbon Nanotube. *Physica E, Low-Dimensional Systems and Nanostructures*.

Joshi Anand, Y., Sharma Satish, C., & Harsha, S. P. (2010d). Dynamic Analysis of a Clamped Wavy Single Walled Carbon Nanotube based Nano Mechanical Sensors. *ASME Journal of Nanotechnology in Engineering & Medicine*, *1*, 031007-1–7.

Kumar, O., Singh, Y., Rao, V. K., & Vijayaraghavan, R. (2008). Carbon nanotubes: Detection of chemical and biological warfare agents. *Defence Science Journal*, *58*(5), 617–625. doi:10.14429/dsj.58.1684

Lassagne, B., Garcia-Sanchez, D., Aguasca, A., & Bachtold, A. (2008). Ultrasensitive mass sensing with a nanotube electromechanical resonator. *Nano Letters*, *8*(11), 3735–3738. doi:10.1021/nl801982v PMID:18939809

Lee, S., & Yoon, D. S. (2007). Self-Assembly of Colloidal Particles on a Patterned Surface with Wettability. *Biochip Journal*, *1*, 193.

Leppard, K., Dimmock, N., & Easton, A. (2007). *Introduction to Modern Virology* (p. 450). Blackwell Publishing Limited.

Li, C., & Chou, T. W. (2003). A structural mechanics approach for the analysis of carbon nanotubes. *International Journal of Solids and Structures*, *40*(10), 2487–2499. doi:10.1016/S0020-7683(03)00056-8

Li, C., & Chou, T.-W. (2003b). Single-walled carbon nanotubes as ultrahigh frequency nano mechanical resonators. *Physical Review B: Condensed Matter and Materials Physics*, *68*, 734051–734053.

Li, C., & Chou, T.-W. (2004). Vibrational behaviors of multi-walled carbon nanotube-based nano mechanical resonators. *Applied Physics Letters*, *84*(1), 121–123. doi:10.1063/1.1638623

Li, C., & Chou, T. W. (2004). Mass detection using carbon nanotube-based nanomechanical resonators. *Applied Physics Letters*, *84*(25), 5246–5248. doi:10.1063/1.1764933

Lin, Y., Taylor, S., Li, H., Fernando, K. S., Qu, L., Wang, W., & Sun, Y. P. (2004). Advances toward bioapplications of carbon nanotubes. *Journal of Materials Chemistry*, *14*(4), 527–541. doi:10.1039/b314481j

Lu, F., Gu, L., Meziani, M. J., Wang, X., Luo, P. G., Veca, L. M., & Sun, Y. P. (2009). Advances in bioapplications of carbon nanotubes. *Advanced Materials*, *21*(2), 139–152. doi:10.1002/adma.200801491

Madigan, M., & Martinko, J. (Eds.), (2005). *Brock Biology of Microorganisms* (11th ed.). Prentice Hall.

Maguina C, Gotuzzo E. (2000). Bartonellosis: New and old. *Infect. Dis. Clin. North Am.*, 14, 1–22. Doi: 10.1016/S0891-5520(05)70215-4

Mahan, G. D. (2002). Oscillations of a thin hollow cylinder: Carbon nanotubes. *Physical Review B: Condensed Matter and Materials Physics*, *65*(23), 235402. doi:10.1103/PhysRevB.65.235402

Mattson, M. P., Haddon, R. C., & Rao, A. M. (2000). Molecular functionalization of carbon nanotubes and use as substrates for neuronal growth. *Journal of Molecular Neuroscience*, *14*(3), 175–182. doi:10.1385/JMN:14:3:175 PMID:10984193

Meirovitch, L. (1997). *Principles and techniques of vibrations* (Vol. 1). New Jersey: Prentice Hall.

Patel, Ajay M., Anand Y Joshi. (2014a). Computational Investigation of mass sensing using defective Double walled Carbon Nanotubes, Procedia. *Materials Science*, *5*, 482–488.

Patel, Ajay M., Joshi A.Y. (2014e). Effect of Waviness on the Dynamic Characteristics of Double Walled Carbon Nanotubes, *Nanoscience Nano technology Letter*, 6, 1-9.

Patel, Ajay M., Anand Y Joshi. (2015b). Evaluating the Vibrational Characteristics of Double Walled Carbon Nanotubes with Pinhole Defects, *Current NanoScience. Bentham Science Publishers*, *11*(3), 371–378.

Patel, A. M., Gosai, N., & Joshi, A. Y. (2015c). A Review on Defects in Carbon Nanotubes. *Applied Mechanics and Materials, Trans Tech Publications, 813-814*, 145–150. doi:10.4028/www.scientific.net/AMM.813-814.145

Patel, A. M., & Joshi, A. Y. (2013). Vibration Analysis of Double Wall Carbon Nanotube based Resonators for Zeptogram level Mass Recognition. *Computational Materials Science, Elsevier Publishers, 79*, 230–238. doi:10.1016/j.commatsci.2013.06.022

Patel, A. M., & Joshi, A. Y. (2014b). Vibration Analysis of Defective Double Walled Carbon Nanotube Based Nano Resonators. *Proceedings of ASME 2014 International Mechanical Engineering Congress and Exposition* (pp. V010T13A024-V010T13A024). American Society of Mechanical Engineers. doi:10.1115/IMECE2014-36454

Patel, A. M., & Joshi, A. Y. (2014c). Investigating the influence of surface deviations in double walled carbon nanotube based nanomechanical sensors. *Computational Materials Science*, *89*, 157–164. doi:10.1016/j.commatsci.2014.03.034

Patel, A. M., & Joshi, A. Y. (2014d). Investigation of Double walled Carbon Nanotubes for Mass Sensing. *Procedia Technology*, *14*, 290–294. doi:10.1016/j.protcy.2014.08.038

Patel, A. M., & Joshi, A. Y. (2015a). Influence of atomic vacancies on the dynamic characteristics of nanoresonators based on double walled carbon nanotube. *Physica E, Low-Dimensional Systems and Nanostructures*, *70*, 90–100. doi:10.1016/j.physe.2015.02.016

Pathak, S. K., Basu, S., Basu, K. K., Banerjee, A., Pathak, S., Bhattacharyya, A., & Basu, J. et al. (2007). Direct extracellular interaction between the early secreted antigen ESAT-6 of Mycobacterium tuberculosis and TLR2 inhibits TLR signaling in macrophages. *Nature Immunology*, *8*(6), 610–618. doi:10.1038/ni1468 PMID:17486091

Qian, D., Wagner, G. J., Liu, W. K., Yu, M. F., & Ruoff, R. S. (2002). Mechanics of carbon nanotubes. *Applied Mechanics Reviews*, *55*(6), 495–533. doi:10.1115/1.1490129

Rivas, G. A., Rubianes, M. D., Rodriguez, M. C., Ferreyra, N. F., Luque, G. L., Pedano, M. L., & Parrado, C. (2007). Carbon nanotubes for electrochemical biosensing. *Talanta*, *74*(3), 291–307. doi:10.1016/j.talanta.2007.10.013 PMID:18371643

Saito, R., Matsuo, R., Kimura, T., Dresselhaus, G., & Dresselhaus, M. S. (2001). Anomalous potential barrier of double-wall carbon nanotube. *Chemical Physics Letters*, *348*(3-4), 187–193. doi:10.1016/S0009-2614(01)01127-7

Sanchez-Portal, D., Artacho, E., Soler, J. M., Rubio, A., & Ordejón, P. (1999). Ab initio structural, elastic and vibrational properties of carbon nanotubes. *Physical Review B: Condensed Matter and Materials Physics*, *59*(19), 12678–12688. doi:10.1103/PhysRevB.59.12678

Smalley, R. E. (2003). *Carbon nanotubes: synthesis, structure, properties, and applications*. Springer Science & Business Media.

Sohlberg, K., Sumpter, B. G., Tuzun, R. E., & Noid, D. W. (1998). Continuum methods of mechanics as a simplified approach to structural engineering of nanostructures. *Nanotechnology*, *9*(1), 30–36. doi:10.1088/0957-4484/9/1/004

Tserpes, K. I., & Papanikos, P. (2005). Finite element modeling of single-walled carbon nanotubes. *Composites. Part B, Engineering*, *36*(5), 468–477. doi:10.1016/j.compositesb.2004.10.003

Wang, J. (2005). Carbon-nanotube based electrochemical biosensors: A review. *Electroanalysis*, *17*(1), 7–14. doi:10.1002/elan.200403113

Wang, X. Y., & Wang, X. (2004). Numerical simulation for bending modulus of carbon nanotubes and some explanations for experiment. *Composites. Part B, Engineering*, *35*(2), 79–86. doi:10.1016/S1359-8368(03)00084-2

Wikipedia. (2015). Togaviridae. Retrieved from https://en.wikipedia.org/wiki/Togaviridae

Wong, S. S., Joselevich, E., Woolley, A. T., Cheung, C. L., & Lieber, C. M. (1998). Covalently functionalized nanotubes as nanometre-sized probes in chemistry and biology. *Nature*, *394*(6688), 52–55. doi:10.1038/27873 PMID:9665127

Wu, W., Palaniapan, M., & Wong, W. K. (2008). Multiwall carbon nanotube resonator for ultra-sensitive mass detection. *Electronics Letters*, *44*(18), 1932–1933. doi:10.1049/el:20081932

Wu, W., Palaniapan, M., & Wong, W. K. (2008). Multiwall carbon nanotube resonator for ultra-sensitive mass detection. *Electronics Letters*, *44*(18), 1932–1933. doi:10.1049/el:20081932

Xu, K. Y., Aifantis, E. C., & Yan, Y. H. (2008). Vibrations of double-walled carbon nanotubes with different boundary conditions between inner and outer tubes. *Journal of Applied Mechanics*, *75*(2), 021013. doi:10.1115/1.2793133

Yakobson, B., Brabec, C., & Bernholc, J. (1996). Nanomechanics of carbon tubes: Instabilities beyond linear response. *Physical Review Letters*, *76*(14), 2511–2514. doi:10.1103/PhysRevLett.76.2511 PMID:10060718

Yang, Y., Callegari, C., Feng, X., Ekinci, K., & Roukes, M. (2006). Zeptogram scale nanomechanical mass sensing. *Nano Letters*, *6*(4), 583–586. doi:10.1021/nl052134m PMID:16608248

Yoon, J., Ru, C. Q., & Mioduchowski, A. (2003). Vibration of an embedded multiwall carbon nanotube. *Composites Science and Technology*, *63*(11), 1533–1542. doi:10.1016/S0266-3538(03)00058-7

Yun, Y., Dong, Z., Shanov, V., Heineman, W. R., Halsall, H. B., Bhattacharya, A., & Schulz, M. J. et al. (2007). Nanotube electrodes and biosensors. *Nano Today*, *2*(6), 30–37. doi:10.1016/S1748-0132(07)70171-8

Zone, V. (2015). ExPASy. Retrieved from https://en.wikipedia.org/wiki/Caliciviridae

Chapter 5
CNS Targeted Nanoparticle Drug Delivery:
CNS Drug Delivery

Dimple Sethi Chopra
Punjabi University, India

ABSTRACT

The idea of formulating brain permeable nanoparticles stems from the need to treat various neurological disorders like Parkinson's disease, Alzheimer's disease, schizophrenia, depression and brain tumors. Neuropeptides, antibiotics, anticancer drugs and many CNS active drugs cannot cross blood brain barrier (BBB). Studies have revealed that when these drugs are loaded on to nanoparticles they not only cross BBB, but also exhibit decreased side effects. The drug can be dissolved, dispersed, encapsulated inside the nanoparticle or attached on to surface of nanoparticles. In 1995, dalargin was the first drug to be delivered across blood brain barrier (BBB) using polysorbate 80 coated nanoparticles. The size of nanoparticles is usually between 10-1000nm. For crossing BBB it should be less than 300 nm.

INTRODUCTION

In the past few years, several high profile global pharmaceutical companies have shutdown major research activities within the neuroscience area. This is because the development of new psychotherapeutic drugs has become a high risk activity with

DOI: 10.4018/978-1-5225-1043-7.ch005

little success. CNS drug development takes considerable amount of time. Out of 5000 potential candidate molecules, only 250 reach pre-clinical evaluation. Out of 250 reaching the pre-clinical stage, 10 reach clinical development. Out of these 10 molecules entering clinical development, only one gets approved by the regulatory authorities and get to the market. The time for development of one drug molecule takes 12-15 years. The cost of developing a new drug is typically US$10-15 billion. Moreover, CNS drugs usually fail at later stage of clinical trials, after significant investment has been made (Wegner & Rujescu, 2013). The new drug molecule should possess desired pharmacological activity, selectivity at receptor site to avoid undesirable side effects, optimal bioavailability, ability to penetrate the blood-brain barrier. The BBB and the efflux transporters are the main cause of failure of most of the CNS drugs under clinical trial.

Ehrlich and his colleagues very the first to describe the existence of blood-brain barrier (BBB) in 1885. They showed that intravenous injection of acidic dye tryptan blue was not able to color brain and spinal cord. This gave an indication of presence of some barrier which separated the CNS from rest of the body (Ehrlich, 1885). The existence of a barrier at the level of cerebral vessels was first postulated by Bield and Kraus (1898) and Lewandowsky (1900) based on the observation that the intravenous injection of cholic acids or sodium ferrocyanide had no pharmacological effects on the CNS, whereas neurological symptoms occurred after intraventricular application of the same substances. After the development of electron microscope, it was found that cerebral endothelial cells were morphologically different from other endothelial cells in the body. The tight junctions between the endothelial cells forms the cellular basis of BBB. The brain capillary endothelial cells are surrounded by astrocytes, microglial cells, pericytes. BBB protect brain from general circulation and toxic substances. Astrocytes are periendothelial structures which play an important role in brain homeostasis. They are involved in maintenance of potassium ion levels, inactivation of neurotransmitters, regulation and production of growth factors and cytokines (Ehrlich,1885). Damage to the BBB or alterations in transport systems may facilitate pathogenesis of many CNS diseases, including HIV-associated CNS dysfunction. HIV-1 infection can result in neuropathologic changes in about one half of infected individuals and also can result in damage to the BBB. HIV-1 and the HIV-1 viral proteins, Tat and gp120, cause alterations in the integrity and function of the BBB through both paracellular and transcellular mechanisms (McRae, 2016).

Mechanisms of Transport Across BBB

The transport of solute molecules across the BBB membrane takes place by four different mechanisms:

1. **Paracellular Diffusion:** This is a non-saturable and non-competitive process occurring between cells. The presence of tight junctions in brain endothelial cells greatly restricts paracellular diffusion. Only small water-soluble molecules can diffuse through the BBB.
2. **Transcellular Diffusion:** It is also non-saturable and non-competitive. And takes place across cells. Molecules which are highly lipophilic and have molecular weight less than 450 Da undergo transcellular diffusion.
3. **Carrier-Mediated Transport:** These carriers are membrane bound and facilitate active transport of specific nutrients across the BBB. Glucose transporters GLUT1 and GLUT3 are of great importance as glucose is main energy source of brain. Monocarboxylate transporters are another important carrier systems for lactate and pyruvate. Similarly specialized carriers exist for amino acids and vitamins. Drugs or prodrugs can cross BBB through this transport mechanism by mimicking nutrients (Ohtsuki & Terasaki, 2007).
4. **Receptor-Mediated Endocytosis:** It facilitates the transport of large endogenous proteins and hormones across the BBB. It is mediated by receptors present on the luminal side of the barrier. Specific receptors have been identified for insulin, insulin-like growth factors, angiotensin II, folates and transferrin. Lipids can also be internalized in the brain in the form of low-density lipoproteins (LDL), which are recognized and endocytosed by the endothelial cells, due to the expression of apolipoproteins on the surface of the LDL (Yoshikawa & Pardridge, 2001).

The massive brain microvasculature has a total surface area of 20 m^2 and a total length of 640 km. The brain capillaries present no fenestrae, few pinocytosis vesicles, and particular tight junctions, also known as zonula occludens, controlling the movement of molecules through the paracellular pathway. This comprises an almost impermeable barrier for drugs administered through the peripheral circulation (Kreuter, 2013). Only small molecules can diffuse through it. BBB prevents transport of many important drugs from blood to brain including anticancer drugs, antibiotics, drugs for treatment of Alzheimer's and Parkinson's disease. It has been a real challenge in developing drugs for CNS disorders (Jain, 2009). A number of methods have been developed to overcome this barrier. Chemical modifications of the drugs, such as increasing their lipophilicity, turn them more prone to be internalized in the brain. Osmotic opening of the tight junctions, direct surgical administration of drugs into the brain, use of prodrugs, use of carrier systems like antibodies or liposomes: Use of molecular tools to bind the drugs such as small immunoglobulins, liposomes or nanoparticles that will act as Trojan horses favouring the drug delivery in brain (Pardridge, 2015).

POLYMERIC NANOPARTICLES

In 1995, nanoparticles were first time used to deliver drugs across the BBB following intravenous administration. The underlying mechanism was found to be receptor-mediated endocytosis followed by transcytosis. Their was no evidence of tight junction opening of brain endothelial cells. Nanoparticles coated with surfactants like polysorbate-80 or poloxamers 188 take advantage of physiological receptor-mediated transport across the BBB. Once nanoparticles enter the blood, they get covered by a corona of biomolecules. These biomolecules are responsible for interaction of nanoparticles with the cellular processes. The physical adsorption of apolipoprotein A-I and /or E mediate interaction with LDL receptors on the BBB. The covalent attachment of apolipoprotein A-I and /or E or of transferring, insulin or antibodies against the respective receptors also enables a similar nanoparticle-mediated drug transport across the BBB (Ohtsuki & Terasaki, 2007).

Apolipoprotein E was chemically linked to human serum albumin nanoparticle (HAS-NP). Loperamide, which does not cross the blood-brain barrier but exerts antinociceptive effects after direct injection into the brain, was used as model drug. This preparation induced antinociceptive effects in the tail-flick test in ICR mice after i.v. injection. In contrast, nanoparticles linked to apolipoprotein E variants that do not recognize lipoprotein receptors failed to induce these effects. These results indicate that apolipoprotein E attached to the surface of nanoparticles facilitates transport of drugs across the blood brain barrier, probably after interaction with lipoprotein receptors on the brain capillary endothelial cell membranes (Michaelis *et al.*, 2006). HAS-NP with different vectors, transferrin, and apoE have been used in two independent series of experiments. Similar results were achieved for the delivery of loperamide to the CNS in both cases (Ulbrich *et al.*, 2009). Transferrin (Tf) is a single glycoprotein synthesized in the liver, and is the main iron-containing protein in blood plasma. Upon combination with Fe^{2+}, its molecular structure changes and it acquires binding affinity for the transferrin receptor (TfR) (Widera et al., 2003; Yang et al., 2009). TfR is widely expressed in capillary endothelial cells of the BBB, and targeted administration *via* the Tf-TfR pathway is a promising method of drug delivery (Xing *et al.*, 2016).

Polymeric nanoparticles are solid colloidal systems of 1-1000nm size, composed of biocompatible copolymers that have low solubility in water (Craparo *et al.*, 2011; Kumar *et al.*, 2010). Synthetic polymers like poly(alkyl cyanoacrylates); polyesters such as polylactides (PLAs); poly (D,L-lactide-co-glycolate) (PLGA) have been used for preparation of nanoparticles. Nanoparticles made of PLA and PLGA are synthesized by emulsion-diffusion and precipitation, whereas poly(alkyl cyanoacrylates) nanoparticles are synthesized by emulsion polymerization and nanoprecipitation method (Craparo *et al.*,2011). Nanoparticles are able to transport substances adsorbed onto their surface or covalently bound or incorporated into their polymer matrix, or encapsulated in them. It is difficult to reveal localization of the incorporated substance, therefore the terms "nanospheres" and "nanocapsules" are not commonly used, instead the term "nanoparticles" is generally used (Denora *et al.*, 2009). Nanoparticles present in the blood are quickly cleared because of opsonization by RES cells. Decreasing the size of nanoparticles increases duration of their circulation in the blood (Pinto *et al.*, 2006). However, adsorption of surfactants on their surface has been shown to be more effective for prolongation of their circulation time in blood (Kreuter, 2001). Modification of the surface of the nanoparticles by adsorption onto their surface or covalent linking of hydrophilic polymers, such as PEG, polysorbate 80, or polysaccharides, leads to an increase in the circulation of nanoparticles in the blood, whereas incorporation of molecules that recognize cellular determinants helps in penetration of nanoparticles through the BBB (Kreuter *et al.*, 1995).

PLGA is a block co-polymer made of polylactic acid (PLA) and polyglycolic acid (PGA) that has been approved by the FDA and European Medicine Agency. PLGA has excellent biocompatibility, in the human body it is hydrolyzed into lactic and glycolic acids, which are naturally-occurring metabolites. PLGA nanoparticles can be easily modified to accommodate a range of functional or targeting options. However, PLGA NPs without modifications also have a number of intrinsic drawbacks. Their negative surface charge inhibits makes cellular uptake. They have a relatively short blood circulation time and they do not pass through the BBB without further modifications (Cai *et al.*,2016).

Polybutylcyanoacrylate nanoparticles were used to facilitate transport of drugs through the BBB (Kreuter *et al.*, 1995). Kreuter and Alyautdin, 2000 performed in vivo experiments that demonstrated the possibility for brain delivery of peptide dalargin using polybutylcyanoacrylate nanoparticles coated with polysorbate 80. It was shown in subsequent experiments with H^3-dalargin that the absence of the polymeric coating of nanoparticles significantly reduced the delivery of the peptide to the brain (Schroeder *et al.*, 2000). The polar hydrophilic drugs tubocurarine (Alyautdin *et al.*, 1998) and neostigmine, P-gp substrates loperamide (Alyautdin *et al.*, 1997) and doxorubicin (Steiniger *et al.*, 2004) and the nerve growth factor (NGF) a protein have also been delivered to the brain through the BBB using this transport system (Kurakhmaeva *et al.*, 2009). These data support the fact that adsorption of polysorbates onto the surface of nanoparticles leads to changes in the characteristics of nanoparticle distribution. Analysis showed that penetration of nanoparticles is provided by stimulation of receptor-mediated endocytosis; primarily, by interaction of nanoparticles with LDL receptors on brain endothelial cells (Wohlfart *et al.*, 2012). The reason for such interaction may be due to adsorption of endogenic LDL from the plasma by polysorbate 80. The role of apoE has been demonstrated by Michaelis *et al.*, 2006. Two types of albumin nanoparticles one coated with polysorbate 80 and other with covalently bound apoE on their surface exhibited approximately equal ability to transport drugs through BBB. It has been found that polysorbate absorbed on the surface of polymeric nanoparticles has three effects: it allows for an increase in the duration of nanoparticle circulation in the blood by decreasing the chance of being captured by RES cells; it indirectly stimulates receptor-dependent endocytosis; and it directly influences Tight junctions of the BBB (Kreuter, 2004; Kreuter *et al.*, 2003; Kreuter *et al.*, 2002).

Rivastigmine, a reversible cholinesterase inhibitor used for treatment of Alzheimer's disease. It was shown that, compared to the free drug, rivastigmine-loaded poly(n-butylcyanoacrylate) (PBCA) nanoparticles (Riv-NP), particularly after being coated with polysorbate 80 (Riv-NP + PS80), accumulated to a greater extent in the brain. It was postulated by Wilson *et al.*, 2008 that this could translate into a better efficacy of rivastigmine. It was found that, compared to Riv-NP, Riv-NP + PS80 accumulates in a significantly lesser amount in the liver, which could potentially limit organ-specific toxicity and drug–drug interactions. Delivery of antineoplastic agents across the BBB has been shown to be improved by utilization of a nanoparticles. Gemcitabine, has proven to have beneficial effect on solid tumors such as non-small-cell lung cancer, pancreatic cancer, and breast cancer. Following incorporation into PBCA and coating with polysorbate 80 (Gem-PBCA + PS80),compared to gemcitabine alone, was found to have significantly improved antiproliferative activity on C6 glioma cells in an *in vitro* assay. The survival time was significantly increased in rats treated with Gem-PBCA + PS80 as compared to those treated with free gemcitabine (Wang *et al.*, 2009). Polymeric nanoparticles coated with polysorbate 80 have been used successfully to transport tacrine (Wilson *et al.*, 2008) and antitumor agent doxorubicin to the brain (Gulyaev *et al.*, 1999). It has been shown both *in vitro* and *in vivo* that polymeric nanoparticles penetrated the BBB, accumulated in the brain parenchyma resulting in strong antitumor effects (Kreuter and Gelperina, 2008). The polymers used as matrices for nanoparticles includes albumin, PLA, PLGA, chitosan, solid lipids, and polyethyleneimines. The main advantages of these polymers are low toxicity, biodegradability, presence of functional groups on the surface, processing stability, and the possibility to influence the excretion rate of drug molecules (Olivier, 2005). The development of novel and smart polymers have led to increase in number of transport vectors that can ferry drug molecules across BBB. Similar to the use of apoE as a vector for receptor-mediated transport, transferrin (OX26) or antibodies to transferrin receptor, insulin and antibodies to insulin, and TAT proteins have been used as targeting agents. Transferrin provides iron transport in the organism; therefore, there is a relatively high concentration of this protein in the plasma, which decreases its transport potential. Antibodies situated on the surface of nanoparticles interact with transferrin receptors in ECs, making it possible to deliver nanoparticle-receptor complex across BBB through transcytosis (Pardridge, 2006). Transferrin has been used to transport albumin nanoparticles containing loperamide through the BBB. The possibility of drug transport to the brain using antibodies to human insulin receptor was first demonstrated by Pardridge in monkeys in 1995. Ulbrich et al demonstrated, in 2011, the possibility for transporting the opioid agonist loperamide through the BBB in experiments using albumin nanoparticles with insulin attached to their surface. Antibodies to the insulin receptor inhibited transport of loperamide to the brain (Ulbrich *et al.*, 2011). TAT peptide is one of the most frequently used vectors for drug transport to the brain. This peptide, via a domain composed of nine to 16 amino acids, may penetrate biological membranes using mechanisms that are independent of transporter- or receptor-mediated transport (Torchilin, 2008) PLA nanoparticles, containing ritonavir and coated with TAT protein, increased the concentration of the drug in the brain 800 times (Rao *et al.*, 2008). Besides this, recent evidence suggests the usage of another cell penetrating peptide (CPP), syncytin-B (SynB), as a vector for drug transport across the BBB (Tian *et al.*, 2012). This group of investigators showed very promising data with nanoparticles that were conjugated to SynB peptide using PEG and gelatin–siloxane. Another protein, gH625, which is a membrane-perturbing domain in the glycoprotein gH of Herpes simplex virus type I, was conjugated onto polystyrene nanoparticles for transport across the BBB. This study was performed in vitro and can be applied to the design of one or more systems of drug delivery to the brain (Guarnieri *et al.*, 2013). Interesting data regarding CPP penetration, which delivered antibodies to the prion protein into mouse brain, were obtained (Skrlj *et al.*, 2013). Placing serotonin on the surface of a nanoparticle led to recognition by the serotonin receptor on the surface of cells, thereby stimulating endocytosis; in the same way, DNA was delivered to the cell (Gopal *et al.*, 2011). Both liposomes (Mora *et al.*, 2002) and polymeric nanoparticles (Nowacek *et al.*, 2009) were successfully used for cell-mediated drug transport. This approach of drug delivery to the brain has several advantages. Targeting of drug to disease sites, resulted in prolonged drug half-lives and time-controlled drug release (Gendelman *et al.*, 2008). The concept of polymeric nanoparticles seems to be conventional, but surface conjugation of these carriers with peptides have improved their targeting capabilities as can be easily see in Table 1. It can be clearly seen that these versatile carriers can be developed as diagnostics for Alzheimer's disease.

Novel Non-Viral Vector for Brain Targeting

The application of viral vectors in drug delivery is restricted by issues such as toxicity, and their is utmost need for safe and effective treatment. Liposomes are a new non-viral vector for the delivery of drugs to the brain, such as doxorubicin,

Table 1. Brain permeable theranostic polymeric nanoparticles

Drug/Bioactive Molecule	Polymer Used/ Stabilizer	Surfactant Coating	Particle Size	Results	References
Dalargin	Polybutyl-cyanoacrylate	Polysorbate-80	260 nm	Significant & Prolonged Anesthesia	Kreuter and Alyautdin, 2000
Tubocurarine (Positively charged)	Polybutyl-cyanoacrylate/ dextran 70 KDa	Polysorbate-80	230 nm	Significant increase in brain concentration	Alyautdi *et al.,* 1998
Loperamide	PLGA	Glycosylated heptapeptide	155 nm	100 times increase in brain concentration	Vergoni *et al.*, 2009
Dopamine	PLGA	Polysorbate-80	215 nm	Significant and sustained increase in antinociceptive effects	Pahuja *et al.*, 2015
Paclitaxel	PLGA	PEG	114 nm	Improved penetration into brain	Nance *et al.*, 2012
(125)I-clioquinol	Polybutyl-cyanoacrylate	--------	200 nm	Promising delivery vehicle For SPECT/PET amyloid imaging (Alzheimer's disease)	Kulkarni *et al.*, 2010
Sulpiride	PEG-PLA	Thiolated cationized bovine serum ablbumin conjugation	329nm	Improved concentration in brain and dose reduction	Parikh *et al.*, 2010
Rifampicin, Isoniazid, Pyrazinamide- ,Ethambutol combination	PLGA	------	186-290 nm	Significant and prolonged drug levels in brain	Pandey & Khuller, 2006
Efavirenz, Lopinavir, Ritonavir antiretroviral combination	PLGA	-------	138.3± 55.4 nm	Significant and suatained levels in brain for 28 days.	Destache *et al.*, 2010
Model drug Coumarin-6	PEG-PLA	Penetratin conjugation	100 nm	Enhanced brain uptake	Xia *et al.,*2012

which can be entrapped in the aqueous phase of the liposome (Schnyder et al., 2005; Xu, 2011; Gao et al., 2013). Brain-derived neurotrophic factor (BDNF) is widely expressed in the central nervous system, especially in the hippocampus and cortex (Bertrand, 2010; Qin, 2011). BDNF and its receptor TrkB play an important role in regulating various biological properties, including neuronal survival, differentiation, growth and plasticity (Ploughman et al., 2009; Tuinstra, 2012; Zhang, 2012). In the developing brain, endogenous BDNF levels are very low; therefore, in cases of brain injury (such as ischemia, degeneration and inflammation) during development, administration of exogenous BDNF is a promising treatment approach. However, as a macromolecular protein, exogenous BDNF cannot directly traverse the blood-brain barrier (BBB) or other biomembranes to exert its therapeutic function (Vijender, 2015). Recently it was investigated whether BDNF could be transported across the blood-brain barrier by tail-vein injection of liposomes conjugated to transferrin (Tf) and polyethylene glycol (PEG), and carrying BDNF modified with cytomegalovirus promoter (pCMV) or glial fibrillary acidic protein promoter (pGFAP) (Tf-pCMV-BDNF-PEG and Tf-pGFAP-BDNF-PEG, respectively). Both liposomes were able to traverse the blood-brain barrier, and BDNF was mainly expressed in the cerebral cortex. BDNF expression in the cerebral cortex was higher in the Tf-pGFAP-BDNF-PEG group than in the Tf-pCMV-BDNF-PEG group. This study demonstrates the successful construction of a non-virus targeted liposome, Tf-pGFAP-BDNF-PEG, which crosses the blood-brain barrier and is distributed in the cerebral cortex (Xing *et al.*, 2016).

Cell-Mediated Drug Transport

Immunocytes including mononuclear phagocytes (dendritic cells, monocytes and macrophages), neutrophils, and lymphocytes) are highly mobile; they have the ability to migrate across impermeable barriers and release their drug cargo at sites of infection or tissue injury. Thus immune cells can be exploited as trojan horses for drug delivery. Using cells as delivery vehicles enables targeted drug transport, and prolonged circulation times, along with reductions in cell and tissue toxicities. For successful cell-mediated drug delivery, high drug loading into cell-carriers, efficient preservation of entrapped therapeutic agents against disintegration and clearance in the host cells, drug triggered release at the site of action, efficient homing of cell-carriers to a disease site, safety of cell-based drug formulations for the whole organism are some of the prerequisite which needs to be achieved. The design of nanocarriers for cell-mediated drug delivery differ from those used for conventional drug delivery systems. Two different approaches are presently utilized in cell-mediated drug delivery: genetically modified cell-carriers producing therapeutically active molecules; and drug loaded cell-carriers used as "Trojan horses" to deliver the drug to the disease site. This disease combating strategy that can be applied to a spectrum of human infectious, cancerous, and degenerative disorders (Batrakova *et al.,* 2011).

To target circulating mononuclear phagocytic cells, negatively-charged nano-sized liposomes were formulated encapsulating serotonin, a BBB impermeable neurological drug. In vitro and in vivo studies confirmed the ability of monocytes to carry nano-sized liposomes loaded with serotonin across the BBB (Afergan *et al.*, 2008). There was also evidence of in vitro delivery of catalase to the brain using a polymeric nanocarrier. Catalase was immobilized on a polyethyleneimine-poly(ethylene glycol) copolymer, and such "nanozymes" were phagocytosed by bone-marrow-derived macrophages (Batrakova *et al.*, 2008). The polymer protected the enzyme from being activated inside the immune cell, while the macrophage delivered the antioxidant enzyme to the brain to reduce oxidative stress and increase survival of dopaminergic neurons in patients with Parkinson's disease.

Efflux Transporters in Brain

Efflux transporters contribute significantly in maintaining brain homeostasis by eliminating toxic endogenous and exogenous substances. The efflux transporters of the ATP-binding cassette (ABC) P-glycoprotein (Pgp) can actively transport large quantities of lipophilic drugs out of the brain capillary endothelial cells. Some of the other ABC efflux transporters are multidrug resistance protein (MRP) and breast cancer resistance protein (BCRP). ABC efflux transporters are beneficial in reducing or preventing adverse effects of drugs and neurotoxic substances. But they also limit distribution of drugs that are valuable in treating CNS disorders. Moreover, neurological disorders such as epilepsy may be associated with overexpression of ABC efflux transporters, resulting in resistance to therapeutic medication. Hence, modulation of ABC efflux transporters forms a novel strategy to enhance the penetration of drugs into the brain (Löscher and Potschka, 2005).

Another very popular ATP-dependent drug transport protein P-glycoprotein (P-gp) is present on the luminal side of BCEC. It is a 170-kDa membrane-associated glycoprotein that actively transports structurally diverse molecules. It has been established, both in vitro and in vivo, that BBB. P-gp can prevent accumulation of drug molecules in the brain, by expelling them from the cells. Inhibition of the P-gp function can be obtained through down-regulation of its expression at transcriptional or translational levels, or by altering membrane targeting after synthesis of the P-gp protein, or through chemical inhibition of transporter functions. The first two approaches have been less developed, whereas chemical inhibition of the transporter functions has been extensively studied for improving permeability across BBB (Tsuji and Tamai,1997; Schinkel, 1999). An alternative approach is use of nanoparticles to carry the drug into the endothelial cells by endocytosis, thereby bypassing the P-gp drug efflux mechanism (Miller *et al.*, 2008).

The activity of P-gp, is increased in pathological conditions like epilepsy, stroke (Spudich *et al.*, 2006; Yang and Rosenberg,2011*)*. This reduces drug efficacy. P-gp actively transport out phenytoin, levetiracetam, lamotrigine, and Phenobarbital ((Denora *et al.,* 2009*)*. In the case of HIV infection, trans-activating transcriptor (TAT) proteins from virion stimulate P-gp activity (Hayashi *et al.*, 2005). Same is true for antitumor drugs like paclitaxel, vincristine and ivermectin. In transgenic mice lacking P-gp, the concentration of vincristine and ivermectin in the brain increased 80–100-fold than that in normal mice. Paclitaxel is effective on tumor cells in vitro, but does not show same activity in vivo because of P-gp efflux. The concentration of paclitaxel can be significantly increased by simultaneous administration of paclitaxel and an inhibitor of P-gp, PSC833 (Denora *et al.,* 2009*)*. The first-generation inhibitors of P-gp, verapamil, diltiazem and cyclosporine, are used to increase brain delivery of viral protease inhibitors, paclitaxel and itraconazole, docetaxel, digoxin, and imatinib. But, they had limitations, like low affinity to P-gp, toxic effects due to high doses of these inhibitors, and simultaneous inhibition of liver microsomal enzymes (CYP3A). A second-generation P-gp inhibitor, an analog of cyclosporine valspodar 10, has great affinity for the transporter, with reduced toxicity and decreased effects on liver microsomal enzymes. A third generation of P-gp inhibitors, including elacridar 11, zosuquidar 12, and tariquidar 13, are devoid of effect on liver microsomal enzymes (CYP3A). Modulation of BCRP/ABCG2 activity also has significant clinical importance due to the possibility of mutual overlapping with P-gp (Laquintana *et al.,*2009*)*. Elacridar 11 and the proton pump inhibitor pantoprazole are known inhibitors of BCRP (Litman *et al.*, 2000). Inhibition of P-gp is desirable for delivery of therapeutic molecules but would at the same time open the BBB to potentially toxic substances. This risk can be acceptable for the treatment of patients with cancer, but should be carefully weighed with long-term therapy of chronic CNS diseases (Alyautdin *et al.*, 2014).

ROLE OF NANOTECHNOLOGY IN TREATMENT OF BRAIN TUMORS

Glioblastomas belong to the most aggressive human cancers with short survival times. It is currently being treated with radiation therapy and surgery. One of the most important problems regarding the treatment of Glioblastomas is the presence of blood brain barrier (BBB) which inhibits the efficient drug delivery into central nervous system (Huse *et al.*, 2013). Nanotechnology can deliver therapeutic drugs into CNS by crossing the BBB. There are different types of nanoparticles (Nps) which can be manipulated for clinical applications as a treatment for CNS-related disorders (Pourgholi *et al.,*2016). Steiniger *et al.*, 2004 have shown that doxorubicin bound to polysorbate-coated nanoparticles crossed the intact blood-brain barrier, thus reaching therapeutic concentrations in the brain. Additionally, binding to nanoparticles may reduces systemic toxicity of doxorubicin. The standard therapy for GBM is surgery followed by radiation and chemotherapy with temozolomide (TMZ). Treatment of GBMs remains a challenge, largely due to the fast degradation of TMZ, inability to deliver an effective dose of TMZ to tumors, and lack of target specificity which may cause systemic toxicity (Zhang *et al.*, 2012). Recently Fang *et al.*, 2015 presented a simple method to synthesize a nanoparticle-based carrier that can protect TMZ from rapid degradation in physiological solutions and can specifically deliver them to GBM cells through the mediation of a tumor targeting peptide chlorotoxin (CTX). This nanoparticle, namely NP-TMZ-CTX, had a hydrodynamic size of less than 100 nm, exhibited sustained stability in cell culture media for up to two weeks, and could accommodate stable drug loading. The NP contains a biotin-functionalized chitosan core that is covalently conjugated with TMZ that were linked to GBM-specific, surface-bound CTX ligands with fast, stable, and stoichiometric biotin-neutravidin interactions (Petros and Desimone, 2010). NP-TMZ-CTX showed higher uptake in GBM cells and was more effective in cell killing than non-targeted NPs and free drug. It was further demonstrated that NP-TMZ-CTX can cross the BBB and deliver TMZ into avascular region of brain. NP-TMZ-CTX offers hope to deliver effective doses of TMZ specifically to GBM cells while minimizing dose-limiting toxicity on other healthy tissues. This NP-based delivery system can be broadly applied to various applications as it can be tailored to incorporate various therapeutics that are otherwise unstable or insoluble in physiological conditions (Fang *et al.*, 2015).

CONCLUSION AND FUTURE PROSPECTS

Biologics such as antibodies because of their large size are unable to cross BBB. They can be made to cross BBB by re-engineering them as an IgG fusion protein. The IgG is a monoclonal antibody against endogenous BBB receptor i.e transferring receptor. So the biologic is ferried into the brain by receptor mediated endocytosis. Many of ligand targeted nanomedicines for brain targeting have progressed into clinical trials. The fusion of classical pharmacology with nanotechnology has opened up many different approaches for new generation of neuroactive drugs to access BBB. An example is targeted drug delivery to brain of BDNF using liposome as non-viral vector. Green synthesized nanoparticles can also be used in discovery of biomarkers and improvement of diagnosis. This forms the basis for new drug development for neurological disorders. It further explores new methods for drug delivery across blood-brain barrier. Nanotechnology based implants are being developed to facilitate regeneration of the nervous system. It is being speculated that femtolasers, nanorobots and other nanodevices will bring advancement in neurosurgery. Finally, the progress of BBB permeability of drugs and biologics are highly dependent on highly precise *in vitro* BBB models which have good prediction power. These models are extremely useful for investigating intracellular sorting mechanism of receptors. This requires microscopy at subcellular resolution. This will also help in studying vesicular compartments involved in transcytosis and finding new transcytosis receptors. All the molecules designed for BBB penetration cannot be tested in animals due to ethical considerations. Many models have been designed to mimic BBB *in vitro*. But none can replace *in vivo* investigations completely. Hence, great care should be exercised in the analysis and interpretation of the results obtained with cultured cells and extrapolating it to the native BBB *in vivo*.

REFERENCES

Afergan, E., Epstein, H., Dahan, R., Koroukhov, N., Rohekar, K., Danenberg, H. D., & Golomb, G. (2008). Delivery of serotonin to the brain by monocytes following phagocytosis of liposomes. *Journal of Controlled Release*, *132*(2), 84–90. doi:10.1016/j.jconrel.2008.08.017 PMID:18805446

Alyautdin, R., Khalin, I., Nafeeza, M. I., Haron, M. H., & Kuznetsov, D. (2014). Nanoscale drug delivery systems and the blood–brain barrier. *International Journal of Nanomedicine*, *9*, 795–811. PMID:24550672

Alyautdin, R. N., Petrov, V. E., Langer, K., Berthold, A., Kharkevich, D. A., & Kreuter, J. (1997). Delivery of loperamide across the blood–brain barrier with polysorbate 80-coated polybutylcyanoacrylate nanoparticles. *Pharmaceutical Research*, *14*(3), 325–328. doi:10.1023/A:1012098005098 PMID:9098875

Alyautdin, R. N., Tezikov, E. B., Ramge, P., Kharkevich, D. A., Begley, D. J., & Kreuter, J. (1998). Significant entry of tubocurarine into the brain of rats by adsorption to polysorbate 80-coated polybutylcyanoacrylate nanoparticles: An in situ brain perfusion study. *Journal of Microencapsulation*, *15*(1), 67–74. doi:10.3109/02652049809006836 PMID:9463808

Batrakova, E. V., Gendelman, H. E., & Kabanov, A. V. (2011). Cell-mediated drugs delivery. *Expert Opinion on Drug Delivery*, *8*(4), 415–443. doi:10.1517/17425247.2011.559457 PMID:21348773

Batrakova, E. V., Li, S., Reynolds, A. D., Mosley, R. L., Bronich, T. K., Kabanov, A. V., & Gendelman, H. E. (2007). A macrophage-nanozyme delivery system for Parkinson's disease. *Bioconjugate Chemistry*, *18*(5), 1498–1506. doi:10.1021/bc700184b PMID:17760417

Bertrand, N., Simard, P., & Leroux, J. C. (2010). Serum-stable long-circulating pH-sensitive PEGylated liposomes. *Methods in Molecular Biology (Clifton, N.J.)*, *605*, 545–558. doi:10.1007/978-1-60327-360-2_36 PMID:20072905

Bield, A., & Kraus, R. (1898). U¨ ber eine bisher unbekannte toxische Wirkung der Gallensaurenauf das Zentralnervensystem. *Zhl Inn Med*, *19*, 1185–1200.

Cai, Q., Wang, L., Deng, G., Liu, J., Chen, Q., & Chen, Z. (2016). Systemic delivery to central nervous system by engineered PLGA nanoparticles. *American Journal of Translational Research*, *8*(2), 749–764. PMID:27158367

Chirio, D., Gallarate, M., Peira, E., Battaglia, L., Muntoni, E., & Riganti, C. et al. (2014). Positive-charged solid lipid nanoparticles as paclitaxel drug delivery system in glioblastoma treatment. *European Journal of Pharmaceutics and Biopharmaceutics*, *88*(3), 746–758. doi:10.1016/j.ejpb.2014.10.017 PMID:25445304

Craparo, E. F., Bond, M. L., Pitarresi, G., & Cavallaro, G. (2011). Nanoparticulate systems for drug delivery and targeting to the central nervous system. *CNS Neuroscience & Therapeutics*, *17*(6), 670–677. doi:10.1111/j.1755-5949.2010.00199.x PMID:20950327

Denora, N., Trapani, A., Laquintana, V., Lopedota, A., & Tropani, G. (2009). Recent advances in medicinal chemistry and pharmaceutical technology-strategies for drug delivery to the brain. *Current Topics in Medicinal Chemistry*, *9*(2), 182–196. doi:10.2174/156802609787521571 PMID:19200004

Destache, C. J., Belgum, T., Goede, M., Shibata, A., & Belshan, M. A. (2010). Antiretroviral release from poly(DL-lactide-co-glycolide) nanoparticles in mice. *The Journal of Antimicrobial Chemotherapy*, *65*(10), 2183–2187. doi:10.1093/jac/dkq318 PMID:20729545

Ehrlich, P. (1885). Oxygen necessity of the organism. A colourful Analytical Study. August Hirschwald, Berlin, Germany.

Fang, C., Wang, K., Stephen, Z. R., Mu, Q., Kievit, F. M., & Chiu, D. T. et al. (2015). Temozolomide Nanoparticles for Targeted Glioblastoma Therapy. *ACS Applied Materials & Interfaces*, *7*(12), 6674–6682. doi:10.1021/am5092165 PMID:25751368

Freskgard, P. O., Niewoehner, J., & Urich, E. (2014). Time to Open the Blood–Brain Barrier Gate for Biologics? *Future Neurology*, *9*(3), 243–245. doi:10.2217/fnl.14.15

Gao, J. Q., Lv, Q., Li, L. M., Tang, X. J., Li, F. Z., Hu, L. Y., & Han, M. (2013). Glioma targeting and blood-brain barrier penetration by dual-targeting doxorubincin liposomes. *Biomaterials*, *34*(22), 5628–5639. doi:10.1016/j.biomaterials.2013.03.097 PMID:23628475

Gendelman, H. E., Kabanov, A., & Linder, J. (2008). The promise and perils of CNS drug delivery: A video debate. *Journal of Neuroimmune Pharmacology*, *3*(2), 58–64. doi:10.1007/s11481-008-9103-9 PMID:18322803

Gopal, V., Xavier, J., Dar, G. H., Jafurulla, M., Chattopadhyay, A., & Rao, N. M. (2011). Targeted liposomes to deliver DNA to cells expressing 5-HT receptors. *International Journal of Pharmaceutics*, *419*(1-2), 347–354. doi:10.1016/j.ijpharm.2011.08.004 PMID:21855617

Guarnieri, D., Falanga, A., Muscetti, O., Tarallo, R., Fusco, S., & Galdiero, M. et al. (2013). Shuttle-mediated nanoparticle delivery to the blood brain barrier. *Small*, *9*(6), 853–862. doi:10.1002/smll.201201870 PMID:23135878

Gulyaev, A. E., Gelperina, S. E., Skidan, I. N., Antropov, A. S., Kivman, G. Y., & Kreuter, J. (1999). Significant transport of doxorubicin into the brain with polysorbate 80-coated nanoparticles. *Pharmaceutical Research*, *16*(10), 1564–1569. doi:10.1023/A:1018983904537 PMID:10554098

Hayashi, K., Pu, H., Tian, J., Andras, I. E., Lee, Y. W., Hennig, B., & Toborek, M. (2005). HIV-Tat protein induces P-glycoprotein expression in brain microvascular endothelial cells. *Journal of Neurochemistry*, *93*(5), 1231–1241. doi:10.1111/j.1471-4159.2005.03114.x PMID:15934943

Huse, J. T., Holland, E., & DeAngelis, L. M. (2013). Glioblastoma: Molecular Analysis and Clinical Implications. *Annual Review of Medicine*, *64*(1), 59–70. doi:10.1146/annurev-med-100711-143028 PMID:23043492

Jain, K. K. (2009). Current status and future prospects of nanoneurology. *Journal of Nanoneuroscience*, *1*(1), 56–64. doi:10.1166/jns.2009.006

Kreuter, J. (2001). Nanoparticulate systems for brain delivery of drugs. *Advanced Drug Delivery*, *47*(1), 65–81. doi:10.1016/S0169-409X(00)00122-8 PMID:11251246

Kreuter, J. (2004). Influence of the surface properties on nanoparticle-mediated transport of drugs to the brain. *Journal of Nanoscience and Nanotechnology*, *4*(5), 484–488. doi:10.1166/jnn.2003.077 PMID:15503433

Kreuter, J. (2013). Mechanism of polymeric nanoparticle-based drug transport across the blood-brain barrier (BBB). *Journal of Microencapsulation: Micro and Nano Carriers*, *30*(1), 49–54. doi:10.3109/02652048.2012.692491 PMID:22676632

Kreuter, J., Alyautdin, R., Kharkevich, D., & Ivanov, A. A. (1995). Passage of peptides through the blood–brain barrier with colloidal polymer particles (nanoparticles). *Brain Research*, *674*(1), 171–174. doi:10.1016/0006-8993(95)00023-J PMID:7773690

Kreuter, J., & Alyautdin, R. N. (2000). Using nanoparticles to target drugs to the central nervous system. In D. J. Begley, M. W. Bradbury, & J. Kreuter (Eds.), *The Blood–Brain Barrier and Drug Delivery to the CNS* (pp. 205–223). New York, NY: Marcel Dekker, Inc. doi:10.1201/9780824741990

Kreuter, J. & Gelperina, S (2008). Use of nanoparticles for cerebral cancer. *Tumori journal* 94, 271–277.

Kreuter, J., Ramge, P., Petrov, V., Hamm, S., Gelperina, S. E., & Engelhardt, B. et al. (2003). Direct evidence that polysorbate-80-coated poly (butylcyanoacrylate) nanoparticles deliver drugs to the CNS via specific mechanisms requiring prior binding of drug to the nanoparticles. *Pharmaceutical Research*, *20*(3), 409–416. doi:10.1023/A:1022604120952 PMID:12669961

Kreuter, J., Shamenkov, D., Petrov, V., Ramge, P., Cychutek, K., Koch-Brandt, C., & Alyautdin, R. (2002). Apolipoprotein mediated transport of nanoparticle-bound drugs across the blood–brain barrier. *Journal of Drug Targeting*, *10*(4), 317–325. doi:10.1080/10611860290031877 PMID:12164380

Kulkarni, P. V., Roney, C. A., Antich, P. P., Bonte, F. J., Raghu, A. V., & Aminabhavi, T. M. (2010). Quinoline-*n*-butylcyanoacrylate-based nanoparticles for brain targeting for the diagnosis of Alzheimer's disease. *WIREs Nanomed Nanobiotechnol*, *2*(1), 35–47. doi:10.1002/wnan.59 PMID:20049829

Kumari, A., Yadav, S. K., & Yadav, S. C. (2010). Biodegradable polymeric nanoparticles based drug delivery systems. *Colloids and Surfaces*, *75*(1), 1–18. doi:10.1016/j.colsurfb.2009.09.001 PMID:19782542

Kurakhmaeva, K. B., Kijindijkhasvili, I. A., Petrov, V. E., Balabanyan, V. U., Voronina, T. A., & Trofimov, S. S. et al. (2009). Brain targeting of nerve growth factor using poly(butyl cyanoacrylate) nanoparticles. *Journal of Drug Targeting*, *17*(8), 564–574. doi:10.1080/10611860903112842 PMID:19694610

Laquintana, V., Trapani, A., Denora, N., Wang, F., Gallo, G. M., & Tripani, G. (2009). New strategies to deliver anticancer drugs to brain tumors. *Expert Opinion on Drug Delivery*, *6*(10), 1017–1032. doi:10.1517/17425240903167942 PMID:19732031

Lewandowsky, M. (1900). Zur Lehre der Zerebrospinalflussigkeit. *Zeitschrift fur Klinische Medizin*, *40*, 480–484.

Litman, T., Brangi, M., Hudson, E., Fetsch, P., Abati, A., & Ross, D. D. et al. (2000). The multidrug-resistant phenotype associated with overexpression of the new ABC half-transporter, MXR (ABCG2). *Journal of Cell Science*, *113*(Pt 11), 2011–2021. PMID:10806112

Löscher, W., & Potschka, H. (2005). Blood-Brain Barrier Active Efflux Transporters: ATP-Binding Cassette Gene Family. *NeuroRx*, *2*(1), 86–98. doi:10.1602/neurorx.2.1.86 PMID:15717060

McRae, M. (2016). HIV and viral protein effects on the blood brain barrier. *Tissue Barriers*. 28;4(1). doi:.10.1080/21688370.2016.1143543

Michaelis, K., Hoffmann, M. M., Dreis, S., Herbert, E., Alyautdin, R. N., & Michaelis, M. et al. (2006). Covalent linkage of apolipoprotein E to albumin nanoparticles strongly enhances drug transport into the brain. *The Journal of Pharmacology and Experimental Therapeutics*, *317*(3), 1246–1253. doi:10.1124/jpet.105.097139 PMID:16554356

Miller, D. S., Bauer, B., & Hartz, A. M. S. (2008). Modulation of P-glycoprotein at the blood–brain barrier: opportunities to improve central nervous system pharmacotherapy, *Pharmacol. Rev.* 60:196–209.

Mora, M., Sagrista, M. L., Trombetta, D., Bonina, F. P., Pasquale, A. D., & Saija, A. (2002). Design and characterization of liposomes containing long-chain N-acylPEs for brain delivery: Penetration of liposomes incorporating GM1 into the rat brain. *Pharmaceutical Research*, *19*(10), 1430–1438.

Nance, E. A., Woodworth, G. F., Sailor, K. A., Shih, T. Y., Xu, Q., & Swaminathan, G. et al. (2012). Dense Poly(Ethylene Glycol) Coating Improves Penetration of Large Polymeric Nanoparticles Within Brain Tissue. *Science Translational Medicine*, *4*(149), 149ra119. doi:10.1126/scitranslmed.3003594 PMID:22932224

Nowacek, A. S., Miller, R. L., McMillan, J., Kanmogne, G., Kanmogne, M., & Mosley, R. L. et al. (2009). NanoART synthesis, characterization, uptake, release and toxicology for human monocyte-macrophage drug delivery. *Nanomedicine (London)*, *4*(8), 903–917. doi:10.2217/nnm.09.71 PMID:19958227

Ohtsuki, S., & Terasaki, T. (2007). Contribution of carrier-mediated transport systems to the blood–brain barrier as a supporting and protecting interface for the brain; importance for CNS drug discovery and development. *Pharmaceutical Research*, *24*(9), 1745–1758. doi:10.1007/s11095-007-9374-5 PMID:17619998

Olivier, J. C. (2005). Drug transport to brain with targeted nanoparticles. *NeuroRx*, *2*(1), 108–119. doi:10.1602/neurorx.2.1.108 PMID:15717062

Pahuja, R., Seth, K., Shukla, A., Shukla, R. K., Bhatnagar, P., & Chauhan, L. K. et al. (2015). Trans-blood brain barrier delivery of dopamine-loaded nanoparticles reverses functional deficits in Parkinsonian rats. *ACS Nano*, *9*(5), 4850–4871. doi:10.1021/nn506408v PMID:25825926

Pandey, R., & Khuller, G. K. (2006). Oral nanoparticle-based antituberculosis drug delivery to the brain in an experimental model. *The Journal of Antimicrobial Chemotherapy*, *57*(6), 1146–1152. doi:10.1093/jac/dkl128 PMID:16597631

Pardridge, W. (2006). Molecular Trojan horses for blood–brain barrier drug delivery. *Current Opinion in Pharmacology*, *6*(5), 494–500. doi:10.1016/j.coph.2006.06.001 PMID:16839816

Pardridge, W. M. (2001). Crossing the blood–brain barrier: Are we getting it right? *Drug Discovery Today*, *6*(1), 1–2. doi:10.1016/S1359-6446(00)01583-X PMID:11165157

Pardridge, W. M. (2015). Blood–brain barrier drug delivery of IgG fusion proteins with a transferrin receptor monoclonal antibody. *Expert Opinion on Drug Delivery*, *12*(2), 207–222. doi:10.1517/17425247.2014.952627 PMID:25138991

Parikh, T., Bommana, M. M., & Squillante, E. III. (2010). Efficacy of surface charge in targeting pegylated nanoparticles of sulpiride to the brain. *European Journal of Pharmaceutics and Biopharmaceutics*, *74*(3), 442–450. doi:10.1016/j.ejpb.2009.11.001 PMID:19941957

Petros, R. A., & Desimone, J. M. (2010). Strategies in the Design of Nanoparticles for Therapeutic Applications. *Nature Reviews. Drug Discovery*, *9*(8), 615–627. doi:10.1038/nrd2591 PMID:20616808

Pinto, R. C., Neufeld, R. J., Ribeiro, A. J., & Veiga, F. (2006). Nanoencapsulation I. Methods for preparation of drug-loaded polymeric nanoparticles. *Nanomed-Nanotechnology*, *2*(1), 8–21. doi:10.1016/j.nano.2005.12.003 PMID:17292111

Ploughman, M., Windle, V., MacLellan, C. L., White, N., Dore, J. J., & Corbett, D. (2009). Brain-derived neurotrophic factor contributes to recovery of skilled reaching after focal ischemia in rats. *Stroke*, *40*(4), 1490–1495. doi:10.1161/STROKEAHA.108.531806 PMID:19164786

Pourgholi, F., Hajivalili, M., Farhad, J. N., Kafil, H. S., & Yousefi, M. (2016). Nanoparticles: Novel vehicles in treatment of Glioblastoma. *Biomedicine and Pharmacotherapy*, *77*, 98–107. doi:10.1016/j.biopha.2015.12.014 PMID:26796272

Qin, Y., Chen, H., Zhang, Q., Wang, X., Yuan, W., & Kuai, R. et al. (2011). Liposome formulated with TAT-modified cholesterol for improving brain delivery and therapeutic efficacy on brain glioma in animals. *International Journal of Pharmaceutics*, *420*(2), 304–312. doi:10.1016/j.ijpharm.2011.09.008 PMID:21945185

Rao, K. V., Reddy, M. K., Horning, J. L., & Labhasetwar, V. (2008). TAT-conjugated nanoparticles for the CNS delivery of anti-HIV drugs. *Biomaterials*, *29*(33), 4429–4438. doi:10.1016/j.biomaterials.2008.08.004 PMID:18760470

Schinkel, A. H. (1999). P-glycoprotein, a gatekeeper in the blood–brain barrier. *Advanced Drug Delivery Reviews*, *36*(2-3), 179–194. doi:10.1016/S0169-409X(98)00085-4 PMID:10837715

Schnyder, A., Krähenbühl, S., Drewe, J., & Huwyler, J. (2005). Targeting of daunomycin using biotinylated immunoliposomes: Pharmacokinetics tissue distribution and *in vitro* pharmacological effects. *Journal of Drug Targeting*, *13*(5), 325–335. doi:10.1080/10611860500206674 PMID:16199376

Schroeder, U., Schroeder, H., & Sabel, B. A. (2000). Body distribution of 3H-labelled dalargin bound to poly(butyl cyanoacrylate) nanoparticles after iv injections to mice. *Life Sciences*, *66*(6), 495–502. doi:10.1016/S0024-3205(99)00619-0 PMID:10794066

Skrlj, N., Drevenšek, G., Hudoklin, S., Romih, R., Šerbec, V. C., & Dolinar, M. (2013). Recombinant single-chain antibody with the Trojan peptide penetratin positioned in the linker region enables cargo transfer across the blood–brain barrier. *Applied Biochemistry and Biotechnology*, *169*(1), 159–169. doi:10.1007/s12010-012-9962-7 PMID:23160949

Spudich, A., Kilic, E., Xing, H., Kilic, U., Rentsch, K. M., & Allenspach, H. W. et al. (2006). Inhibition of multidrug resistance transporter-1 facilitates neuroprotective therapies after focal cerebral ischemia. *Nature Neuroscience*, *9*(4), 487–488. doi:10.1038/nn1676 PMID:16565717

Steiniger, S. C., Kreuter, J., Khalansky, A. S., Skidan, I. N., Bobruskin, A. I., & Smirnova, Z. S. et al. (2004). Chemotherapy of glioblastoma in rats using doxorubicin-loaded nanoparticles. *International Journal of Cancer*, *109*(5), 759–767. doi:10.1002/ijc.20048 PMID:14999786

Tajes, M., Fernández, E. R., Jiang, X. W., Morató, M. B., Guivernau, B., & Pichot, A. E. et al. (2014). The blood-brain barrier: Structure, function and therapeutic approaches to cross it. *Molecular Membrane Biology*, *31*(5), 152–167. doi:10.3109/09687688.2014.937468 PMID:25046533

Tian, X. H., Wei, F., Wang, T. X., Wang, P., Lin, X. N., & Wang, J. et al. (2012). In vitro and in vivo studies on gelatin-siloxane nanoparticles conjugated with SynB peptide to increase drug delivery to the brain. *International Journal of Nanomedicine*, *7*, 1031–1041. PMID:22403486

Torchilin, V. P. (2008). Tat peptide-mediated intracellular delivery of pharmaceutical nanocarriers. *Advanced Drug Delivery Reviews*, *60*(4–5), 548–558. doi:10.1016/j.addr.2007.10.008 PMID:18053612

Tsuji, A., & Tamai, I. (1997). Blood–brain barrier function of P-glycoprotein. *Advanced Drug Delivery Reviews*, *25*(2-3), 287–298. doi:10.1016/S0169-409X(97)00504-8

Tuinstra, H. M., Aviles, M. O., Shin, S., Holland, S. J., Zelivyanskaya, M. L., & Fast, A. G. et al. (2012). Multifunctional multichannel bridges that deliver neurotrophin encoding lentivirus for regeneration following spinal cord injury. *Biomaterials*, *33*(5), 1618–1626. doi:10.1016/j.biomaterials.2011.11.002 PMID:22130565

Ulbrich, K., Hekmatara, T., Herbert, E., & Kreuter, J. (2009). Transferrin-and transferrin-receptor-antibody-modified nanoparticles enable drug delivery across the blood–brain barrier (BBB). *European Journal of Pharmaceutics and Biopharmaceutics*, *71*(2), 251–256. doi:10.1016/j.ejpb.2008.08.021 PMID:18805484

Ulbrich, K., Knobloch, T., & Kreuter, J. (2011). Targeting the insulin receptor: Nanoparticles for drug delivery across the blood–brain barrier (BBB). *Journal of Drug Targeting*, *19*(2), 125–132. doi:10.3109/10611861003734001 PMID:20387992

van der Meel, R., Vehmeijer, L. J., Kok, R. J., Storm, G., & van Gaal, E. V. (2013). Ligand-targeted particulate nanomedicines undergoing clinical evaluation: Current status. *Advanced Drug Delivery Reviews*, *65*(10), 1284–1298. doi:10.1016/j.addr.2013.08.012 PMID:24018362

Vergoni, A. V., Tosi, G., Tacchi, R., Vandelli, M. A., Bertolini, A., & Costantino, L. (2009). Nanoparticles as drug delivery agents specific for CNS: In vivo biodistribution. *Nanomedicine: NBM*, *5*, 369–377. PMID:19341816

Vijender, S., & Bhavuk, G. (2015). Neuroplasticity and its relevance to psychiatric disorders. *Current Psychopharmacology*, *3*(3), 151–157. doi:10.2174/2211556004666150313234631

Wang, C. X., Huang, L. S., Hou, L. B., Jiang, L., Yan, Z. T., Wang, Y. L., & Chen, Z.-L. (2009). Antitumor effects of polysorbate-80 coated gemcitabine polybutylcyanoacrylate nanoparticles *in vitro* and its pharmacodynamics in vivo on C6 glioma cells of a brain tumor model. *Brain Research*, *1261*, 91–99. doi:10.1016/j.brainres.2009.01.011 PMID:19401168

Wegner, G., & Rujescu, D. (2013). The current development of CNS drug research. *The International Journal of Neuropsychopharmacology*, *16*(07), 1687–1693. doi:10.1017/S1461145713000345 PMID:23651558

Widera, A., Norouziyan, F., & Shen, W. C. (2003). Mechanisms of TfR-mediated transcytosis and sorting in epithelial cells and applications toward drug delivery. *Advanced Drug Delivery Reviews*, *55*(11), 39–66. doi:10.1016/j.addr.2003.07.004 PMID:14597140

Wilson, B., Samanta, M. K., Santhi, K., Kumar, K. P., Paramakrishnan, N., & Suresh, B. (2008). Targeted delivery of tacrine into the brain with polysorbate 80-coated poly(n-butylcyanoacrylate) nanoparticles. *European Journal of Pharmaceutics and Biopharmaceutics*, *70*(1), 75–84. doi:10.1016/j.ejpb.2008.03.009 PMID:18472255

Wilson, B., Samanta, M. K., Santhi, K., Kumar, K. P., Paramakrishnan, N., & Suresh, B. (2008). Poly(n-butylcyanoacrylate) nanoparticles coated with polysorbate 80 for the targeted delivery of rivastigmine into the brain to treat Alzheimer's disease. *Brain Research*, *1200*, 159–168. doi:10.1016/j.brainres.2008.01.039 PMID:18291351

Wohlfart, S., Gelperina, S., & Kreuter, J. (2012). Transport of drugs across the blood–brain barrier by nanoparticles. *Journal of Controlled Release*, *161*(2), 264–273. doi:10.1016/j.jconrel.2011.08.017 PMID:21872624

Xia, H., Gao, X., Gu, G., Liu, Z., Hu, Q., & Tu, Y. et al. (2012). Penetratin-functionalized PEG-PLA nanoparticles for brain drug delivery. *International Journal of Pharmacology*, *436*(2), 840–850. doi:10.1016/j.ijpharm.2012.07.029 PMID:22841849

Xing, Y., Wen, C., Li, S., & Xia, Z. (2016). Non-viral liposome-mediated transfer of brain-derived neurotrophic factor across the blood-brain barrier. *Neural Regeneration Research*, *11*(4), 617–622. doi:10.4103/1673-5374.180747 PMID:27212923

Xu, S. L., Liu, Y., Tai, H. C., Zhu, J., Ding, H., & Lee, R. J. (2011). Synthesis of transferrin (Tf) conjugated liposomes via Staudinger ligation. *International Journal of Pharmaceutics*, *404*(1-2), 205–210. doi:10.1016/j.ijpharm.2010.10.053 PMID:21056642

Yang, X., Koh, C. G., Liu, S., Pan, X., Santhanam, R., & Yu, B. et al. (2009). Transferrin receptor-targeted lipid nanoparticles for delivery of an antisense oligodeoxyribonucleotide against Bcl-2. *Molecular Pharmaceutics*, *6*(1), 221–230. doi:10.1021/mp800149s PMID:19183107

Yang, Y., & Rosenberg, G. A. (2011). Blood–brain barrier breakdown in acute and chronic cerebrovascular disease. *Stroke*, *42*(11), 3323–3328. doi:10.1161/STROKEAHA.110.608257 PMID:21940972

Yoshikawa, T., & Pardridge, W. M. (2001). Delivery of peptides and proteins through the blood–brain barrier. *Advanced Drug Delivery Reviews*, *46*(1-3), 247–279. doi:10.1016/S0169-409X(00)00139-3 PMID:11259843

Zhang, J., Stevens, M. F., & Bradshaw, T. D. (2012). Temozolomide: Mechanisms of Action, Repair and Resistance. *Current Molecular Pharmacology*, *5*, 102–114. doi:10.2174/1874467211205010102 PMID:22122467

Zhang, L., Xu, T. Y., Wang, S., Yu, L. Q., Liu, D. X., Zhan, R. Z., & Yu, S. Y. (2012). Curcumin produces antidepressant effects via activating MAPK/ERK-dependent brain-derived neurotrophic factor expression in the amygdala of mice. *Behavioural Brain Research*, *235*(1), 67–72. doi:10.1016/j.bbr.2012.07.019 PMID:22820234

Chapter 6
Silver Oxide-Copper Oxide Nanocomposite Preparation and Antimicrobial Activity as a Source for the Treatment of Fish Diseases:
Silver Oxide-Copper Oxide Nanocomposite Preparation and Antimicrobial Activity

Sayed Reza Shaffiey
Novin Pak Nanotech Co. Ltd., Mazandaran Science and Technology Park, Iran

Sayedeh Fatemeh Shaffiey
Novin Pak Nanotech Co. Ltd., Mazandaran Science and Technology Park, Iran

DOI: 10.4018/978-1-5225-1043-7.ch006

ABSTRACT

Aeromonas hydrophila *is a heterotrophic, gram negative bacterium which is primary or secondary cause of ulcers, fin rot, tail rot, and hemorrhagic septicaemia in fish. The treatments for this infection are only restricted to some antibiotics. So, novel materials are being searched for combating with bacterial infections and the resulting consequences. In this chapter,* Ag_2O/CuO *nanocomposites were synthesized chemically and characterized by X-ray diffraction (XRD), scanning electron micrograph (SEM) and transmission electron microscope (TEM). The antimicrobial activities of* Ag_2O/CuO *nanoparticles (NPs), was checked by both well diffusion and turbidometric (spectrophotometric) method. Synthesized nanoparticles exhibited their antimicrobial efficacy in both the standard inhibitory assays; these results thus provide a scope for further research on the application of* Ag_2O/CuO *nanoparticles as disinfectant and/or antibiotic in the fishery industry.*

INTRODUCTION

Fish disease is one of the major threats to the sustainable development of aquaculture causing loss of millions of dollars annually. *Aeromonas hydrophila* is a gram negative bacterium that can grow both in aerobic and anaerobic condition and causes a variety of diseases in both fish and human populations. There are contradicting views on whether the microbe is a primary cause of diseases or an opportunistic one causing diseases to hosts that are immune compromised and stressed (Nielsen et al., 2001). However, this pathogen has been found to cause ulcers, fin rot, tail rot, and hemorrhagic septicemia and epizootic ulcerative syndrome (EUS) in fishes (Austin et al., 1996; Roberts, 1997). The ubiquitous nature of the bacterium in aquatic environments provides significant opportunity for animals, mainly fish and amphibians to contact and ingest organisms (Seshadri et al., 2006). The treatment for the bacterial infection is limited to very few antibiotics like Tetracycline and Remet-30®. Moreover, the bacteria are developing resistance to the available antibiotics because of prolonged use and new compounds for combating bacteria-borne diseases are being searched for.

Nanotechnology is an emerging scientific field and considered to have potential to generate new and innovative materials. Nanotechnology provides the ability to engineer the properties of materials by controlling their size and this has driven research towards a multitude of potential uses for nanomaterials (Saifuddin et al., 2009). Nanoparticles exhibit distinct optical, thermal, chemical and physical properties from that of the bulk material due to their higher surface area to volume ratio.

The synthesis of nanoparticles from metals possesses various biological processes through co-enzymatic systems. The interaction of these nanoparticles with biologically active ligand in the animal system through chelation (Johari et al., 2009). Due to the increase in the outbreak of bacterial diseases in the aquaculture industry and the development of bacterial resistance, new antibacterial agents are required. Silver nanoparticles have proved to be one of the most effective metallic nanoparticles and good antibacterial activity against some bacterial pathogens (Gong et al., 2007) and fish pathogens (Soltani et al., 2009).

At the present time, antibacterial properties of nano metal oxides have found as novel of antimicrobial agent. Also, the majority of scientists have proposed that the nano metal oxide ions, such as silver or copper have useful to controlling of infectious microorganisms (Bustos-Martinez et al., 2006; Lin et al., 1996; Lin et al., 1998; Shaffiey et al., 2015; Shaffiey et al., 2014).

Guogang Ren achieved so widespread experience on the antibacterial effects of nano copper oxides against of *Escherichia coli* and meticillin-resistant *Staphylococcus aureus* (Ren et al., 2009). Jayesh et al. (2007) showed that copper nanocrystales have great promise as antimicrobial agent against *Escherichia coli, Bacilus subtilis* and *Staphylococcus aureus*. They also, assume that copper nanoparticles could be have greater affinity to surface active groups of *Bacilus subtilis*, which may be have led to its greater bactericidal effect. However, limited information on the possible antimicrobial activity of CuO and Ag_2O nanoparticles are available and the mechanism of action of the copper nanocrystales is not yet fully established but nowadays, we know that the bactericidal effect of metal nanoparticles can be attributed to their small size, photocatalystic of activity and high surface to volume ratio, which allows them to interact closely with microbial membranes and is not merely due to the release of metal ions in solution (Morones et al., 2005).

In the current study, we wish to synthesis, characterization and evaluation of Ag_2O/CuO nanoparticles (NPs) as a new class drug against gram negative bacteria, *Aeromonashydrophila* ATCC 7966T, and compare new drug effects with Tetracycline as reference antibacterial drug.

MATERIALS AND METHODS

Chemicals

All chemicals used in the experiment were analytical reagent grade. Silver nitrate ($AgNO_3$, 99.99%), copper nitrate, ethanol, oxalic acid was purchased from Merck, Germany. De-ionized water was used throughout the experiment. Nutrient agar and peptone water medium were purchased from E. Merck Co; Darmstadt Germany. Tetracycline antibiotic was purchased from Fluka Germany.

Synthesis of Ag_2O/CuO Nanocomposites

Silver nitrate and copper nitrate were added to ethanol in a two neck flask giving a 0.3 M blue solution. The temperature was raised to 60 °C and after 30 min of continuous stirring, oxalic acid (OA) was rapidly added to the solution. The molar ratio of Ag:Cu:OA was 2:1:1. The system was kept at 60 °C under reflux for 2h and a blue precipitate was obtained; then the resulting viscous gel was dried at 80 °C overnight. The dried Ag_2O/CuO oxalate was ground and calcined at 400 °C for 2h.

Characterization of Ag_2O/CuO NPs

Synthesized NPs were characterized by transmission electron microscopy (TEM). TEM analysis was carried out using a 200 kV JEOL transmission electron microscope (JEOL Ltd., Tokyo, Japan). Crystalline, structure, and crystallite size of NPs were determined by X-ray diffraction (XRD) technique using a Rigaku-Miniflex X-ray diffractometer (Rigaku Corporation, Tokyo, Japan) with Cu-Kα radiations (λ = 0.15406 nm). Surface morphology of product was characterized by using a scanning electronic microscopy (SEM, Cam Scan MV2300) with an accelerating voltage of 30 kV.

Disk Diffusion Test

Antimicrobial activity of the synthesized Ag_2O/CuO NPs were determined using Gram-negative bacteria (*Aeromonas hydrophila* ATCC 7966T) following a modified Kirby Bauer disc diffusion method (Bauer et al., 1966). A lawn of bacterial culture was prepared by spreading 100 μl culture broth, having1.5×10^8 CFU/ml of test organism on solid nutrient agar plates. The plates were allowed to stand for 10–15 minutes, to allow for culture absorption. After culture, four wells were made at equal distance by using a sterile steel cork borer (8mm diameter). Wells were sealed with 1 ml of molten agar (0.8% nutrient agar) to prevent leakage from the bottom of the plate. The bacteria were plated onto solid nutrient agar plates. Using a micropipette, 20-100% (V/V) of 100 μl (50 μg) of the nanoparticles solution sample was poured into each of four wells on all plates. After incubation at 22-25$\pm$ 2 °C for 24 hours, the size of the zone of inhibition was measured with a ruler with up to1 mm resolution. Each experiment was repeated three times, and the resulting bacterial growth on three plates corresponding to a particular sample were averaged and reported ($p<0.05$). A solvent blank was run as a negative control whereas the antibiotic (tetracycline) was used as a positive control.

Determination of Minimum Inhibitory Concentration (MIC) and Minimum Bactericidal Concentration (MBC)

The minimum inhibitory concentration (MIC), defined as the lowest concentration of material that inhibits the growth of an organism (Shaffiey et al., 2014), was determined based on batch cultures containing varying concentration of Ag_2O/CuO nanoparticles in suspension (0–300 μg/ml). Sterile Erlenmeyer flasks (500 ml), each, containing 100 ml peptone water medium were sonicated for 10 min after adding the nanoparticles to prevent aggregation of the nanoparticles. Subsequently, the flasks were inoculated with 1 ml of the freshly prepared bacterial suspension in order to maintain initial bacterial concentration 1.5×10^8 CFU/ml.

Bacterial growth was measured as increase in absorbance at 625 nm determined using a spectrophotometer (Thermo Spectronic, Helios Epsilon, USA). The experiments also included a positive control (flask containing nanoparticles and peptone water medium, devoid of inoculums) and a negative control (flask containing inoculums and peptone water medium, devoid of nanoparticles). The negative controls indicated the microbial growth profile in the absence of nanoparticles. The absorbance values for positive controls were subtracted from the experimental values (flasks containing peptone water media, inoculums and nanoparticles). All the experiments were carried out in triplicate. Ag_2O/CuO nanoparticles were tested for bactericidal effect using the microbial culture selected for the study.

The minimum bactericidal concentration (MBC) (Shaffiey et al., 2014), the lowest concentration of nanoparticles that kills 99.9% of the bacteria was also determined from the batch culture studies. For growth inhibitory concentration (MIC) the presence of viable microorganisms was tested and the lowest concentration causing bactericidal effect was reported as MBC.

RESULTS

The XRD Analysis

The XRD pattern of Ag_2O/CuO nanoparticles (Figure 1) were compared and interpreted with standard data of International Centre of Diffraction Data (ICDD). The average crystallite sizes (C.S) of the nanoparticles were calculated using the Debye-Scherer Equation from the major diffraction peaks

$$\left(d = \frac{0.9\lambda}{\beta\cos\theta}\right).$$

Figure 1. XRD pattern of Ag_2O/CuO nanoparticles

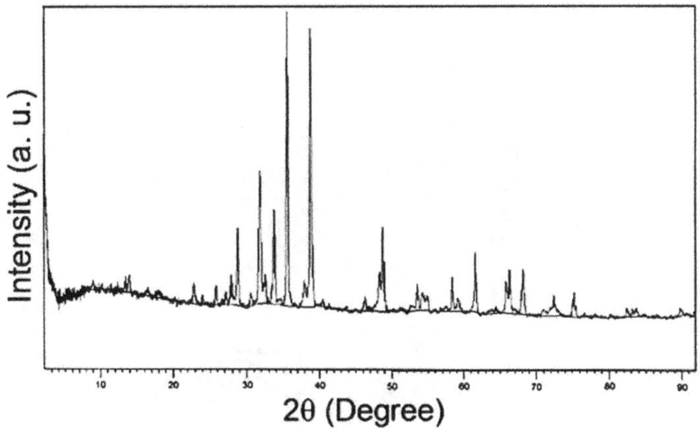

where d is the mean diameter of the nanoparticles, λ is wavelength of X-ray radiation source, β is the angular FWHM of the XRD peak at the diffraction angle θ (Cullity, 1978). The mean size of nanoparticles estimated by XRD is 48 nm.

Figure 2 shows the SEM image of prepared Ag_2O/CuO NPs. It shows that the Ag_2O/CuO NPs are in spherical shape.

TEM micrograph of Ag_2O/CuO NPs is shown in Figure 3, which demonstrates spherical shape and narrow particle size distribution. The mean particle size estimated by TEM is 23-60 nm, which compares well with the size estimated from XRD data.

Figure 2. Scanning electron micrograph (SEM) of Ag_2O/CuO nanocomposite

Figure 3. TEM image of prepared Ag_2O/CuO nanoparticles

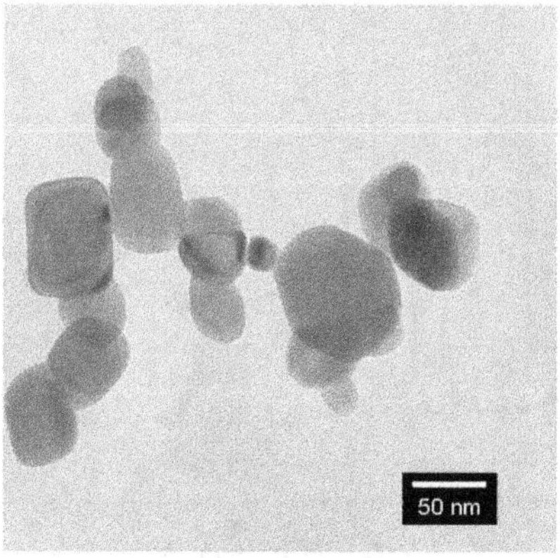

Antimicrobial Properties

In batch studies, a greater lag phase and lower maximum absorbance (at 625 nm) were observed as the concentration of nanoparticles increased. As concentration of nanoparticles increased to MIC of the respective strain, no growth was observed in the flask. The bactericidal effect of nanoparticles is dependent on the concentration of nanoparticles and the initial bacterial concentration (Ruparelia et al., 2008). The initial bacterial concentration was almost constant at 1.5×10^8 CFU /ml irrespective of nanoparticles concentration and microbial strain.

The MIC observed for Ag_2O/CuO NPs is 70 µg/ml for *Aeromonas hydrophila* ATCC 7966T and MBC value for Ag_2O/CuO NPs is 280 µg/ml (Table 1).

Table 1. The diameter of inhibition zone (DIZ) and MBC of Ag_2O/CuO nanoparticles (annealed at 400 °C temperature) against Aeromonas hydrophila

MBC (Log CFU/ml)	DIZ (mm)	Ag_2O/CuO Nanoparticles Concentration (µg/ml)
6.7±0.0	0	Negative Control[a]
5.2±0.0	3	60
4.9±0.0	5	120
2.1±0.0	7	180
1.3±0.0	8	240
NCD[b]	11	300

[a]Negative Control: flask containing inoculums and peptone water medium, devoid of nanoparticles
[b]NCD = No culturable cells detected

Figure 4. Zone of inhibition of Ag_2O/CuO nanocomposite

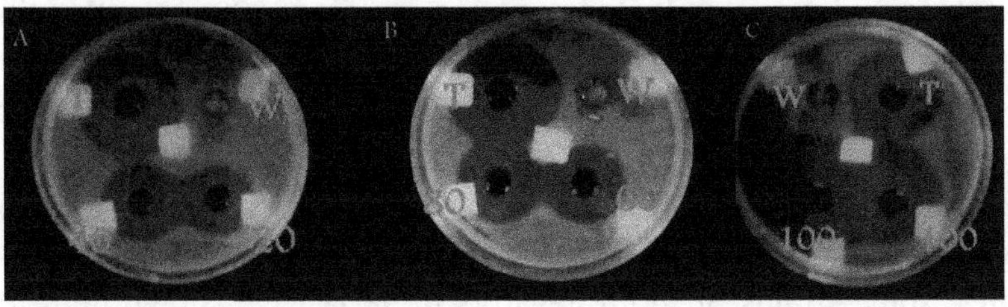

Figure 4 exhibit the zone of inhibition of Ag_2O/CuO NPs synthesized and positive control, a known antibiotic tetracycline, against gram negative bacteria (*Aeromonas hydrophila*).

DISCUSSION

The fish sector contributes a major role in the aquaculture industry worldwide. These resources are expected to have high demand in national and international levels. Disease outbreaks in aquaculture as an important limiting factor in production and trade. Chemotherapeutics are drugs which are capable of affecting or killing micro-organisms, especially bacteria in the fish culture. Several chemicals *viz.*, formalin, malachite green, methyl blue, copper sulphate and potassium permanganate have been used to cure the bacterial fish diseases. But, these chemicals produced undesirable effects in the water as well as organisms (Yanong, 2003).

Various studies suggest that silver ions react with -SH group of proteins. Silver nanoparticles uncouple respiratory electron transport from oxidative phosphorylation, which inhibits respiratory chain enzymes or interferes with membrane permeability to protons and phosphates (Duran et al., 2010). Schreurs and Rosenberg, (1982) reported that Ag^+ inhibits phosphate uptake and exchange in *Escherichia coli* and causes efflux of accumulated phosphate as well as of mannitol, succinate, glutamine, and proline.

Binding of silver nanoparticle depends on the available surface area of the nanoparticles and smaller nanoparticles with high surface area to volume ratio are more lethal due to availability of wide surface area for interaction (Guzman et al., 2008).

Regarding that the diameter of inhibition zone (DIZ), reflects the sensitivity of the organism, strains of sensitive, show larger DIZ and the resistant strains show smaller DIZ. Investigators proved antibacterial properties of copper oxide nanoparticles. While so far has been reported only few studies on the antibacterial properties of copper nanoparticles. Cioffi, (2005) and Shaffiey, (2014) researches showed that copper oxide nanoparticles are considered as an important antibacterial agent. Also during research Guogang Ren et al. in (2009) proved antibacterial properties of copper nanoparticles against *Escherichia coli* and methicilin resistant *Staphylococcus aureus*. It also Guogang proved that copper oxide nanoparticles, is effective on a wide range of nosocomial infection.

Guogang Ren reported Ag-Cu combined nanoparticles as strong antibacterial agents against methicillin resistant *Staphylococcus aureus*. According to Guogang Ren studies, some strains of gram negative bacteria have shown high relatively sensitivity to nano copper oxide combined whith nano silver oxide.

CuO nanoparticles were effective in killing a range of bacterial pathogens involved in hospital-acquired infections. However, in comparison with nano Ag and nano Cu, higher concentrations of nano CuO were required to achieve a bactericidal effect.

Preliminary research on the antiviral (versus influenza A and SARS viruses) capabilities of nanomaterials has shown that nanoparticles such as Ag and Cu release Ag^{+} and Cu^{2+} ions that cause local pH and conductivity changes. This liberation of metal ions into solution will then have the capability to inactivate or kill viruses. Cu^{2+} and Ag^{+} ions are also small enough to disrupt bacterial cell membranes and gain entry in order to disrupt enzyme function. Indirect effects through changes in the surrounding charge environment may also impact on the effectiveness of nanoparticulate metals against microorganisms (Sondi et al., 2004).

CONCLUSION

Fish diseases and the consequent financial loss has become a threat for the sustainable development of fisheries and aquaculture industry. Moreover, development of antibiotic resistance among the bacteria has complicated the situation.

This study reports the successful synthesis of Ag_2O/CuO nanocomposite. The antimicrobial screening studies were also performed in the study. The antimicrobial screening suggests that synthesized Ag_2O/CuO NPs exhibited moderate activity toward *Aeromonas hydrophila*.

REFERENCES

Austin, B., & Adams, C. (1996). Fish pathogens. In B. Austin, M. Altwegg, P. J. Gosling, & S. Joseph (Eds.), *The genus Aeromonas*. Chichester: John Wiley & Sons.

Bauer, A. W., Kirby, W. M., Sherris, J. C., & Turck, M. (1966). Antibiotic susceptibility testing by a standardized single disk method. *American Journal of Clinical Pathology*, *45*(4), 493–496. PMID:5325707

Bustos-Martinez, J. A., Hamdan-Partida, A., & Gutierrez-Cardenas, M. (2006). *Staphylococcus aureus*: La reemergencia de un patógeno en la comunidad. *Rev Biomed*, *17*, 287–305.

Cioffi, N., Torsi, L., Ditaranto, N., Tantillo, G., Ghibelli, L., Sabbatini, L., & Traversa, E. et al. (2005). Copper nanoparticle/polymer composites with antifungal and bacteriostatic properties. *Chemistry of Materials*, *17*(21), 5255–5262. doi:10.1021/cm0505244

Cullity, B. D. (1978). *Elements of X-ray Diffraction* (2nd ed.). USA: Edison-Wesley Publishing Company Inc.

Duran, N., Marcato, P. D., Conti, R. D., Alves, O. L., Costa, F. T. M., & Brocchi, M. (2010). Potential use of silver nanoparticles on pathogenic bacteria, their toxicity and possible mechanisms of action. *Journal of the Brazilian Chemical Society*, *21*(6), 949–959. doi:10.1590/S0103-50532010000600002

Gong, P., Li, H., He, X., Wang, K., Hu, J., Tan, W., & Yang, X. et al. (2007). Preparation and antibacterial activity of FeO @Ag nanoparticles. *Nanotechnology*, *18*(28), 604–611. doi:10.1088/0957-4484/18/28/285604

Guzman, M. G., Dille, J., & Godet, S. (2008). Synthesis of silver nanoparticles by chemical reduction method and their antibacterial activity. *World Acad Sci Eng Technol*, *43*, 357–364.

Jayesh, P., Chatterjee, A. K. P., Duttagupta, S., & Mukherji, S. (2007). Strain specificity in antimicrobial activity of silver and copper nanoparticles. *Nanomedicine (London)*, *3*, 95–101.

Johari, R., Kumar, G., Kumar, D., Singh, S. (2009). Synthesis and antibacterial activity of m (ii) schiff base complex. *J Ind Coun chem.*, 26(1), 23-27.

Lin, Y. E., Vidic, R. D., Stout, J. E., Mccartney, C. A., & Yu, V. L. (1998). Inactivation of Mycobacterium avium by copper and silver ions. *Water Research*, *32*(7), 1997–2000. doi:10.1016/S0043-1354(97)00460-0

Lin, Y. E., Vidic, R. D., Stout, J. E., & Yu, V. L. (1996). Individual and combined effects of copper and silver ions on inactivation of *Legionella pneumophila*. *Water Research*, *30*(8), 1905–1913. doi:10.1016/0043-1354(96)00077-2

Morones, J. R., Elechiguerra, J. L., Camacho, A., Holt, K., Kouri, J. B., Ramirez, J. T., & Yacaman, M. J. (2005). The bactericidal effect of silver nanoparticles. *Nanotechnology*, *16*(10), 2346–2353. doi:10.1088/0957-4484/16/10/059 PMID:20818017

Nielsen, M. E., Høi, L., Schmidt, A. S., Qian, D., Shimada, T., Shen, J. Y., & Larsen, J. L. (2001). Is *Aeromonas hydrophila* the dominant motile Aeromonas species that causes disease outbreaks in aquaculture production in the Zhejiang Province of China? *Diseases of Aquatic Organisms*, *46*, 23–29. doi:10.3354/dao046023 PMID:11592699

Ren, G., Hu, D. W. C., Cheng, E. A., Vargas-Reus, M., Reip, P. P., & Allaker, R. (2009). Characterisation of copper oxide nanoparticles for antimicrobial applications. *International Journal of Antimicrobial Agents*, *33*(6), 587–590. doi:10.1016/j.ijantimicag.2008.12.004 PMID:19195845

Roberts, R. J. (1997). Epizootic ulcerative syndrome (EUS): progress since 1985. In T. W. Flegel & I. H. MacRae (Eds.), *Diseases in Asian aquaculture III*. Manila: Asian Fish Soc.

Ruparelia, J. P., Chatterjee, A. K., Duttagupta, S. P., & Mukherji, S. (2008). Strain specificity in antimicrobial activity of silver and copper nanoparticles. *Acta Biomaterialia*, *4*(3), 707–716. doi:10.1016/j.actbio.2007.11.006 PMID:18248860

Saifuddin, N., Wong, C.W., Nur Yashumira, A.A. (2009). Rapid Biosynthesis of Silver nanoparticles using culture supernatant of bacteria with Microwave irradiation. *e-J Chem*, 6(1), 61-70.

Schreurs, W. J. A., & Rosenberg, H. (1982). Effect of silver ions on transport and retention of phosphate by *Escherichia coli*. *Journal of Bacteriology*, *152*, 7–13. PMID:6749823

Seshadri, R., Joseph, S. W., Chopra, A. K., Sha, J., Shaw, J., Graf, J., & Heidelberg, J. F. et al. (2006). Genome Sequence of *Aeromonas hydrophila* ATCC 7966T: Jack of All Trades. *Journal of Bacteriology*, *188*(23), 8272–8282. doi:10.1128/JB.00621-06 PMID:16980456

Shaffiey, S. F., Shaffiey, S. R., Ahmadi, M., & Azari, F. (2014). Synthesis and evaluation of bactericidal properties of CuO nanoparticles against *Aeromonas hydrophila*. *Nanomecicine Journal*, *1*(2), 199–205.

Shaffiey, S. R., Shaffiey, S. F., & Ahmadi, M. (2015). Synthesis and evaluation of bactericidal properties of Ag_2O nanoparticles against *Aeromonas hydrophila*. *Int. J. Nano Dimens*, *6*(3), 263–269.

Soltani, M., Ghodratnema, M., & Ahari, H., Ebrahimzadeh mousavi, H.A., Atee, M., Dastmalchi, F., Rahmanya, J. (2009). The inhibitory effect of silver nanoparticles on the bacterial fish pathogens, *Streptococcus iniae, Lactococcus garvieae, Yersinia ruckeri* and *Aeromonas hydrophila. J Vet Res*, *3*(2), 137–142.

Sondi, I., & Salopek-Sondi, B. (2004). Silver nanoparticles as an antimicrobial agent: A case study on *E. coli* as a model for gram-negative bacteria. *Journal of Colloid and Interface Science*, *275*(1), 177–182. doi:10.1016/j.jcis.2004.02.012 PMID:15158396

Yanong, R. P. F. (2003). Fungal diseases of fish. In M.P. Jones (Ed.), Fungal Diseases, veterinary clinics of north America, exotic animal practice (Vol. 6). Philadelphia: W.B. Saunders Co.

Chapter 7
Performance Analysis of FET-Based Nanoiosensors by Computational Method

Keka Talukdar
Nadiha High School, India

Anil Shantappa Malipatil
Guru Nanak Dev Engineering College – Bidar, India

ABSTRACT

Human body has a hierarchy of structures. There are many organs which are affected much earlier by a disease when it is detected. Most modern sensors can detect anomalies when its concentration in the body fluid reaches to millimolar range. But more sensitive biosensors should detect disease from much lower concentration like femtomolar range. So, extremely sensitive biosensors are needed for early detection of fatal diseases at their early stage. It should detect the target molecule from a very low concentration of analyte. Also, molecules which we often need to detect are too small in size. So Nanotechnology and biotechnology should shake hand to detect nanosized particles from an extremely low concentration solution. Hence we are in a real need of a biosensor. Here we are interested in charged biomolecules and will discuss the performance of Field- Effect Transistor based biosensosrs by computational method.

DOI: 10.4018/978-1-5225-1043-7.ch007

INTRODUCTION

Recalling Moor's law, it is the real picture that an IC chip now contains billions of transistors. But simultaneous idea of lab on a chip is the real challenge of a nanobiosensor so that it can monitor our whole physiological process as we require. Also, it should direct us for personalised medicine. Now people are trying to develop sophisticated and intelligent nanobiosensors. Nanowire, nanosphere sensors, nano cantilever sensors, field- effect transistor (FET) type sensors, array sensors are some of the models of newly developed sensors.

There are several processes to detect biomolecules from solution depending on their mass, charge, redox potential and even optical index. A biosensor is an analytical sensing device that converts a biological response into an electrical signal to sense materials like antibody, protein, enzyme etc. Detection of disease, quality of food estimation, monitoring environmental condition, agriculture, and pharmaceutical uses are some of the essential applications of biosensor. There are some types of biosensors like potentiometric, amperometric and calorimetric paper biosensor which are commonly used nowadays. With the progress of nanoscience nanotechnology, a tremendous breakthrough can be expected in near future. A sensor must be accurate, sensitive, have very low response time and should be reproducible and it is must be free from electrical noise. Besides, it should be cheap, portable and easy to use.

In every aspect of biosensing, the label-free approach using FET nanobiosensors are proving their superiority regarding target specificity, cost effectiveness, the ease of use and portability than the existing label-based enzyme-linked immunosorbent assay (ELISA) bosensing technique. As the various studies on biosensors are still in the developing stage, so simulation studies now have greater importance. Here we have done a detail analysis of various level free biosensors like flexure FET sensor, double gate FET pH sensor and extended gate FET pH sensor. Many of their important characteristics are studied in detail.

BACKGROUND

The following review briefly describes the progress of biosensing technology until now including both experimental and computational contributions.

Experimentally there are many literatures which address the development of various nanobiosensors which are based on diverse principles of detection and produce different types of devices. Nanoparticles, carbon nanotubes (CNTs), naowire, magnetic particles, metallic platelets and graphene sheets are some of the candidates found to be used for this purpose. Not only in nanomedicine, biosensors are used in finding food pathogens, environment pollutants and in bioterrorism. Lee and Fauchet

(2007) Synthesized high sensitive plane photonic crystal micro cavity biosensor on silicon insulator wafer. The authors presented their method both theoretically and experimentally. This methodology is helpful to detect extremely little mass of the order 2.5fg by getting the resonance redshift from the binding of glutaraldehyde and bovine serum albumin. The sensor can measure the dehydrated protein, so it has a different importance in the field of biosensing. The well agreement between experimental data and theoretical results points towards the high performance of such sensor. By the improvement of Q factor and proper positioning of the biological molecule, quality of the sensor can be improved. Zhu, Du and Fu (2009) developed rhombic silver nanoparticles for the localized surface Plasmon resonance of biosensor. The detection is improved from the triangular particles to rhombic nanoparticles biosensor. By this method, one can sense the refractive index sensitivity till 330 nm with the help of Discrete Dipole Approximation by algorithm designed method. To diagnose the genetic diseases in early stage, it demands the detection of the opposite base pairs in DNA. The silicon nanowire FET biosensors have wide importance because of its capacity to detect protein established on frequency domain electrical measurement (Zheng, Gao & Leiber, 2010). The measurement of antibodies of silicon nanowire bisosensor is done with the help of frequency domain and power spectral density of voltage. The frequency and general time arena measured with this setup with antigen concentration given more detection sensitivity for the frequency domain. The present work confirms the measurements of frequency domain are free set about general time domain measurements which leads to finding the high sensitive electrical proteins.

Many investigators have investigated the performance of biosensor for the detection of DNA. The electrochemical biosensor reported by Wei, Lillehoj and Ho (2010) is a successful one in getting a specific DNA and convert it into proper signal transduction. This method gives more impact for the rapid and multiplexed detection of DNA for the clinical applications. Supervision of adhering phylogenetic relation and dynamics of protein interactions is required for the diagnosis and development of drug to get new remedial candidates. Surface Plasmon resonance is the wonderful way to in this regard but limited by low sensitivity as well as low output. Duan et al. (2012) synthesized silicon nanowire FET biosensor to get the kinship of protein–ligand binding and dynamics for femtomolar concentrations sensitivity. Hafaiedh, Baccar, ktari and Abdelghani (2012) developed Mulitwalled Carbon nanotubes by labeling gold for the purpose of peroxidose biosensor by the pressure vapor deposition technique. The labeled MWCNT can be used for Streptavidin-HRP Immobilization by the help of thiol-acid; this can be a stimulating process with EDS/NHS for the selected protein. If the cross sectional area of the synthesized multiwalled carbon nanotube is large, then the reception of biosensor is higher and can be reproduced with greater stability. Suaifan, Esseghaier and Zourob (2013) developed a sensible and exclusive calorimetric technique for the protease diagnosis. Their method is an attempt for the wonderful sensitivity, selectivity and it is also successful in differentiating protease targets. Silicon nanowire which is used as bistable resistor comprises npn metal oxide semiconductor without gate (Moon et al., 2013). This comprises three basic parts source (n), floating body (p)-which is uncovered as sensing area and Drain (n). The metal-oxide semiconductor biosensor gives new hope as a chip biosensor due to its compact in size and its low cost. The photogene rated charge carriers can be regulated with the help of redo reactions of molecules on the surface of the electrode. A glucose sensor was synthesized by Tang et al. (2014). The advantages of the photoelectron chemical biosensor are, it operates with minimal voltage and it can operate with solar energy. This type of biosensor is very helpful to diagnose disease related problems as well as for the biomarker detection and it provides significant chances for the treatment of diabetes and many other diseases due to glucose. To diagnose enzyme that catalyses the splitting of proteins is mainly depends on the valuable machines and it also takes lot of time. In the former study the authors gave some idea of getting such a low cost biosensor.

Label free biosensor is now getting popular for technical experts. The process uses chemical or biological receptors for sensing biomolecules and efficiently gives the selectivity and affinity. Cancer biomarkers can be detected in a label-free way by using the target binding activity between antigens and antibody (Lee, Nguyen & Sim, 2015). By proper choosing of bioreceptors or antibodies, it is possible to detect various cancer biomarkers using the multiple lebel-free nanoplasmonic biosensor platforms. By the method of one–pot solvothermal novel phenol biosensor based on laccase was prepared (Mei et al., 2015) to improve analytical performances of electrochemical biosensors. In this study reduced graphene oxide supported on palladium-copper alloyed nanocages was used as a platform for enzyme immobilization. It shows improved catalytic performances as compared with other biosensors. This type of sensors can be used as highly sensitive phenolic biosensor. To avoid the infections occurred in hospital due to the methicillin resistant staphylococcus aureus. Nawattanapaiboon et al. (2015) prepared selective DNA biosensor by LAMP-surface Plasmon resonance. By using self- assembled monolayer surface, they detected Lamp amplicons from MRSA with the help of immobilized streptvidin- biotinylated probes. Using this system multi targets analysis can be done without any labeling. The broad range of pathogens can be detected by creating DNA primer and probe sequences.

MicroRNAs (MiRNAs) have great importance as biomarkers to detect cancer in early stage. Gold nanoparticles with graphene FET bio sensor were synthesized by Cai et al. (2015) which is highly sensitive and capable of label free detection. Due to its high sensitivity and selectivity, the biosensor can be used to detect the miRNA in serum samples which helps to detect gene related diseases. Silicon nanowire biosensor has been proved to be successful to detect disease-related biomarkers, but to find multiple aspects of disease-related detection is yet to be solved. Zhang et al. (2015) synthesized silicon nanowire FET biosensor by the top down lithography technique to get manifold sensing for IL-8 and TNF-alpha for saliva. The method has great importance because of its specific and manifold sensing capability. It can also be used for label free detection and it's a non-destructive too. The biosensor may also be helpful to study the early stage of oral squamous cell carcinoma.

Garcia et al. (2015) developed a nanobiosensor by using 4-hydrosyphenylpyruvate dioxygenase to detect the mesotrione. In the present situation more herbicide are used in the agricultural field to get more production, due to which environmental pollution is increasing and it is also affecting the human health. The sensor may help to monitor the herbicides used in agriculture. By using molecular docking and molecular dynamics simulation, the interaction between HPPD and mesotrione was revealed. The atomic force microscope and fluorescence microscopy give clear evidence that HPPD freeze in automatic -tacked single layer. This nanobiosensor can be more useful in the detection of mesotrione molecules. According to the authors, their sensors have a wide importance due it is dependability, accuracy and applicability in wide range of fields.

The first transistor based label free detection of antibody-antigen reaction was reported in 1989. With proper biasing, many nanobiosensors came to news for their outstanding performances. Zheng, Patolsky, Cui, Wang and Lieber (2005), Patolsky, Zheng and Lieber (2006), Zhang et al. (2010), Cai et al. (2014), Colombelli, Manera, Taurino, Catalano and Convertino (2016) reported such high performance nanobiosensors. The new generation of label free biosensors is more accurate and requires short time to sense than the previous biosensors. These small devices have the ability to get the multiple targets within the interfering molecules. Nair and Alam (2010) has given the methodology for such biosensor. The minimum time to detect the surface functionalization to discover the major constituents of unspecified adsorption can be found by considering the surface exclusion, diffusion limited transport and void distribution function. In the literature some experimental techniques are available for the selectivity that gives optimization, but no theory is available for the same. The theory of selectivity gives wide importance for the better sensitivity and selectivity for multi label free biosensors.

Biologically active field – effect transistors are mainly made up of semiconductor devices. A recent result confirms the generation of a field effect by these biosensors. They got wide importance due to its ultra-speed in the detection, dependability and cost effectiveness. Heitzinger, Mauser and Ringhofer (2008) formulated 2D and 3D multi ordered models for planar structures and nanowire sensor. The prominent change in the feature length of the biosensors is essential in the multi ordered models. By using electrostatic potential, the theory of such biosensor can be connected to any of the charge transport model of biosensor. The variation of diameters by using conductance simulation for nanowires gives the great prominence for the dipole moment with the inclusion of surface charge. The porting conditions are mainly depending on the dipole moment density and surface charge density of biosensor layer.

Optimization of the side gate voltage and oxide thickness to enhance the conductance and sensitivity of a Si nanowire (NW) biosensor was demonstrated analytically by Karger (2008) using diffusion-drift model and nonlinear Poisson-Boltzmann equation. The theoretical calculation agreed well with the experimentally obtained values for NW with rectangular cross-section. Often experimental results are taken into consideration by the theoreticians to get a way of better result.

A computational approach was developed by Vico et al. (2011) for observing the conduction sensitivity for nanobio FET sensor, which is connected to the specific target molecule. The process demonstrated basically works on two models namely screening model of surface charge sensors in liquids and PROPKA method. The later method is used to find the charge of protein and can be used as input data for SMSCSL. This approach gives the dependence of conduction sensitivity on many parameters for the ideal wire and explains the experimental setup. The presented methodology gives best and accurate results and also gives the accurate values for charge carrier density of nanowire. This is the first model for computational analysis of nanobiosensor made by FET. It confirms the dependence of ionic strength for conductance sensitivity. It was reported in their study that conductance sensitivity is more when ionic strength is low.

Chen and Lee (2010) analyzed a biosensor with $3 \times N_a \times N_p$ degrees of freedom by multiscale atomistic/continuum theory. Here N_a and N_p are the number of atoms per unit cell and number of representative unit cells, respectively. The authors highlighted beam theory, classical finite element method, generalized atomistics finite element method (GAFEM) and molecular dynamics (MD) simulation method for resonant analysis. This work demonstrated the two main issues; those are resonance analysis for varying the amount of target analyte of specific biosensor and the highest resonance frequency. They validated the Courant- Friedrichs- lewy condition in classical finite element modelling but due to non-local effect the condition could not be validated by wave propagation theory in GAFEM. However, the presented work is helpful to get the operation of micor/nano biosensors from atomistic computational approach. The Atomistic field theory with GAFEM gives the accurate results within low computational expenses. The first fundamental resonant frequency can be accurately predicted by this theory for a nanobiosensor. According to them, MD simulation also gives precise results, but it is too expensive to use.

In the capacitive model of Ahn et al. (2011), the size dependent sensitivity of a double gate nanowire FET (DG-NW FET) biosensor was investigated. According to them the sensitivity of the biosensor was dependent on two coupling ratios; one between the channel and the biomolecules and other between channel and gates. The biasing condition as well as sensor sizes was found to be responsible for better sensitivity issue. According to Sorensen et al. (2007) the change in conductance with the surface charge density directly is given as,

$$\frac{\Delta G}{G_0} = -\frac{2}{\mathrm{Re}\, p_0} \Gamma\left(\Gamma_l \sigma_b + \sigma_s\right) \tag{1}$$

Here R is the radius of the nanowire, e is the electronic charge, p_0 the hole density, σ_b and σ_s are the surface charge density on the surface and at a distance l. Γ is the actual sensitivity of the sensor and Γ_l is the actual sensitivity of the nanowire. The important assumption of this model is that a single charge can describe the change in electrostatic potential due to binding of target molecule. The above parameter in Equation 1 is the conductance sensitivity which comes out to be 0.258 in their work. Adequate explanation for the uncertainty of the measurement of the conductance sensitivity is given by the authors and the main reason of which is the coverage of the receptors in the nanowire.

Heitzinger et al. (2010) with the help of stochastic linearized poission–Boltzmann equation observed the fluctuation in biofunctionalized boundary layers of nanowire field effect biosensors. In the model based on stochastic partial differential equations (PDEs) for fluctuations, stochastic elliptic PDE for electrostatic potential in the FET biosensor is considered. Mainly the effect of charge concentration in the boundary layer on the current through the transducer is carefully observed. The optimized conditions, as reported by the authors can be obtained by choosing suitable surface charge, oligomer length and ionic strength of the electrolyte. At the same time homogenization method was employed by them to find the scaling law for the variance and covariance of the electrostatic potential. A lot of information about current and potential in the vicinity of the boundary is obtained in their study. Both single and double stranded DNA oligomers were considered for observing the fluctuation of biofunctionalized layer by changing the surface charge, oligomer length and electrolytic ionic strength.

Label free technologies often require target capture capacity of the sensor in a tiny volume of fluid. Sometimes for critical investigations, a small quantity of sample may be needed to examine several times. Microcantilevers, electrochemical sensors and FET biosensors can do the same and detect target molecule from femtomolar range concentration. Sensors having microfluidic channels can achieve remarkable automation (Meolin & Quake, 2007).

Talukdar, Bhushan, Viswanath and Mitra (2013) have reported about biologically sensitive FET by Monte Carlo simulation. Silicon nanowire (SiNW) sensor was reported by them and performance was analyzed. SiNW biosensor was found to have considerably high performance when the dimension of the NW reaches about 10nm. SiNW biosensor was also taken up for various studies by many researchers (Tang, Bansaruntip, Nakayama, Yenilmez, Chang & Wang, 2006; Gao, Lu, Wang & Li, 2016). Simulation studies on SINW for the detection of biotin/streptavidin binding was carried out by Wand and Yaci (2010) and an elaborative study on SiNW biosensor was presented by Li, Wang and Tao (2013).

Biochemical sensors are developed by using photonic crystal resonators. Hsiao et al. (2011) gave the computational study for the nanophotonic biosensors by using nano resonators. The photonic crystal structure contains holes maintained in hexagonal manner on silicon block. Nanoring is generated by taking few holes along the hexagonal structure. Nnanoring resonator is placed between the two photonic crystal waveguides created by ejection of two lines of holes. When protein comes near the functionalized surface, a sensing hole is created which gives rise to change in the resonant wavelength. Hsiao and Lee (2011) describes single and dual nanorings. Double ring gives two different resonant modes and directions of propagation are antiparallel to each other. The single nanoring has more wavelength shift than the dual nanoring and the stability of single nanoring is less than dual nanoring resonator. The authors suggest that the dual nanoring is the wonderful sensing mechanism and can sense the two different molecules.

In brief, the progress of FET based biosensor is tabulated in Table 1.

Table 1. Some theoretical and experimental studies and their major results

Authors	Year	Aim of Their Work	Main Results
Liu and Cui	2007	Experimental work on ISFET pH sensor with multilayer film	Mobility is 35.68 cm^2/V
Chen and Lee	2010	Atomistic analysis with different amount of target analyte and study of dependence of resonant frequency on finite element mesh	Atomistic theory with generalized atomistic finite element model gave accurate result
Go et al.	2012	Theoretical study with heterogeneous nanowire_ nanoplate planar transistor sensors	Got extremely high sensitivity >10V/pH
Ahn et al.	2013	Theoretical study on double-gate Silicon nanowire FET	Sensitivity obtained as 68mV/pH
Lee et al.	2013	Experimental study of EGFET pH sensor with CZTSe nanoparticles	Voltage sensitivity found as 9mV/pH
Lee at al.	2014	Experimental work EGFET pH sensor with annealed TiO_2 sensing membrane	Voltage sensitivity found as 58.71mv/pH
Das et al.	2014	Experimental work on PdO based EGFET pH sensor	Reported Voltage sensitivity of 62.87/pH
Jang and Cho	2014	Theoretical and experimental study for dual gate ISFET with ultra-thin body	425.89mV/pH sensitivity was observed.

NEED FOR MORE COMPUTATIONAL WORK

The above review signifies that, in spite of the ongoing research on the FET based label free detection, the experimental result does not show any effectual advantage in the medical or pharmaceutical industry due to the low output signal of such biosensor in reality. Amplification required for practical use often associated with noise induced by parasitic molecules. Only one literature, reported by Jang and Cho (2014) has put some focused idea of sensor-amplifier concept into a biomolecular assay platform which amplifies the signal on spot of sensing. But after that no massive progress regarding the expected sensitivity >10 V/pH, or at least of the order of several hundreds of millivolt/pH is reported which can be used in practice. So there is a real need of more and more computational work before using these sensors in-vivo. If the experimental work would be sufficient, then there would be appreciable advancement until now. The current state-of art diagnostics biosensors are basically of two types; ELISA, or amplification of a sample by polymerase chain reaction (PCR). Both require large quantity of sample and much time. But advancement in diagnostic technology is essential in the current fourth generation of civilization. The sensors should be multitasking, low cost, ultra-fast and ultra-sensitive and unimpeachable. Here the present authors have described a computational method to analyze the performance of some special types of FET based biosensors. It is noteworthy that nanoelectronics and biosensing technology have merged in one place so that we can get the most valuable results theoretically which may focus some idea for the future technology experts to device super sensitive biosensors. The structure of each sensor, computational methodology for their performance and results of computation are elaborated.

SENSOR STRUCTURE

A biosensor essentially contains three parts; Bioreceptor, transducer and detector. The biological recognition element binds the specific biomolecule. A signal processing unit is also required which produces a signal on the occurrence of a specific biological process. The transducer converts the biochemical reaction into an electrical signal. Finally, a signal processor to convert the electronic signal to a meaningful parameter is required along with the display machine. The progress in biosensing is now manifold for the advancement of transistor biosensing mechanism. Much shorter response time with amplified sensitivity and simplicity of such types of nanobiosensors are the importance of today's fast growing world.

Figure 1. Simple diagram of a flexure FET nanobiosensor

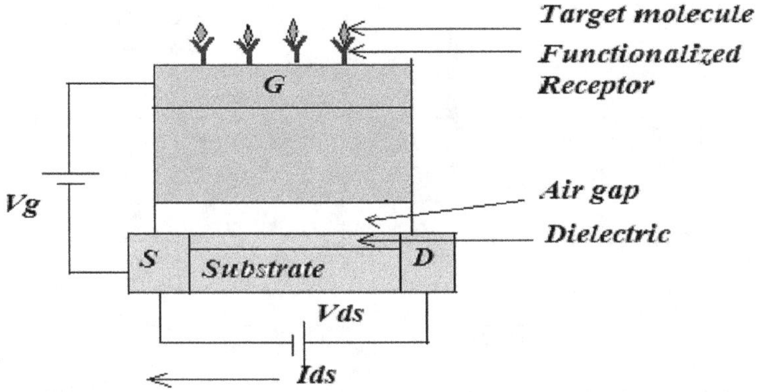

A flexure FET biosensor contains a suspended gate having low leakage current. The gate is a cantilever with two ends fixed. Over the substrate a thin layer of dielectric is placed to separate gate and channel. On the application of gate voltage, electromechanical movement induced by change of drain current is observed. With movement of the suspended gate, the thickness of the air gap changes which changes the capacitance of the device. The structure of a flexure FET biosensor is given in Figure 1.

An extended gate FET (EGFET) pH sensor is a recent addition in the field of sensors. It contains a part used as a pH sensitive membrane and a metal interconnecting layer structure is easier to prepare than that of an ion-sensitive FET(ISFET). In EGFET the FET part is isolated from the chemical environment and there is an extension of the gate electrode at the end of which the chemically sensitive membrane is deposited. Different parts of such a sensor are given in Figure 2. A double gate FET (DGFET) contains two gate regions; one is fluid gate or front gate and other is back gate. Front gate means an electrode which is immersed in the electrolyte. The reference electrode sets the driving voltage and the back gate amplifies the surface potential induced by the pH of solution. Figure 3 depicts the structure of a DHFET pH sensor.

Modeling Method

Here we are interested in charged biomolecules like protein and DNA which have specific charge. DNAs are always negative. But protein molecules are positive or negative depending on the environment surrounding it. However, we know the charge of the specific protein from the very beginning. By electrostatic interaction of the charged target molecules with the functionalized receptors of known identity

Figure 2. Structure of an EGFET biosensor

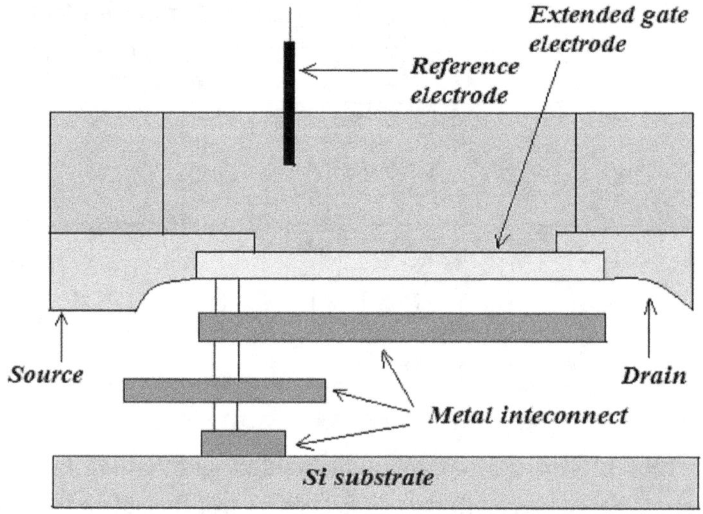

Figure 3. Structure of a DGFET pH sensor

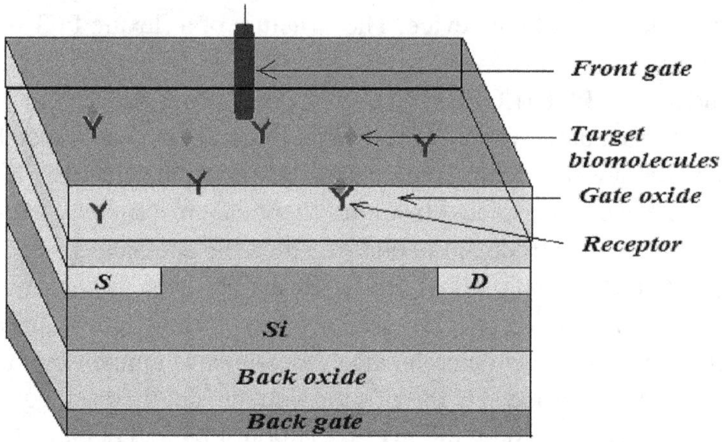

detects the presence of the target biomolecule. The excess charge after binding of the target and receptor molecule governs the conductivity of the FET carrier electrons. The diffused biomolecules in the sensor volume is thus detected by the change of conductivity of the electrons. Here the performance of three types of sensors will be discussed; FFET, EGFET pH sensor and DGFET pH sensor.

The surface of the sensor is functionalized with receptor molecules of known identity. When the target molecules, fed in the sensor structure are captured by the receptors, the excess charge on capturing, changes the conductivity of the FET channel carriers. Thus the presence of any charged target molecule can be assured by the simple process. Voltage applied in between the source and drain regulates the current flow. Here we consider *n* type channel and so, a current is produced in the *n* channel by mobile electrons between drain and source. In the vertical direction an electric field is generated on the application of gate voltage. The vertical field regulates the charge of the channel and hence the conductivity is also regulated by it. Inspired by this basic idea, FET can be used as a biosensor by functionalizing the gate region by suitable receptors which when captures a target charged biomolecule, the current of the gate region is changed. The drain-source current of the bioFET sensor is given as follows

$$I_{DS} = \frac{1}{2}\mu C\left(\frac{W}{L}\right)\left(V_{GS} - V_{th}\right)^2 \text{..... at saturation region } \left(\mathrm{V}_{DS} \geq V_{GS} - V_{th}\right) \quad (2)$$

$$= \mu C\left(\frac{W}{L}\right)\left[\left(V_{GS} - V_{th}\right)V_{DS} - \frac{1}{2}V_{DS}^2\right] \text{..... at triode region } \left(\mathrm{V}_{DS} < V_{GS} - V_{th}\right) \quad (3)$$

$$\mathrm{C} = \left[\Sigma \frac{di}{\left(\varepsilon_0 \varepsilon_i\right)}\right]^{-1} \quad (4)$$

μ is the mobility of electron in the channel, W and L are the width and length of the channel respectively. In between gate and channel a charge accumulation takes place and hence a capacitance equal to C is generated. Dielectric thickness is given by di and ε_0 and ε_i are the dielectric constant of vacuum and the material deposited. However, depending on the sensor structure, these values are modified.

The settling time of such sensor is much less than that of the ELISA model. The theoretical model is based on self-consistent solutions of Diffusion-Capture model. (for the settling time response), Poisson-Boltzmann and Drift-Diffusion Equations (for electrolyte screening and conductance modulation) and the statistical properties of biomolecule adsorption (Selectivity).

Drift Diffusion Model

In FET based biosensor, a functionalized layer of receptor biomolecules is used instead of gate. As discussed earlier, the conductance of the transducer changes after capturing of target molecules by receptors. Here targets are DNA molecules, may be single stranded or double standed, but having definite negative charge. When the target molecules move around the solution, it's movement is called diffusion. There has no regular path of the molecules, rather it moves randomly and the time for the molecule to reach the sensor surface can be calculated statistically and is proportional to the square of the distance. A depetion zone of size x is created when targets bind the receptor

$$x \sim \sqrt{Dt} \tag{5}$$

When the thickness of the depletion zone is comparable to the sensor size, the zone becomes flat. It enters the channel radially and then again grows as before.

The kinetics of absorption of biomolecule on sensor surface is given by the following two very known equations.

$$\frac{d\rho}{dt} = D\nabla^2\rho \tag{6}$$

$$\frac{dN}{dt} = k_F(\mathrm{N}_0 - N)\rho_s - k_R N \tag{7}$$

Equation 1 gives the diffusion of target molecules in the sensor surface. Diffusion process is a random walk in presence of many other water molecules. Here ρ is the concentration of the analyte molecules and D is its diffusion coefficient. The second equation represents the capture of biomolecules on the functionalised receptors. N is the density of conjugated receptorgs, N_0, the density of receptors on the sensor surface and k_F , k_R and ρ_s are the capture constant, dissociation constant and concentration of analyte particles at the surface of the sensor.

Diffusion distance of the molecule is a dirac delta function at time t=0. But after a certain time they diffuse symmetrically on both directions. The average displacement of the molecule is given by Equation 1. The solution of Equation 1 and 2 are

$$-\frac{\kappa_F N_0 + E - \kappa_F N_{equi}}{\kappa_F \rho_0 + \kappa_R} \log\left(1 - \frac{N}{N_{equi}}\right) + \frac{\kappa_F}{\kappa_F \rho_0 + \kappa_R} N = Et \tag{8}$$

Also the minimum time of response is given by Go and Alam (2009)

$$t = t_0 \left(\frac{\rho_0}{\rho} \right)^{\alpha_k} (k{=}1,2,...) \tag{9}$$

Selectivity of the sensor can be defined as

$$SNR = \frac{S}{\eta} \tag{10}$$

Signal and noise component both are derived to find the signal Noise ratio (SNR).

Considering charge neutrality on the sensor surface, the sensitivity of the sensor according to Nair and Alam (2008) can be given as

$$S = \frac{2\varepsilon_{ox}\Psi_0\left(N\left(t\right)\right)}{qa^2N_D \log\left(1 + \frac{t_{ox}}{a}\right)} \tag{11}$$

Here a is the radius of the nanowire and t_{ox} is the thickness of the oxide. N_D represents the doping density and Ψ_0 is the potential at the sensor surface.

Solution of Equation 1 and Equation 2 gives the kinetic response of the sensors or the time evolution of the capturing of the target molecules. Steady state solution is can be achieved at infinite time. At the saturation time, we get the scaling law for nanobiosensor.

Posson-Boltzmann Equation

When biomolecules in an ionic solution is concerned, Poisson-Boltzmann equation must be solved. Debye-Huckel theory (Debye and Huckel, 1923) was proposed for continuum model of molecules in ionic solutions which mains builds on Poisson-Boltzmann equation. Olivares and McQuarrie (1976) solved the equation for getting the effect of electrostatic screening of ions on capturing of biomolecules.

The ISFET and EGFET has nearly same structure except in the later structure the sensor surface is isolated from the transducer by an interconnect. nearly same In this model the device parameters are directly taken from the experimental results.

ssDNA are immobilised on the FET type sensor. The target DNA with known charges are injected into the FET channel. The mean field Poisson equation is the basic equation to describe the potential which is given below in cylindrical coordinates

$$-\nabla.(\varepsilon(r,\phi,\mathrm{z})\nabla)\,\mathrm{V}(r,\phi,\mathrm{z}) =$$

$$-\left(\frac{1}{r}\partial_r\left(\varepsilon r\partial_r\right)+\frac{\varepsilon}{r^2}\partial_{\phi\phi}+\frac{\varepsilon_\phi}{r^2}\partial_\phi+\partial_z\left(\varepsilon\partial_z\right)\right)V(r,\phi,\mathrm{z}) = n(r,\phi,\mathrm{z}) \quad (12)$$

The Poisson equation is meant for the domain $\Omega \subseteq \mathbb{R}^3$ which includes sensor and the region containing analyte. where *V* is the electrostatic potential, ε is the permittivity and *n* is the charge distribution. The right hand side of the equation is computed self consistently from the transport equations of the carriers and the additional charges produces by the receptor and the DNA molecules (Heitzinger & Klimeck, 2007). Approximation is that *n* is assumed to be a function of the exponential of the potential V without loss of generality.

Non-linear Poission-Boltzmann equation is given as:

$$-\nabla^2\Psi\left(r\right)+\frac{\kappa^2}{\beta}\sinh\left(\beta\Psi\right)=\frac{q}{\varepsilon_w}\sum_i Z_i\delta\left(r-r_i\right) \quad (13)$$

where Ψ is the electrostatic potential, inverse of κ is the Debye Huckel screening length and r is the spatial coordinate. z_i and r_i are the partial charge and position of the reference atom in the biomolecule. $\beta = \frac{q}{k_B T}$, where *q* is the electronic charge, κ^{-1} is called the Debye screening length. When the electrostatic potential is very small, the first term of the Taylor series expansion of the potential function can be used and then we get linearized Poisson-Boltzmann equation.

The whole sensor volume can be divided into three parts:

R_1: The region when the biomolecule exists.
R_2: The region outside the biomolecule, where no ions of the solution is present.
R_3: The region contains the solution with ions.

For the three regions Poisson-Boltzmann equation is solved applying suitable boundary conditions at the interfaces.

The compact model of conventional MOSFET by Taur and Nin (2009) is used for calculation. For EGFET or DGFET Poisson-Boltzmann equation is to be solved separately for different regions of the sensor. These are top gate oxide, bottom gate oxide, interface region, Silicon body and bottom oxide and gate interface. However, to obtain maximum sensitivity the DEFET sensor is operated in inversion-inversion region.

FFET Sensor

The static behavior of the FFET is governed by the theory of spring-mass model of Kam, Lee, Howe and King (2005) and Nathanson, Newell and Wickstrom (1967). The static behavior is given by the equation following Jain, Nair and Alam (2012)

$$k\left(y_0 - y\right) = \frac{1}{2}\varepsilon_0 E_{air}^2 A \tag{14}$$

Here k is the stiffness, E_{air} is the electric field in the air, W is the area of the gate electrode, ε_0 is the free space permittivity, y_0 and y are the air gap and position of the gate electrode respectively.

The drain source current is given as

$$I_{DS} = \mu_n L Q_i \frac{V_{DS}}{W} \tag{15}$$

L is the length of the gate electrode, μ_n is the mobility of the electrons in the channel, Q_i is the inversion charge density, V_{DS} is the drain source voltage and W is the width of the gate electrode. If I_{DS1} and I_{DS2} are the drain source current before and after the capturing of biomolecules, then according to Jain et al. (2012) the sensitivity S of the sensor is given as

$$S = \frac{I_{DS1}}{I_{DS2}} \approx \exp\left(\gamma_1 \sqrt{N_s} - \gamma_2 N_s\right) \tag{16}$$

γ_1 and γ_2 are some constants dependent on sensor geometry.

For modeling and simulation open source software LAMMPS (Plimpton, 1995) and online simulation tool of NanoHub.org are used.

RESULTS AND DISCUSSION

Considerable research progress in the field of modeling and simulation has been noticed while taking about applications of nanotechnology in medicine. Different sensors have their own limitations uptil now when used for in vitro studies. Till now the so called ELISA technique is very popular though sometimes it kills a lot of time. Here simulation approach to introduce nanoparticles as sensor materials to bind and detect charged DNA from analytes is presented. Here three types of sensors are taken up for investigation; all of them are FET type sensor. FFET sensor, EGFET pH sensor and DGFET pH sensor are discussed separately.

Flexure FET Sensor

The response of a nanobiosensor can be characterized by different parameters. These are sensitivity, settling time, selectivity etc. The surface of the sensor is functionalized with some receptors of known identity. When they find complementary biomolecules in its vicinity it captures it and the overall transfer characteristics are changed. This gives the sensitivity. So the change of the drain source current on capturing of a target molecule gives the sensitivity of the sensor. Figure 4(a) and 4(b) show the sensitivity of the FFET nanobiosensor at various lengths at a width 10 nm and 100nm respectively.

Figure 4. (a): Sensitivity of a FFET sensor for various lengths at a fixed width of (a) 10nm (b) 100nm of the channel

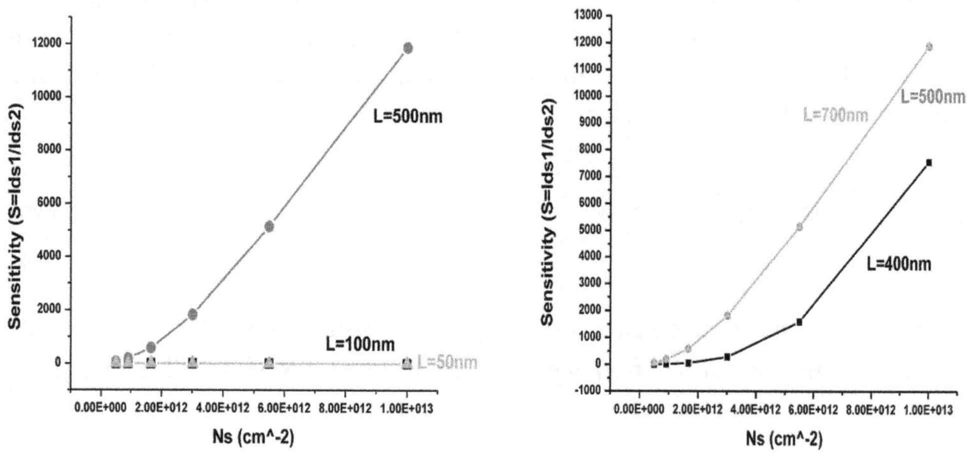

Sensitivity of the FFET depends on the sensor geometry (Figure 4, 5). Maximum sensitivity is obtained for a length of 500nm and width of 100nm of the gate electrode. But the sensitivity is lost for very low length and width. Remarkable sensitivity of almost 12000 is reported in this study which signifies that the sensor is supersensitive with this geometrical configuration. So this type of sensor can detect fatal diseases like cancer in its much earlier stage. When a gate bias V_G is applied to the cantilever, the gates moves towards the dielectric layer and the increases the gate capacitance. Obvious increase of drain source current, I_{DS} is observed. I_{DS} can be calculated from Equation 3 in the steady state. In a limit close to pull in and in sub threshold regime, FFET sensor shows highest sensitivity. It can detect both charged and uncharged molecule as it combines the effect of two sensors; one is a mechanical one and other is an electrical.

After capture, I_{DS} decreases again as at that time the separation between the gate and the dielectric is increased again. So biomolecules can be targeted either by their mass and size or by their charge. Figure 5 and 6 depict the signal-noise ratio (SNR) with respect to receptor density and variation of density of captured molecules over time respectively. For calculating these two properties the length and the width of the gate electrode are taken as 400nm and 100nm respectively. Here the other parameters of simulation are listed below:

Figure 5. Signal noise ratio (SNR) of the biosensor

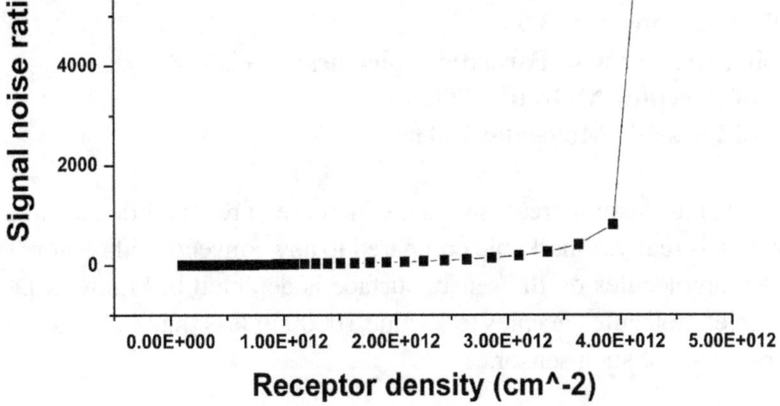

Figure 6. Variation of the density of target molecules the presence of parasitic molecules over time

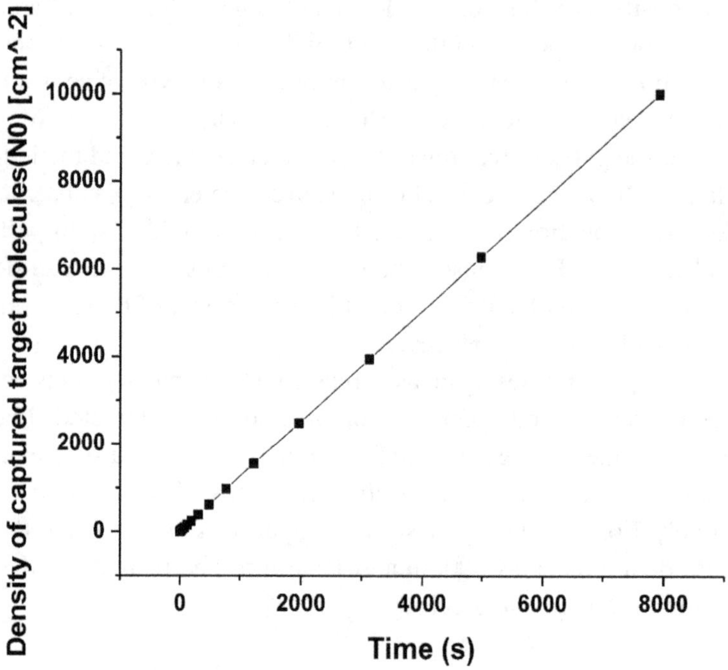

- **Thickness of Gate Electrode:** 40nm.
- **Thickness of Air Gap between Gate and Dielectric Medium:** 100nm.
- **Dielectric Thickness:** 100 nm.
- **Lowe Value of Analyte Concentration:** 1e-15 M.
- **Upper Value of Analyte Concentration:** 1e-6 M.
- **Minimum Number of Molecules:**
- **Drain to Source Voltage:** 0.5 V.
- **Dielectric Constant:** 3.9.
- **Doping Density with Parasitic Molecules:** 6e+22/m^3.
- **Size of Receptor Molecule:** 200nm.
- **Size of Parasitic Molecule:** 100 nm.

SNR of the biosensor increases with the increase of receptor density and reaches to 9113 which is really remarkable compared to any conventional sensor. Transient capture of biomolecules on the sensor surface is depicted in Figure 6. Density of captured target molecules reaches to a value 10000 in less than 2.5 hour, indicating high performance of such sensor.

Figure 7. Current voltage relationship of the FFET biosensor

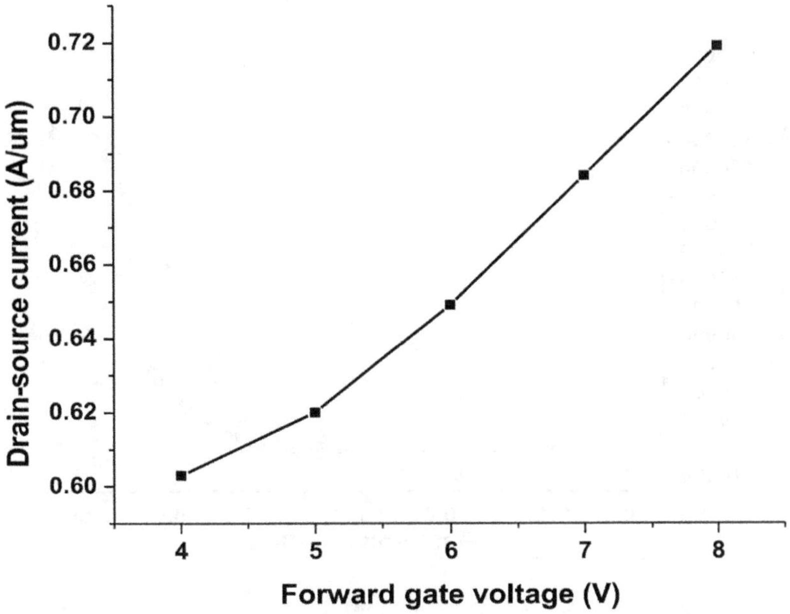

One important characteristic of a FET is its current-voltage relationship. For the single gate FFET the drain current increases with the increase in pull in gate voltage (Figure 7). Due the high sensitivity of the sesor, the drastic increase of current is observed with the increase of pull in gate voltage.

Extended Gate FET pH Sensor

An EGFET is a device that isolates FET from chemical environment. pH response in this EGFET sensor is obtained in the pH range of 4-8. pH measurement is done by combining the effect of a metal oxide silicon field effect transistor (MOSFET) and a chemically sensitive membrane instead of the gate electrode. These two parts serve as an ion selective field effect transistor (ISFET). Modern sensors based on EGFET reveal new methods of pH or nitrate measurements in much easier way (Fernandes et al., 2016; Minami et al. (2016). Figure 8 (a) and (b) shows the current-voltage relationship of the sensor for different values of channel length and width. The simulation parameters used here are as follows:

- **Top Oxide Thickness:** 400nm.
- **Back Oxide Thickness:** 150nm.
- **Silicon Layer Thickness:** 800nm.

Figure 8. Voltage-current characteristics of the EGFET sensor for (a) channel length 500nm and width 10nm (b) channel length 1µm and width 500nm

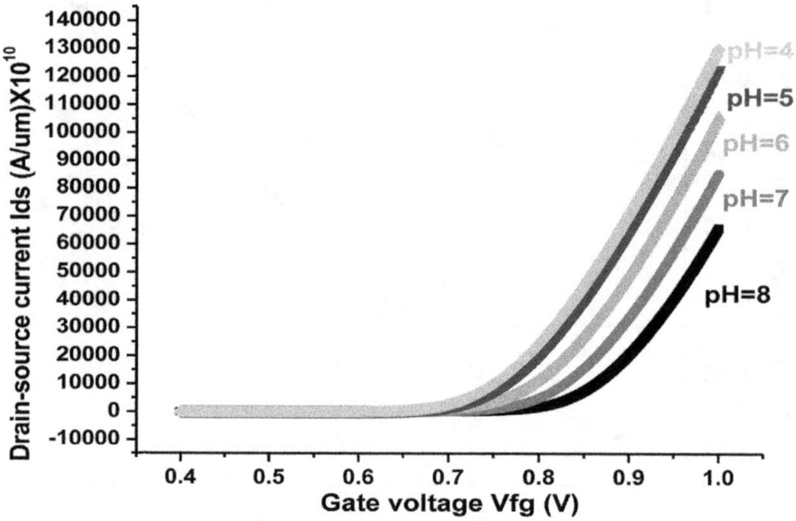

(a)

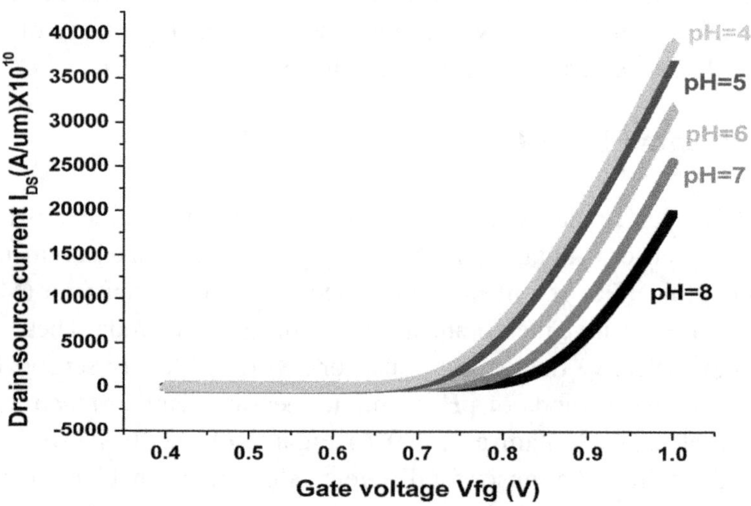

(b)

- **Doping Density:** 1e15.
- **Size of Parasitic Molecules:** 100nm.
- **Size of Receptor Molecules:** 200nm.
- **Concentration of Target Molecules:** 1e-12 M.
- **Concentration of Parasitic Molecules:** 1e-6 M.
- **Electrolyte Concentration:** 0.1M.
- **Surface Density:** e14.

The drain-source current I_{DS}, depends on the pH of the buffer solution. Hence variation of I_{DS} indicates the sensor sensitivity. Keeping the drain-source voltage V_{DS} fixed, variation of I_{DS} is measured with changing gate voltage V_G. In Figure 8 different colored lines represent curves for different pH value of solution. As time elapses, the drain I_{DS} increases gradually for the same pH value. Here V_{DS} is fixed as 0.1 V. The curves indicate a threshold voltage shift which depends on concentration of H^+ ions. The threshold voltage shifts towards right for higher pH value.

On increasing the pH of solution, the reference electrode voltage increases almost linearly. From the graphs Figure 9 (a) -(d), we can calculate pH sensitivity for fixed I_{DS} by calculating small change of voltage with respect to small change in pH.

Voltage sensitivity, S_V thus can be written as

$$S_V = \left.\frac{\Delta V_R}{\Delta pH}\right|_{I_{DS}}$$

Here ΔV_R is the small change in reference voltage and ΔpH is the small change in the pH value. Here I_{DS} is 0.9 mA. From the four curves average sensitivity is found as 63.59 mV/pH. Maximum ISFET or EGFET studies reported much lower pH sensitivity in μV/pH (Lee et al., 2013; Chang and Yang, 2012). Some studies (Lee et al., 2014; Yang et al., 2013) report the sensitivity around the Nernst limit of 59 mV/pH for EGFET biosensor. Very recent study reveals about breaking of the Nernst limit like Al-Hardan et al. (2016). They got a value of 66mV/pH which is a super Nernstian value by using porous silicon as an EGFET. Our value is very much similar to them.

DGFET pH Sensor

In order to cross the Nernst limit of sensitivity DGFET instead of an ISFET has an ideal structure. It is a low noise differential amplifier for the pH signal. Hence it has always better response than that of a conventional ISFET. According to Go

Figure 9. Threshold voltage variation with pH for an extended gate FET pH sensor for (a) L=500nm, W=100nm and V_{DS}=0.1 V (b) L=700nm, W=500nm, V_{DS}=0.1 V, (c) L= 100nm, W= 100nm, V_{DS}=0.2 V, (d)L=500 nm, W=10 nm, V_{DS}=0.2 V

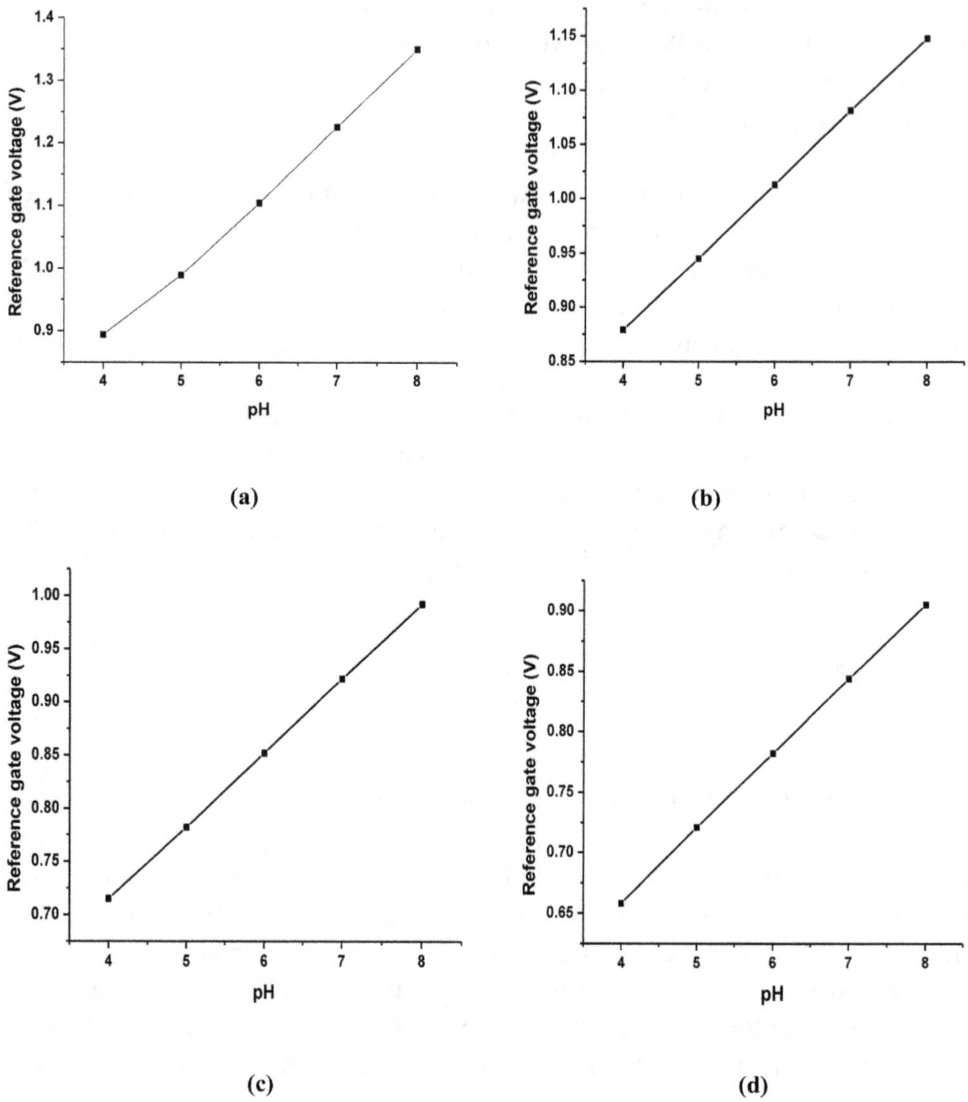

et al. (2012), maximum pH sensitivity for DGFET sensor is achieved in inversion-inversion regime and SNR will be higher when fluid gate is biased in the depletion region. Transfer characteristics are obtained for this sensor keeping the fluid gate bias fixed and varying the forward gate bias. The authors here have analyzed the

Figure 10. Transfer characteristics of the DGFET sensor for different pH values in front gate operation mode (a) for L= 1μm, W=900nm, V_{DS}=0.1 V (b) L= 500 nm, W= 100nm, V_{DS}=-0.1 V

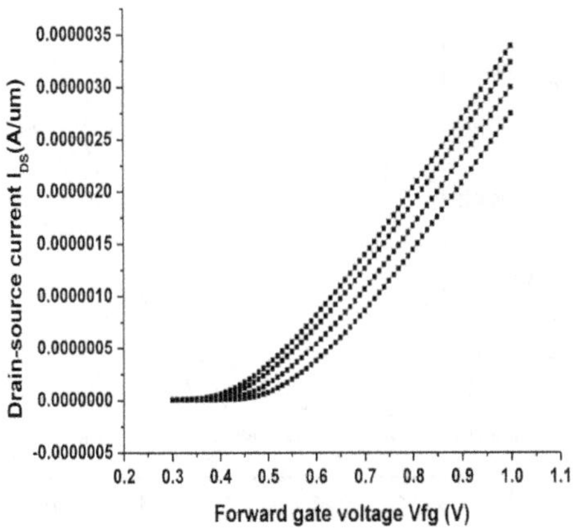

(a)

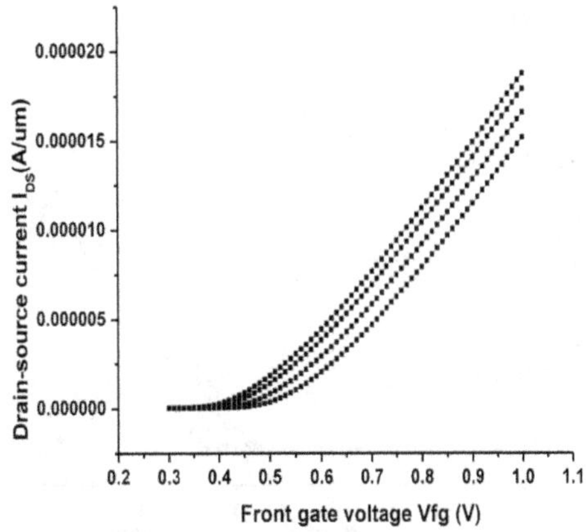

(b)

effect of changing the length, width and forward gate voltage on the current-voltage relationship of the sensor in Figure 10 (a) and (b) The steep rise of the curves indicates the high sensitivity of this structure.

With little negative bias of gate source region, and length and width of the device in the nanometer regime, steeper curves are obtained. With the increase of device length or width, sensitivity is gradually lost. The back gate bias is fixed at 0.0 V. Here the device is operated in nMOSFET mode and the device simulated between the pH ranges 4-7. Transfer characteristics are shown in Figure 11 (a) and (b).

Calculating the pH sensitivity in the front gate mode, nothing interesting result is obtained. But when it is operated in the back gate mode, high sensitivity is reported. Threshold voltage vs pH of the sensor is depicted in Figure 12.

The slope of the curve gives the ph sensitivity with respect to voltage change. Remarkable sensitivity is obtained for this sensor which is found to be around 200 mV/pH. In the depicted figure below, it comes out to be 203.49mV/pH. Here back gate voltage V_{fg} is set to 0.1 V and front gate voltage is set to 0.0 V. Noise is also amplified with the signal, but by properly choosing the channel material, its noise can be set to minimum label. The idea of supportive bias of one gate for the amplification of the shift of threshold voltage and the biasing of other main gate to increase current, worked nicely for a DGFET. It also improves instrument limited pH resolution capability of the device.

Figure 11. Transfer characteristics of the DGFET sensor for different pH values in back gate operation mode (a) for L= 1μm, W=1μm, V_{DS}=0.1 V (b) L= 500 nm, W= 100nm, V_{DS}=-0.1 V

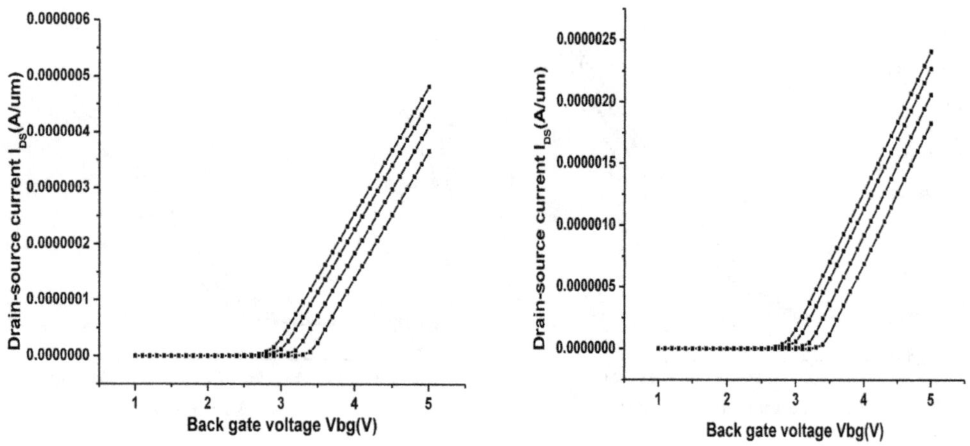

Figure 12. Variation of threshold voltage with pH of solution for L= 500nm, W=100nm, V_{DS}=0.1 V

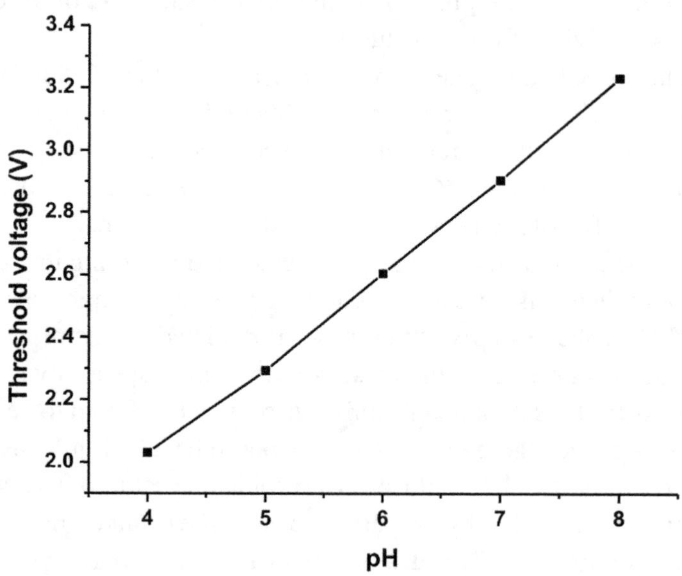

CONCLUSION

A detailed study of the performance of a FFET sensor, EGFET pH sensor and DGFET pH sensors is presented in this work. Maximum sensitivity is obtained for DGFET sensor. For the flexure FET sensor, maximum sensitivity is obtained for a device of length 500nm and 700nm and width 100nm. But decreasing the width to a much lower value, sensitivity is again decreased. For this sensor SNR is considerably high and the value obtained is 9113. Compared to the conventional and most popular ELISA technique, it can detect about 10000 target molecules/cm^2 on the surface from a concentration of 1X10^{-15} M of the analyte. High current is also reported in the operation of FFET sensor. EGFET pH sensor shows high sensitivity of about 63.59mV/pH which is slightly higher than the Nernst limit. Transfer characteristics of the sensor clearly indicates threshold voltage shift towards right for higher pH value. Due to the chemically sensitive layer, the threshold voltage depends on the concentration of H^+ ions in the solution. High current is observed for a device length of 500nm and width of 10nm. Front gate and back gate operation both are studied. When front gate bias is set to zero with a length of 50nm and width of 10nm of the device, maximum drain-source current is found to flow. DGFET structure is reported to have the maximum sensitivity of 203.49mV/pH when operated in the back gate mode. As DGFET is a low noise differential amplifier for pH, hence it may be used with great success in future for rapid detection of fatal disease and from a very low concentration of the analyte.

REFERENCES

Ahn, J.-H., Choi, S.-J., Han, H. W., Park, T. J., Lee, S. Y., & Choi, Y.-K. (2011). Investigation of size dependence on sensitivity for nanowire FET biosensors. *IEEE Transactions on Nanotechnology*, *6*, 1405–1411.

Al-Hardan, , Abdul Hamid, M., Ahmed, N., Jalar, A., Shamsudin, R., & Othman, N. et al. (2016). High Sensitivity pH Sensor Based on Porous Silicon (PSi) Extended Gate Field-Effect Transistor. *Sensors (Basel, Switzerland)*, *16*(6), 839. doi:10.3390/s16060839 PMID:27338381

Cai, B., Huang, L., Zhang, H., Sun, Z., Zhang, Z., & Zhang, G.-J. (2015). Gold nanoparticles-decorated graphene field-effect transistor bio-sensor for femtomolar microRNA detection. *Biosensors & Bioelectronics*, *74*, 329–334. doi:10.1016/j.bios.2015.06.068 PMID:26159152

Cai, B., Wang, S., Huang, L., Ning, Y., Zhang, Z., & Zhang, G.-J. (2014). Ultrasensitive label-free detection of PNA-DNA hybridization by reduced graphene oxide field-effect transistor biosensor. *ACS Nano*, *8*(3), 2632–2638. doi:10.1021/nn4063424 PMID:24528470

Chang, S-P., Yang, T.-H. (2012). Sensing performance of EGFET pH sensors with Cuo nanowires fabricated on glass substrate. *International Journal of Electrochemical Sensing*, 7(6), 5020-5027.

Chen, J. & Lee., J.E. (2010). Atomistic analysis of nano/micro biosensors. *Interaction and Multiscale mechanics*, 3(2), 111-121.

Chen, J., Fang, Z., Lie, P., & Zeng, L. (2012). Computational lateral flow biosensor for proteins and small Molecules: A new class of strip logic gates. *Analytical Chemistry*, *84*(15), 6321–6325. doi:10.1021/ac301508b PMID:22780691

Clemens, H., Liu, Y., Mauser, N. J., Ringhofer, C., & Dutton, R. W. (2010). Calculation of fluctuations in boundary layers of nanowire Field effect biosensors. *Journal of Computational and Theoretical Nanoscience*, *7*, 1–7.

Colombelli, A., Manera, M.G., Taurino, A., Catalano, M., Convertino, A. Au nanoparticles decoration of silica nanowires for improved optical bio-sensing. *Sensor actuators B: Chemical*, 226, 589-597.

Das, , Ko, D. H., Chen, C.-H., Chang, L.-B., Lai, C.-S., & Chu, F.-C. et al. (2014). Highly sensitive palladium oxide thin film extended gate FETs as pH sensor. *Sensors and Actuators. B, Chemical*, *205*, 199–205. doi:10.1016/j.snb.2014.08.057

De Vico, L., Sørensen, M. H., Iversen, L., Rogers, D. M., Sørensen, B. S., & Brandbyge, M. et al. (2010). Quantifying signal changes in nano-wire based biosensors. *Nanoscale*, *3*(2), 706–717. doi:10.1039/C0NR00442A PMID:21173975

Duan, X., Li, Y., Rajan, N. K., Routenberg, D. A., Modis, Y., & Reed, M. A. (2012). Quantification of the affinities and kinetics of protein interactions using silicon nanowire biosensors. *Nature Nanotechnology*, *7*(6), 401–407. doi:10.1038/nnano.2012.82 PMID:22635097

Fernandes, J. C., Raphael, R. A. S., & Mulato, M. (2016). Effects of measurements conditions on an extended gate FET used as pH sensor. *Materials Research*, *19*(1), 267–272. doi:10.1590/1980-5373-MR-2015-0248

Gao, A., Lu, N., Wang, Y., & Li, T. (2016). Robust ultrasensitive tunneling-FET biosensor for point-of care diagnostics. *Scientific Reports*, *2*, 22554. doi:10.1038/srep22554 PMID:26932158

Garcia, P. S. et al. (2015). A Nanobiosensor based on 4-hydrosyphenylpyruvate dioxygenase Enzyme for mesotrione detection. *IEEE Sensors Journal, 15*(4), 2106–2113. doi:10.1109/JSEN.2014.2371773

Go, J., & Alam, M. A. (2009). Statistical interpretation of "femtomolar" detection. *Applied Physics Letters, 95*(3), 033110. doi:10.1063/1.3176017 PMID:19690630

Hafaiedh, I., Baccar, H., Ktari, T., & Abdelghani, A. (2012). Electrochemical chacterization of Streptavidin-HRP Immobilized on Multiwall carbon Nanotubes for Biosensor Applications. *Journal of Biomaterials and Nanobiotechnology, 3*(01), 31–36. doi:10.4236/jbnb.2012.31005

Heitzinger, C., Kennell, R., Klimeck, G., Mauser, N., Mclennan, M., & Ringhofer, C. (2008). Modeling and simulation of field effect biosensors and their deployment on the nanoHUB. *Journal of Physics: Conference Series, 107*, 012004. doi:10.1088/1742-6596/107/1/012004

Heitzinger, C., & Klimeck, G. (2007). Computational Aspects of the three dimensional feature-scale simulation of silicon nanowire field-effect sensors for DNA detection. *Journal of Computational Electronics, 6*(1-3), 387–390. doi:10.1007/s10825-006-0139-x

Heitzinger, C., Mauser, N. J., & Ringhofer, C. (2010). Multiscale modeling of planar and nanowire field-effect biosensors. *SIAM Journal on Applied Mathematics, 70*(5), 1634–1654. doi:10.1137/080725027

Hsiao, F.-L., & Lee, C. (2011). Nanophotonic biosensors using hexagonal nanoring resonators: Computational study. Journal of micro/nanolithography. *MEMS and MOEMS, 10*(1), 013001. doi:10.1117/1.3532834

Hung, S.-C., Cheng, N. J., Yang, C.-F., & Lo, Y.-P. (2014). Investigation of extended-gate field-effect transistor pH sensors based on different-temperature-ennealed bi-layer MWCNTs-In_2O_3 films. *Nanoscale Research Letters, 9*(1), 502. doi:10.1186/1556-276X-9-502 PMID:25288911

Jain, A., Nair, P. R., & Alam, M. A. (2012). Flexure-fet biosensor to break the fundamental sensitivity limits of biosensors using nonlinear electromechanical coupling. *Proceedings of the National Academy of Sciences of the United States of America, 109*(24), 9304–9308. doi:10.1073/pnas.1203749109 PMID:22623527

Jang, H.-J., & Cho, W.-J. (2016). Performance enhancement of capacitive-coupling dual-gate ion-sensitive field-effect transistor in ultra-thin-body. *Scientific Reports, 4*, 5284. PMID:24923751

Kam, H., Lee, D. T., Howe, R. T., & King, T.-J. (2005). A new nano-electro-mechanical field effect transistor (NEMFET) design for low-power electronics.*Proceedings of IEEE International Electron Device Meeting* (pp. 463-466).

Karger, A. (2008). Theoretical investigation of silicon nanowire pH sensor. *Proceedings of the 2nd IEEE RAS & EMBS International Conference on Biomedical Robotics and Biomechatronics* (pp. 765-769). Retrieved from http://lammps.sandia.gov/

Lee, I.-K., Bae, T.-E., Lau, H.-C., Lim, J.-O., & Cho, W.-J. (2014). A study on a differential-measurement biosensor using high performance extended-gate field-effect transistors with TiO_2 sensing membranes. *Sensor Letters*, *12*(6), 1003–1007. doi:10.1166/sl.2014.3193

Lee, J.U., Nguyen, A.H. & Sim, S.J. (2015). A nanoplasmonic biosensor for label-free multiplex detection of cancer biomarkers. *Biosensors and Bioelectronics*, 74, 341-346.

Lee, M., & Fauchet, P. M. (2007). Two dimensional silicon photonic crystal based biosensing platform for protein detection. *Optics Express*, *15*(8), 4530. doi:10.1364/OE.15.004530 PMID:19532700

Lee, P.-Y., Chang, S.-P., Kuo, P. J., & Shei, S. C. (2013). Sensing performance of EGFET pH sensors with CZTse nanoparticles fabricated on glass substrates. *International Journal of Electrochemical Sensing*, *8*, 3866–3875.

Li, G., Wang, Y., & Tao, Q. (2013). Modeling and simulation of silicon nanowire-based biosensors. In *Modeling and control for micro/nano devices and systems*. London: Taylor and Francis.

Mao, Y., Shin, K.S., Wang, X., Ji, Z., Meng., H. & Chui, C.O. (2016). Semiconductor electronic label-free assay for predictive toxicology. *Scientific Reports*, *6*, 24982. doi:10.1038/srep24982

Mei, L.-P., Feng, J.-J., Wu, L., Zhou, J.-Y., Chen, J.-R., & Wang, A. J. (2015). Novel phenol biosensor based on laccase immobilized on reduced graphene oxide supported palladium–copper alloyed nanocages. *Biosensors & Bioelectronics*, *74*, 347–352. doi:10.1016/j.bios.2015.06.060 PMID:26159155

Meolin, J., & Quake, S. R. (2007). Microfluidic large scale integration: The evolution of design rules for biological automation. *Annual Review of Biophysics and Biomolecular Structure*, *36*(1), 213–231. doi:10.1146/annurev.biophys.36.040306.132646 PMID:17269901

Minami, T., Sasaki, Y., Minamiki, T., Wakida, S., Kurita, R., Niwa, O., & Tokito, S. (2016). Selectrive nitrate detection by an enzymatic sensor based on an extended-gate type organic field-effect transistor. *Biosensors & Bioelectronics*, *81*, 87–91. doi:10.1016/j.bios.2016.02.036 PMID:26921557

Moon, D., Peycelon, M., Kim, J.-Y., Ahn, J.-H., Park, T. J., & Choi, Y. K. (2013). A biristor based on a floating-body silicon nanowire for biosensor applications. *Applied Physics Letters*, *102*(4), 043701. doi:10.1063/1.4789904

Mu, L., Chang, Y., Sawtelle, S. D., Wipf, M., Duan, X., & Reed, M. (2015). Silicon nanowire field effect transistors-A versatile class of potentiometric nanobiosensors. *Access, IEEE*, *3*, 287–302. doi:10.1109/ACCESS.2015.2422842

Nair, P. R., & Alam, M. A. (2006). Performance limits of nanobiosensors. *Applied Physics Letters*, *88*(23), 233120. doi:10.1063/1.2211310

Nair, P. R., & Alam, M. A. (2007). Dimensionally frustrated diffusion towards fractal absorbers. *Physical Review Letters*, *99*(25), 256101. doi:10.1103/PhysRevLett.99.256101 PMID:18233533

Nair, P. R., & Alam, M. A. (2007). Dimensionally frustrated diffusion towards fractal adsorbers. *Physical Review Letters*, *99*(25), 256101. doi:10.1103/PhysRevLett.99.256101 PMID:18233533

Nair, P. R., & Alam, M. A. (2008). Screening limited response of nanobiosensors. *Nano Letters*, *8*(5), 1281–1285. doi:10.1021/nl072593i PMID:18386914

Nair, P. R., & Alam, M. A. (2010). Theory of selectivity of label-free nanobiosensors: A geometro-physical perspective. *Journal of Applied Physics*, *107*(6), 064701. doi:10.1063/1.3310531 PMID:20428486

Nair, P. R., Go, J., Landells, G. J., Pandit, T. R., Alam, M., Jin, X., et al. (2014). BioSensorLab. Retrieved from https://nanohub.org/resources/senstran DOI:10.4231/D3000014H

Nathanson, H. C., Newell, W. E., Wickstrom, R. A., & Davis, J. R. Jr. (1967). Resonant gate transistor. *IEEE Transactions on Electron Devices*, *14*(3), 117–133. doi:10.1109/T-ED.1967.15912

Nawattanapaiboon, K., Kiatpathomchai, W., Santanirand, P., Vongsakulyanon, A., Amarit, R., & Somboonkaew, A. et al. (2015). SPR-DNA array for detection of methicillin-resistant Staphylococcus aureus (MRSA) in combination with loop-mediated isothermal amplification. *Biosensors & Bioelectronics*, *74*, 335–340. doi:10.1016/j.bios.2015.06.038 PMID:26159153

Olivares, W., & McQuarrie, D. A. (1976). A vibrational approach to the theory of ionic solutions. *The Journal of Chemical Physics*, *65*(9), 3604. doi:10.1063/1.433545

Patolsky, F., Zheng, G., & Lieber, C. M. (2006). Fabrication of silicon nanowire devices for ultrasensitive, label free, real-time detection of biological and chemical species. *Nature Protocols*, *1*(4), 1711–1724. doi:10.1038/nprot.2006.227 PMID:17487154

Plimpton, S. (1995). Fast Parallel Algorithms for Short-Range Molecular Dynamics. *Journal of Computational Physics*, *117*(1), 1–19. doi:10.1006/jcph.1995.1039

Talukdar, K., Bhushan, M., Viswanathc, A. K., & Mitrad, A. K. (2013). Simulation study of the performance of a biologically sensitive field effect transistor. *International Journal of Nano Dimension*, *4*, 85–89.

Sørensen, M. H., Mortensen, N. A., & Brandbyge, M. (2007). Screening model for nanowire surface-charge sensors in liquid. *Applied Physics Letters*, *91*, 102105. doi:10.1063/1.2779930

Suaifan, G. A., Esseghaier, C., Ng, A., & Zourob, M. (2013). Ultra-rapid colorimetric assay for protease detection using magnetic nanoparticle-based biosensors. *Analyst (London)*, *138*(13), 3735–3739. doi:10.1039/c3an36881e PMID:23676852

Tang, J., Wang, Y., Li, J., Da, P., Geng, J., & Zheng, G. (2014). Sensitive enzymatic glucose detection by TiO2 nanowire photoelectrochemical biosensors. *Journal of Materials Chemistry A*, *2*(17), 6153–6157. doi:10.1039/C3TA14173J

Taur, Y., & Ning, T. H. (2009). *Fundamentals of modern VLSI devices* (2nd ed.). Cambridge University Press.

Vico, et al. (2011). Quantifying signal changes in nanowire-based biosensors. *Nanoscale*, *3*(2), 706–717. doi:10.1039/C0NR00442A PMID:21173975

Wang, Y., Dostalek, J., & Knoll, W. (2011). Magnetic Nanoparticle-enhanced biosensor based on Grating-coupled surface Plasmon resonance. *Analytical Chemistry*, *83*(16), 6202–6207. doi:10.1021/ac200751s PMID:21711037

Wei, F., Lillehoj, P. B., & Ho, C. M. (2010). DNA Diagnostics: Nanotechnology-enhanced electrochemical detection of nucleic acids. *Pediatric Research*, *67*(5), 458–468. doi:10.1203/PDR.0b013e3181d361c3 PMID:20075759

Yang, C.-M., Wang, J.-C., Chiang, T. W., Lin, Y.-T., Juan, T.-W., & Chen, T.-C. et al. (2014). Hydrogen ion sensing characteristics of IGZO/Si electrode in EGFET. *International Journal of Nanotechnology*, *11*(1/2/3/4), 15–26. doi:10.1504/IJNT.2014.059806

Zhang, M.-L., Fan, X., Zhou, H.-W., Shao, M.-W., Zapien, J. A., Wong, N.-B., & Lee, S.-T. (2010). High efficiency surface-enhanced Raman scattering substrate based silicon nanowires array decorated with silver nanoparticles. *Journal of Physical Chemistry*, *114*, 1969–1975.

Zhang, Y., Chen, R., Xu, L., Ning, Y., Xie, S., & Zhang, G. J. (2015). Silicon Nanowire Boisensor for highly Sensitive and Multiplexed detection of oral squamous cell carcinoma biomarkers in saliva. *Analytical Sciences*, *31*(2), 73–78. doi:10.2116/analsci.31.73 PMID:25746803

Zheng, G., Gao, X. P. A., & Leiber, C. M. (2010). Frequency domain detection of biomolecules using silicon nanowire biosensors. *Nano Letters*, *10*(8), 3179–3183. doi:10.1021/nl1020975 PMID:20698634

Zheng, G., Patolsky, F., Cui, Y., Wang, W. U., & Lieber, C. M. (2005). Multiplexed electrical detection of cancer markers with nanowire sensor arrays. *Nature Biotechnology*, *23*(10), 1294–1301. doi:10.1038/nbt1138 PMID:16170313

Zhu, S., Du, C., & Fu, Y. (2009). Fabrication and characterization of rhombic silver nanoparticles for biosensing. *Optical Materials*, *31*(6), 769–774. doi:10.1016/j.optmat.2008.07.014

Chapter 8
Self-Setting Calcium Phosphate Bone Cement Preparation, Characterization and Drug Delivery for Skeletal System:
Self-Setting Calcium Phosphate Bone Cement

Sayed Reza Shaffiey
Novin Pak Nanotech Co. Ltd., Mazandaran Science and Technology Park, Iran

Sayedeh Fatemeh Shaffiey
Novin Pak Nanotech Co. Ltd., Mazandaran Science and Technology Park, Iran

DOI: 10.4018/978-1-5225-1043-7.ch008

ABSTRACT

The reactivity of acid base cements forming hydroxyapatite (HA) such as, β-tricalcium phosphate (β-TCP) and dicalcium phosphate dehydrate (DCPD). Amorphous calcium phosphates, prepared by precipitation from supersaturated solutions, can also react to form apatitic cements since they are thermodynamic unstable with respect to HA and have a setting reaction more independent of particle size. Calcium phosphate cement containing an antibiotic can be used for filling bone defects and to ensure local antibiotherapy. Therefore, in the present chapter proposal, cement paste was prepared by combining cement liquids comprised of Na_2HPO_4 with cement powders. Gentamicin sulfate was also loaded on the cements and its in vitro release was evaluated over a period of time. The cement setting times were compared before and after drug addition. According to results, the initial and final setting times of samples increase after drug addition.

INTRODUCTION

Calcium phosphate materials are widely applied as a variety of bone substitute materials in dental and medical fields (Shaffiey & Shaffiey, 2016). Brown and Chow in 1986 were firstly developed calcium phosphate cements (CPCs). Calcium phosphate cement is a group of calcium phosphate ceramic which has excellent biocompatibility and bioactivity properties. It is suitable to be used only for non- or low load-bearing applications such as bone filler (Friedman et al., 1998; Kamerer et al., 1994; Constantz et al., 1995) bone augmentation (Horstmann et al., 2003; Strauss & Egol, 2007; Liverneaux, 2004; Welch, Zhang & Bronson, 2003; Aral et al., 2008) and drug delivery applications (Ginebra, Traykova & Planell, 2006).

CPCs exhibit various benefits comparing other calcium phosphate based materials. Being injectable and easy to shape let them to be used to fill geometrically complicated bone defects (Del Real, Wolke, Vallet-Regí, & Jansen, 2002). Reactants for HA cements are crystalline calcium phosphates like tetracalcium phosphate (TTCP) or α/β-tricalcium phosphate (α/β-TCP) combined with slightly acidic compounds like dicalcium phosphate anhydride (DCPA) or dicalcium phosphate dehydrate (DCPD). On account of distinctive rates of disintegration of reactants, setting responses will just happen if the dynamic solubilities of all segments are harmonious; subsequently, this is ordinarily balanced by means of the particle size/particular surface range of the reactant particles. The powder stage after blending with a fluid stage like water or a watery arrangement shapes a glue which in light of the solidifying response can frame HA (apatite cements) or DCPD (brushite cements) in the wake of being connected inside of the body (Julien et al., 2007), (Kokubo, 2008), (Van den Vreken et al., 2010), (Ginebra, Traykova & Planell, 2006), (Yang, Troczynski & Liu, 2002).

Cements formed at pH<4.2, which are obtained by using acidic calcium phosphates or phosphorous acid, form brushite (DCPD, $CaHPO_4 . 2H_2O$) (Mirtchi, Lemaitre & Terao, 1989; Mirtichi, Lemaitre & Munting, 1989) and above 4.2, hydroxyapatite (HA) is formed.

Osteomyelitis is an intense or perpetual provocative procedure of the bone and its structures optional to disease with pyogenic life forms. Osteomyelitis may be limited or it may spread through the marrow, periosteum, cortex, and cancellous tissue. The bacterial pathogen differs on the premise of the persistent's age and the component of contamination. Bacterial reasons for intense and direct Osteomyelitis incorporate group related methicillin-safe *Staphylococcus Aureus* (MRSA), Kingella Kingae and others. Treatment for osteomyelitis includes the start of intravenous anti-infection agents that enter bone and joint depressions, referral of the patient to an orthopedist or general surgeon and conceivable restorative irresistible illness meeting (King & Kulkarni, 2010).

Couple of confusions in reconstructive orthopedics represents a more difficult hindrance to clinical accomplishment than disease, and bacterial osteomyelitis remains an imperative and overwhelming orthopedic and clinical issue. An assortment of prophylactic and helpful systems has been intended to decrease the occurrence and effect of disease. Routine treatment utilizing systemic anti-microbials is costly, inclined to complexities and frequently unsuccessful. In this way, neighborhood treatment is craved to treat or avert osteomyelitis (Lew & Waldvogel, 2004; Ciampolini & Harding, 2000).

Gentamicin is an aminoglycosidic antibiotic for bacterial infections caused by staphylococcus which are sensitive to Gram-negative bacteria (Doadrio et al., 2004) and of particular interest in orthopaedic surgery where it can be joined with bone graft materials to keep from contamination after surgery.

In this research, we prepared gentamicin sulfate loaded Calcium phosphate bone cement and carried out an investigation of the cement injectability in a constant injection velocity; then, the effect of medication addition on the cement properties and its injectability was investigated; as a result, we obtained an injectable medication loaded calcium phosphate bone cement which did not contain any polymeric additives to improve injectability.

MATERIALS AND METHODS

Synthesis and Characterization of β-TCP Powder

The sample (β-TCP) was prepared by semi-wet stirring method. Raw materials were stirred with ethanol for 3h at room temperature and dried at 80 °C for 24h, respectively. The mixtures were heat-treated at 1500 °C and then quenched after holding for

24h. Where DCPD and Calcium hydroxide ($Ca(OH)_2$), (Aldrich Co. USA) was used for β-TCP as a raw material. The mixing molar ratio of samples the β-TCP was 1:1.

TG/DTA Analysis

TG/DTA (TG-1280, Rigaku, Japan) analysis was performed from 30 °C to 1450 °C with the heating rate of 10 °C/min in air.

X-Ray Diffraction

All of the synthesized β-TCP powders and set cement were analyzed by X-ray diffractometer (MPD-Pro, PANalytical Co., Netherland) operated at 40 kV and 30 mA, respectively and uses Cu-Kα radiation (1.5406Å). For qualitative analysis, XRD diagrams were recorded in the interval $10° \leq 2\theta \leq 40°$ at a scan speed of 2°/min giving a step size 0.02° and the step time 1s. The phase creation was checked by means of JCPDS reference designs for β-TCP (PDF Ref. 09-0169) and HA (PDF Ref. 09-0432).

Cementing Properties

Each β-TCP powder and DCPD were milled with 1:1 molar ratio in ethanol for 24h and then dried. 100 mg CPC powder were immersed into gentamicin (Aldrich Co. USA) solution (concentration 40 mg/ml), were shaken at 80 rpm and 37 °C for 2 h. After centrifugation, the supernatant liquid dried in a desiccator. After centrifugation the aqueous supernatant was analyzed spectrophotometrically using calibration curves on a Hitachi UV U3010 spectrophotometer. Molar disodium phosphate dihydrate ($Na_2HPO_4.2H_2O$) (Merck, Germany) was used as stock cement liquid. The cement powder is mixed with a 1.55 P/L ratio for 1min on a glass plate at room temperature. The cement mixture placed into a plastic mold (15 mm D × 15 mm H) and stored 95% relative humidity chamber at 37 °C for cement-setting. The setting time of the cement was measured using a Vicat apparatus according to the ASTM C187-98. The setting temperature of the samples was also measured with setting time. The samples for compressive strength were prepared by using a plastic syringe mold (15 mm D × 10 mm H) with a pressure of 2 MPa. The specimens were held in a 95% relative humidity chamber at 37 °C for 1 day and then were placed in a drying oven at 40 °C for 24 h. The compressive strength was measured using the Universal Testing Machine (R&B Co. Korea) at a cross head speed of 0.5 mm/min and a load speed of 0.5 kgf/min with 500 kgf load cell. The crystalline phases of hardened bone cements were characterized using XRD. The hardened specimen was then removed from the mold and stored in 30 mL of a physiological-like solution (PLS) for 7 days at 37 °C with the solution being replaced daily. PLS has a composition ([$CaCl_2$] = 1.15 mmol/L, [KH_2PO_4] = 1.20 mmol/L, [NaCl] = 133 mmol/L, [HEPES] = 50 mmol/L, and pH = 7.4) similar to that of serum in terms of calcium, phosphate and electrolyte contents (Eidelman, Chow & Brown, 1987).

Particle Size Distribution Analysis

Particle size distribution of the powder sample was determined by the Laser Particle Sizer Analysette 22 (Fritsch GmbH, Germany) recording the scattering of a He-Ne laser beam (wavelength 632.8 nm) caused by the solid material suspended in de-mineralized water. Disintegration of the solid powders was upheld by an ultrasound shower with stirrer. The lowest particle size accessible is ca. 0.6 μm due to the wavelength used for measurement.

Surface Area Analysis

The BET method was employed to measure the surface area of samples with a Micromeritics Gemini 2360 analyzer, whereby nitrogen gas molecules are adsorbed onto a solid surface, which allows measuring of the surface area of a material. The samples were first prepared using a Gemini Vac Prep degasser. Then, dry and degassed samples were analyzed using the single/multipoint adsorption method for surface area assessment.

Injectability

The injectability was defined as the ability of cement to move through a syringe–catheter device. This test was examined through drawing extrusion curves of the cement paste. The cement paste was filled into an insulin syringe (without needle) which was placed between the compression plates of a computer-controlled Universal Testing Machine (Zwicki, Zwick/Roell, Ulm, Germany).

Scanning Electron Microscopy

The samples were coated with a thin layer of gold (Au) by sputtering (EMITECH K450X, England) and then the shape and morphology of prepared cement were observed on a scanning electron microscope (SEM; Stereoscan S 360 Cambridge) that operated at the acceleration voltage of 25 kV.

Table 1. Absorbance of gentamicin solution of different concentration at 290 nm

Concentration of Gentamicin Solution (mg/ml)	Absorbance
0.1	0.120
0.5	0.540
1.0	1.099

In Vitro Drug Release Studies

Preparation of Phosphate-Buffered Saline solution (PBS) 7.4 was based on British Pharmacopoeia. 2.38 g of disodium hydrogen orthophosphate, 0.19 g of potassium dihydrogen orthophosphate, and 8.0 g of sodium chloride were dissolved in sufficient amount of distilled water to produce 1000 ml. Distilled water can be added slowly to increase the pH of the solution until it reaches pH 7.4 with the help of pH meter (Metrohm 692, Switzerland). Determination of in vitro drug release profile the CPC-gentamicin implants were immersed into vials with 10 ml of PBS 7.4. Each set of implant were placed into incubator shaker set at 37 °C and 40 rpm for a better prediction of in vivo release. Two milliliters of each sample were withdrawn at specified time points up to 14 days. 2 ml of fresh phosphate buffer were replaced for each 2 ml withdrawn to maintain sink condition.

Figure 1. Calibration curve graph of gentamicin sulfate

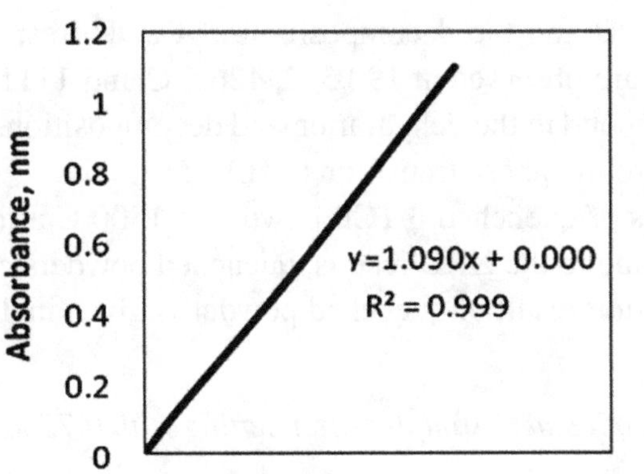

Calibration Curve of Gentamicin

A calibration curve of gentamycin sulfate was prepared. Gentamicin solution with the concentrations of 0.1%, 0.5% and 1.0% were prepared. The model drug solution concentrations were analyzed *via* UV-Visible Spectrophotometer (Hitachi UV U3010) at 290 nm peak absorbency to obtain the respective absorbance values. The wavelength at 290 nm was determined as the peak absorbency wavelength as maximum absorbency at this particular wavelength was observed. The equation of the linear curve shown in Table 1 and Figure 1 produced with the three points of different concentrations was used to quantify the concentrations of gentamicin in the withdrawn sample solutions. Drug release profiles for the different sets of implants up to 14 days were obtained.

RESULTS

Particle Size Distribution

Particle size distribution results showed a broad particle size distribution and the mean particle diameter (d50) of 12 μm, which is shown in Figure 2. According to BET technique, specific surface area was approximately 6 m^2/g. As that is illustrated in Figure 2, the powder component had a broad particle size distribution and moderate particle diameter. In fact, the particle size of the powder plays an important role in the extent of powder reactivity. Specific surface area of powder increases with decreasing particle size; as a result, increases cement hydration and reactivity. Also, broad particle size distribution will help to better cement injectability (Bohner and Baroud, 2005).

TG-DTA curve in Figure 3 shows twice weight loss peaks at the temperature range of 170-200 and 380-460°C, which are related to the water evaporation/decomposition of DCPD and the decomposition of CaOH, respectively. And three endothermic peaks are observed at 190.5°C, 426.5°C and 1111.7°C. The first and second peaks correspond to the dehydration and decomposition. The last endothermic peak is related to the phase transition to HA.

The XRD results of quenched β-TCP powder at 1500°C after 24h were shown in Figure 4. According to the XRD results, quenched powders contains β-TCP and HA. The quantification result of quenched powder is given in Table 2.

Table 2. HA content of cement samples after setting with 0.75 M Na_2HPO_4 solution for 24 h

Grinding Time (h)	2	4	8	12	24
HA content of cement (%)	25.2	42.3	60.2	86.1	93.4

Figure 2. The weight percent and cumulative volume on the Y axis is plotted versus the particle diameter in microns on the X axis

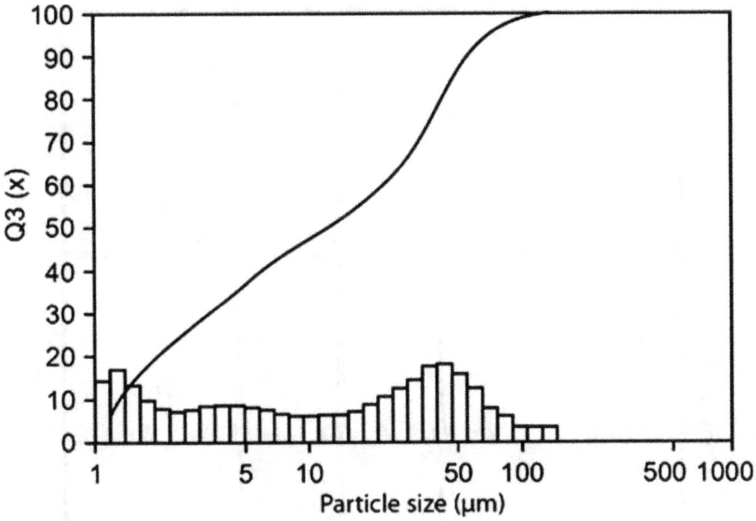

Figure 3. TG-DTA patterns of mixed raw materials heated from 30°C to 1450°C with heating rate 10°C/min

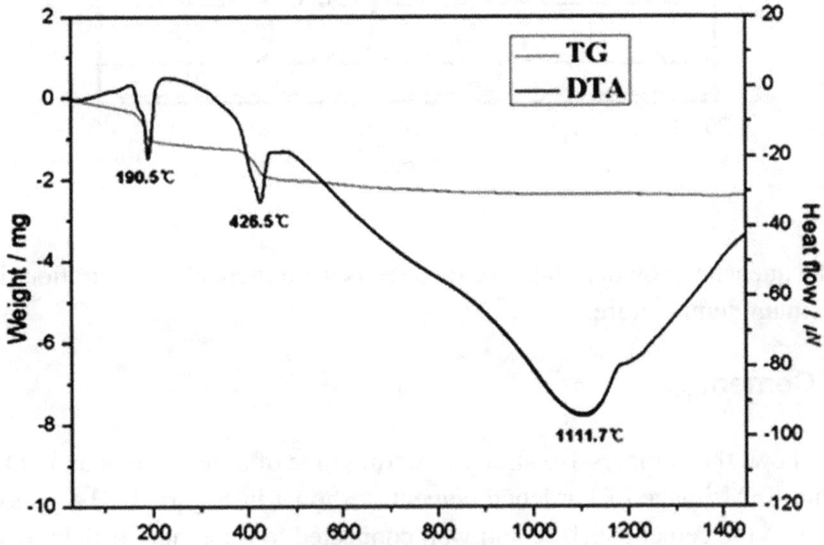

Figure 4. XRD patterns of synthetic β-TCP powder quenched after 24h duration at 1500°C

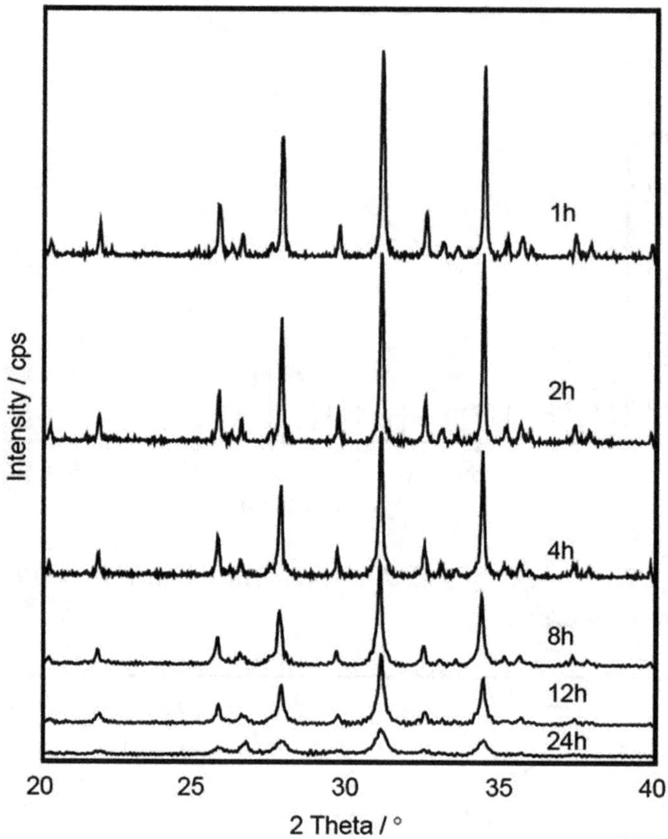

In the quenched powder, the β-TCP contents were increased as duration time at the maximum temperature.

Bone Cement

Table 3 shows the compressive strength, setting time of bone cement and optimum P/L ratio. SEM image of hardened cement is shown in Figure 5. HA phase was observed on the cement surface and well connected to the other particle. It shows precipitation reactions between β-TCP and DCPD with hardener. It is an evidence of phase conversion from two starting materials to HA.

With increasing the drug dose to the cement samples, the setting times increased which could be attributed to the change of sulfate ion concentration (Bohner, Lemaitre & Ring, 1996) in calcium phosphate bone cements (Table 4).

Table 3. Compressive strength, setting time and optimum P/L ratios

Cement	Hardener	Optimum P/L (Weight/ Volume Ratio)	Setting Time (min)	Compressive Strength (MPa)
β-TCP with DCPD	De-ionized water	1.55	23.8±0.2	12.91
	Na_2HPO_4, 0.5 M	1.55	17.5±0.2	28.12
	Na_2HPO_4, 0.75 M	1.55	9.3±0.2	46.78
	Na_2HPO_4, 1 M	1.55	6.1±0.2	52.67

Table 4. Compressive strength, setting time and optimum P/L ratios for gentamisin loaded samples

Cement	Hardener	Optimum P/L (Weight/ Volume Ratio)	Setting Time (min)	Compressive Strength (MPa)
β-TCP with DCPD+ gentamisin sulfate	De-ionized water	1.55	27.0±0.3	11.27
	Na_2HPO_4, 0.5 M	1.55	21.0±0.2	24.43
	Na_2HPO_4, 0.75 M	1.55	10.1±0.2	39.48
	Na_2HPO_4, 1 M	1.55	7.8±0.2	48.65

Figure 5. SEM images of hardened bone cements; β-TCP with DCPD +Na_2HPO_4, 1M solution

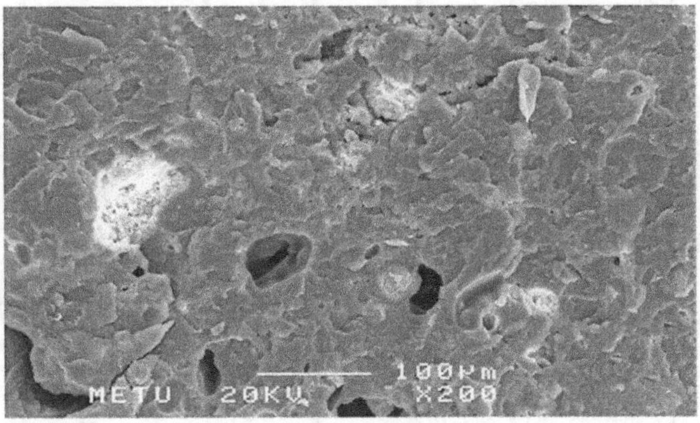

Figure 6 shows the XRD patterns of hardened cements. HA and β-TCP phases have just existed in bone cement body. It shows all DCPD phase conversed to HA from reaction with β-TCP and hardener. Figure 7 shows scanning electron micrograph of cement after 7 days in PLS at 37 °C.

Table 5 shows the crystal phase quantification result of hardened stored 95% relative humidity chamber at 37°C for 24h then stored 40°C dry oven for 24h. All of the hardened cement phases are HA and β-TCP. In this result, all most of β-TCP and DCPD were hydrated by water to form HA.

As shown in Figure 8, the drug release from CPC implants under in vitro was measured by UV spectrophotometer. The results showed two stages: The main stage happened amid the initial 10 days which was compared to the quick medication discharge from the surface of grid. Thereafter in the second stage, the discharge rate

Figure 6. XRD patterns of bone cements hardened in a 95% relative humidity chamber at 37 °C for 24h; β-TCP with DCPD

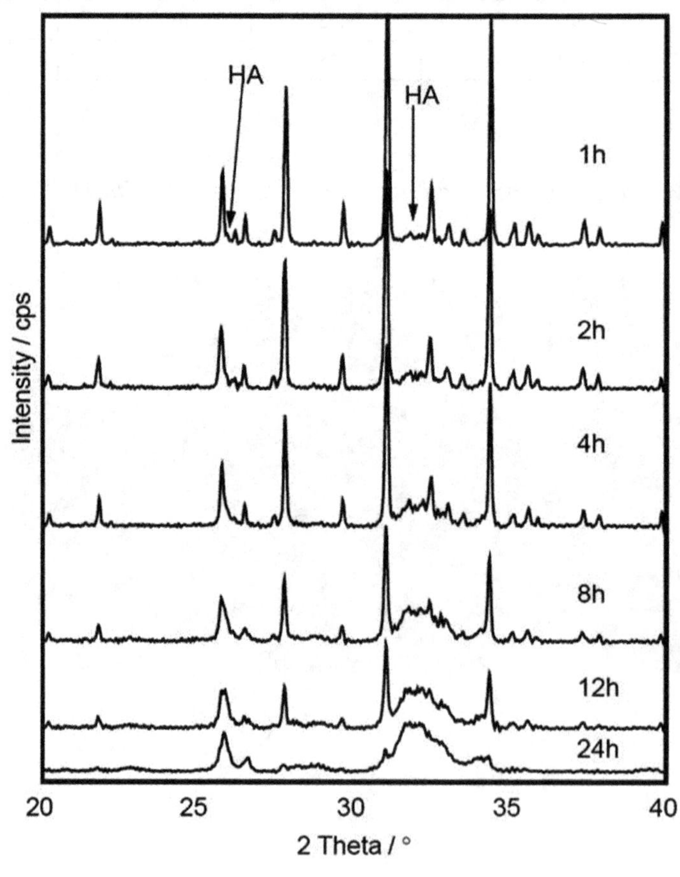

Table 5. The quantitative analysis (wt%) of crystalline phases of bone cement hardened in a 95% relative humidity chamber at 37 °C for 24h and then dried 40 °C for 24h

Hardener	De-Ionized Water	Na_2HPO_4, 0.5 M	Na_2HPO_4, 0.75 M	Na_2HPO_4, 1 M
β-TCP	1.5	1.1	0.6	0.2
DCPD	1.3	0	0	0
HA	97.2	98.9	99.4	99.8

Figure 7. Scanning electron micrograph of cement after 7 days in PLS at 37°C

notably backed off to zero between days 10 and 16 because of medication discharge from pores and splits introducing in the cement structure. Initial burst-release was about 3% in 24 h; followed by a very slow release rate a nearly constant release rate was found from days 10 and 15 for 95% drug release.

DISCUSSION

Setting time correlation of the medication stacked and without medication tests recommended that the medication expansion could come about an increment in setting time. Be that as it may, the setting time expanded with expanding the medication is added to vicinity of sulfate particle focuses in gentamicin sulfate of the cement. In this study, we balanced the measure of medication in bond with its setting time; then we kept it consistent in other steps.

Figure 8. The release of gentamicin loaded CPC, a continuous gentamincin release form CPC was observed for up to 14 days

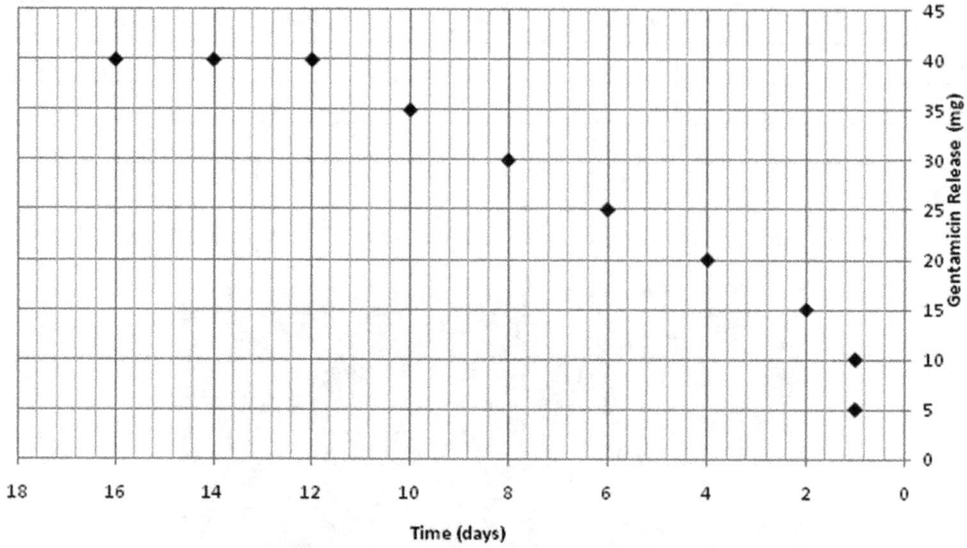

To draw a discharge profile, in this research, we examined the medication arrival of the consistent measure of medication in diverse hatching times. The medication discharge results revealed two discharge steps. In the initial step, the profile has direct association with the time. This high introductory medication focus discharge could give a decent hindrance against contamination in the adjoining bone which it proceeded with gradually until second step (like other anti-infection discharging from bone cements). Compared with other medication stacked CPCs in which the medication discharges were finished amid 7 days. Diverse considers, for example, distinction bond arrangement and microstructure, diverse dosages of medication and distinctive strategy for estimation could change the outcomes essentially. Clearly, the injectability of CPC is critical in clinical applications that include deformities with restricted availability or tight pits, when there is a requirement for exact situation of the glue to fit in with an imperfection zone, or when utilizing insignificantly obtrusive surgical methods. It has been accounted for by specialists that calcium phosphate bonds are inadequately injectable, consequently, a few analysts added polymeric added substances to defeat this issue. In this study creators attempted to get injectable calcium phosphate concrete without including polymers which could abstain from repudiating medication properties. On the other hand, the increment of infusion rate (from 0.5 to 2 mm/min) would prompt an increment in plunger development and a diminishing in injectability. Consequently, the lower rate of infusion brought about an increment in injectabilty. Likewise, medicate expansion could move the expulsion bends towards less plunger development. Subsequently, it is found that medication expansion did not aggravate cement injectability, as well as could enhance it. Our outcomes demonstrated this bond can possibly be conveyed through a syringe or a tool of practices which include insignificantly obtrusive systems, and filling in constrained expert.

The quality of concrete depends upon different elements for instance particle size of the bond powder, nature of added substances and wetting proportion. The impact of grouping of quickening agent in the medium was examined keeping in mind the end goal to advance the bond.

CONCLUSION

In conclusion, we prepared biphasic drug loaded calcium phosphate bone cement consisting of β-TCP and DCPD phases. Moreover, this cement could release gentamicin sulfate within the time in the surgical site and could be injected through a syringe or a surgical gun. The high gentamicin-loaded and injectable CPC have successfully revealed prepared to achieve drug sustaining release for up to 14 days.

REFERENCES

Aral, A., Yalçin, S., Karabuda, Z. C., Anil, A., Jansen, J. A., & Mutlu, Z. (2008). Injectable calcium phosphate cement as a graft material for maxillary sinus augmentation: An experimental pilot study. *Clinical Oral Implants Research*, *19*, 612–617. doi:10.1111/j.1600-0501.2007.01518.x PMID:18474064

Bohner, M., & Baroud, G. (2005). Injectability of calcium phosphate pastes. *Biomaterials*, *26*(13), 1553–1563. doi:10.1016/j.biomaterials.2004.05.010 PMID:15522757

Bohner, M., Lemaitre, J., & Ring, T. A. (1996). Effects of sulfate, pyrophosphate and citrate ions on the physico chemical properties of cements made of beta tricalcium phosphate – phosphoric acid – water mixtures. *Journal of the American Ceramic Society*, *79*(6), 1427–1434. doi:10.1111/j.1151-2916.1996.tb08746.x

Brown, W. E., & Chow, L. C. (1986). US Patent, 4612053, 16 September.

Chow, L. C., Takagi, S., & Ishikawa, K. (1994). Formation of hydroxyapatite in cement systems. In W. E. Brown & E. Constanz (Eds.), *Hydroxyapatite and related materials* (pp. 127–137). Boca Raton, FL: CRC Press.

Ciampolini, J., & Harding, K. G. (2000). Pathophysiology of chronic bacterial osteomyelitis. Why do antibiotics fail so often? *Postgraduate Medical Journal*, *76*(898), 479–483. doi:10.1136/pmj.76.898.479 PMID:10908375

Constantz, B. R., Ison, I. C., Fulmer, M. T., Poser, R. D., Smith, S. T., VanWagoner, M., & Rosenthal, D. et al. (1995). Skeletal repair by in situ formation of the mineral phase of bone. *Science*, *267*(5205), 1796–1799. doi:10.1126/science.7892603 PMID:7892603

Del Real, R. P., Wolke, J. G. C., Vallet-Regí, M., & Jansen, J. A. (2002). A new method to produce macropores in calcium phosphate cements. *Biomaterials*, *23*(17), 3673–3680. doi:10.1016/S0142-9612(02)00101-1 PMID:12109693

Doadrio, A. L., Sousa, E. M., Doadrio, J. C., Pérez Pariente, J., Izquierdo-Barba, I., & Vallet-Regí, M. (2004). Mesoporous SBA-15 HPLC evaluation for controlled gentamicin drug delivery. *Journal of Controlled Release*, *97*(1), 125–132. doi:10.1016/j.jconrel.2004.03.005 PMID:15147810

Eidelman, N., Chow, L. C., & Brown, W. E. (1987). Calcium phosphate saturation levels in ultrafiltered serum. *Calcified Tissue International*, *40*(2), 71–78. doi:10.1007/BF02555708 PMID:3105836

Friedman, C. D., Costantino, P. D., Takagi, S., & Chow, L. C. (1998). Bone Source hydroxyapatite cement: A novel biomaterial for craniofacial skeletal tissue engineering and reconstruction. *Journal of Biomedical Materials Research*, *43*(4), 428–432. doi:10.1002/(SICI)1097-4636(199824)43:4<428::AID-JBM10>3.0.CO;2-0 PMID:9855201

Ginebra, M. P., Traykova, T., & Planell, J. A. (2006). Calcium phosphate cements: Competitive drug carriers for the musculoskeletal system. *Biomaterials*, *27*(10), 2171–2177. doi:10.1016/j.biomaterials.2005.11.023 PMID:16332349

Ginebra, M. P., Traykova, T., Planell, J. A. (2006). Calcium phosphate cements as bone drug delivery systems: a review. *Journal of controlled release*, 113, 102–110.

Horstmann, W. G., Verheyen, C. C. P. M., & Leemans, R. (2003). An injectable calcium phosphate cement as a bone-graft substitute in the treatment of displaced lateral tibial plateau fractures. *Injury*, *34*(2), 141–144. doi:10.1016/S0020-1383(02)00105-5 PMID:12565022

Julien, M., Khairoun, I., LeGeros, R. Z., Delplace, S., Pilet, P., Weiss, P., & Guicheux, J. et al. (2007). Physico-chemical mechanical and in vitro biological properties of calcium phosphate cements with doped amorphous calcium phosphates. *Biomaterials*, *28*(6), 956–965. doi:10.1016/j.biomaterials.2006.10.018 PMID:17123598

Kamerer, D. B., Hirsch, B. E., Snyderman, C. H., Costantino, P., & Friedman, C. D. (1994). Hydroxyapatite cement: A new method for achieving watertight closure in transtemporal surgery. *The American Journal of Otology*, *15*, 47–49. PMID:8109630

King, R. W., & Kulkarni, R. (2013). Osteomyelitis in Emergency Medicine. *Emedicine*. Retrieved from http://emedicine.medscape.com/article/785020-overview#showall

Kokubo, T. (2008). Bioceramics and their clinical applications. Cambridge: Woodhead Pub. and Maney Pub.

Lew, D. P., & Waldvogel, F. A. (2004). Osteomyelitis. *Lancet*, *364*(9431), 369–379. doi:10.1016/S0140-6736(04)16727-5 PMID:15276398

Liverneaux, P. A. (2004). Osteoporotic distal radius curettage-filling with an injectable calcium phosphate cement. A cadaveric study. *European Journal of Orthopaedic Surgery & Traumatology*, *15*(1), 1–6. doi:10.1007/s00590-004-0182-x

Mirtchi, A. A., Lemaitre, J., & Terao, N. (1989). Calcium phosphate cements: Study of the beta-tricalcium phosphate-monocalcium phosphate system. *Biomaterials*, *10*(7), 475–480. doi:10.1016/0142-9612(89)90089-6 PMID:2804235

Mirtichi, A. A., Lemaitre, J., & Munting, E. (1989). Calcium phosphate cements: Action of setting regulators on the properties of the beta-tricalcium phosphate-monocalcium phosphate cements. *Biomaterials*, *10*(9), 634–638. doi:10.1016/0142-9612(89)90120-8 PMID:2611315

Shaffiey, S. R., & Shaffiey, S. F. (2016). Surface enamel remineralization by biomimetic nano hydroxyapatite crystals and fluoride ions effects. *Journal of Ceramic Processing Research*, *17*(2), 109–112.

Strauss, E. J., & Egol, K. A. (2007). The management of ankle fractures in the elderly. *Injury*, *38*(3), 52–59. PMID:17723786

Van den Vreken, N. M. F., Pieters, I. Y., Declercq, H. A., Cornelissen, M. J., & Verbeeck, R. M. H. (2010). Characterization of calcium phosphate cements modified by addition of amorphous calcium phosphate. *Acta Biomaterialia*, *6*(2), 617–625. doi:10.1016/j.actbio.2009.07.038 PMID:19654057

Welch, R. D., Zhang, H., & Bronson, D. G. (2003). Experimental tibial plateau fractures augmented with calcium phosphate cement or autologous bone graft. *J. Bone Joint Surg.*, *85*, 222–231. PMID:12571298

Yang, Q., Troczynski, T., & Liu, D. M. (2002). Influence of apatite seeds on the synthesis of calcium phosphate cement. *Biomaterials*, *23*(13), 2751–2760. doi:10.1016/S0142-9612(02)00010-8 PMID:12059025

Chapter 9
Mineralized Nanofibers for Bone Tissue Engineering

Ozan Karaman
Izmir Katip Celebi University, Turkey

ABSTRACT

The limitation of orthopedic fractures and large bone defects treatments has brought the focus on fabricating bone grafts that could enhance ostegenesis and vascularization in-vitro. Developing biomimetic materials such as mineralized nanofibers that can provide three-dimensional templates of the natural bone extracellular-matrix is one of the most promising alternative for bone regeneration. Understanding the interactions between the structure of the scaffolds and cells and therefore the control cellular pathways are critical for developing functional bone grafts. In order to enhance bone regeneration, the engineered scaffold needs to mimic the characteristics of composite bone ECM. This chapter reviews the fabrication of and fabrication techniques for fabricating biomimetic bone tissue engineering scaffolds. In addition, the chapter covers design criteria for developing the scaffolds and examples of enhanced osteogenic differentiation outcomes by fabricating biomimetic scaffolds.

INTRODUCTION

Bone defects, are currently being treated by using auto- and allograft procedures. These techniques have limitations in their clinical usage such as immune response, donor-site morbidity, and lack of availability (Damien & Parsons, 1991; Kretlow,

DOI: 10.4018/978-1-5225-1043-7.ch009

Young, Klouda, Wong, & Mikos, 2009; Scheller, Krebsbach, & Kohn, 2009). Therefore, the interest in tissue engineering applications for bone graft procedures has been growing rapidly. Typical tissue engineering models for bone graft applications require a cell supporting scaffold in order to maintain a 3-dimensional substrate for cells to function during the formation of bone tissue (Burg, Porter, & Kellam, 2000; Frohlich et al., 2008). Among the 6 million fractures occurring every year in the United States, there are more than 0.5 million skeletal injuries that require bone graft procedures (Greenwald, Boden, & Barrack, 2010).

Biomimetic bone tissue engineering scaffolds need to meet several criteria in order to be a successful candidate for bone graft procedures. An ideal biomimetic scaffold should completely integrate with the surrounding bone tissue, exhibit sufficient mechanical strength, and function as a three dimensional framework for osteoprogenitor cells to fabricate bone matrix. In order to provide adequate 3D structure and allow for full integration into the site of the bone defect, a biomimetic scaffold should fulfill the following requirements:

1. It should mimic the physical and chemical composition of the natural bone tissue,
2. It should be biocompatible,
3. It should be porous to allow bone tissue organization and vascular formation,
4. It should have appropriate surface chemistry and mechanical strength characteristics to support cellular attachment, proliferation, and differentiation, and
5. It should degrade simultaneously with the newly grown bone tissue (Karageorgiou & Kaplan, 2005; (Stevens, Yang, MohandaS, Stucker, & Nguyen, 2008).

Bone tissue is a composite matrix mainly composed of approximately 70 wt% inorganic crystals (mainly hydroxyapatite) (NF) and 30 wt% of organic matrix (mainly Type I collagen) (Rho, Kuhn-Spearing, & Zioupos, 1998). Mineralization of collagen NF is a well-regulated process mediated by many extracellular matrix proteins, such as bone sialoprotein (BSP) and osteonectin (ON). These non-collagenous proteins contain a large amount of glutamic acid (Glu) residues, which act as frameworks for depositing apatite crystals on collagen fibers in natural bone (Fujisawa, Wada, Nodasaka, & Kuboki, 1996; Sarvestani, He, & Jabbari, 2008, 2007). The apatite crystals provide mechanical strength for bone formation (Fantner et al., 2004), while the collagen fibers provide structural support and locations for cell adhesion and regulate cell functions such as cell adhesion, proliferation, differentiation, and mineralization (Beachley & Wen, 2010). The simple fractures in bone tissue can be treated by standard conservative or surgical therapy. However, in some cases, such as extended bone defects due to trauma or cancerous resection, bone grafting procedures are being performed with engineered biomimetic materials (Farhadi et al., 2006; Logeart-Avramoglou, Anagnostou, Bizios, & Petite, 2005). Consequently, biomimetic bone scaffolds aim to mimic the hydroxyapatite/collagen nanofibrous structure observed in natural bone; in this chapter the structure of the materials and the fabrication techniques used to develop mineralized nanofibrous biomimetic scaffolds for bone tissue engineering will be reviewed.

FABRICATION OF NANOFIBERS BASED SCAFFOLDS FOR BONE TISSUE ENGINEERING

In order to mimic the 3D structure of natural bone, biomimetic scaffolds have to be fabricated using different methodologies to facilitate the cell distribution and to guide their growth into three-dimensional space. Collagen nanofibers in natural bone tissue have a fibrous structure with a diameter range of 50 nm to 500 nm (Elsdale & Bard, 1972; Ma, 2008). The ideal biomimetic scaffold fabrication techniques should allow for the accurate duplication of the nanofibrous structure of collagen in natural bone. Several methods have been developed to fabricate nanofibers, such as self-assembly (Hong, Legge, Zhang, & Chen, 2003), phase separation (Ma & Zhang, 1999), and electrospinning (Fang, Wang, Niu, Lin, & Wang, 2010; (Xue et al., 2009).

Electrospinning

The electrospinning technique for scaffold fabrication uses electrostatic force for the production of polymeric fibers ranging from nanoscale to microscale. This process is controlled by a high intensity electric field formed between a syringe containing the polymer solution, and a collecting plate. The force generated by the high voltage generator between the needle and the collecting plate overcomes the surface tension of the polymer solution and stretches the accelerating jet as fibers toward the collector, while the solvent evaporates during the process. Electrospinning has been used to produce nanofibers (NFs) (Xu, Inai, Kotaki, & Ramakrishna, 2004) from natural biopolymers, like collagen and chitosan, as well as synthetic polymers, such as poly(L-lactide) (PLLA) and poly(ε-caprolactone) (PCL) (Huang, Zhang, Kotaki, & Ramakrishna, 2003; Li, Xie, Yuan, & Xia, 2008; Whited, Whitney, Hofmann, Xu, & Rylander, 2011; Zhang et al., 2010). Among these fabrication techniques, electrospinning has been regarded as the most promising approach to fabricate nanofibers with adjustable fiber diameter from nanometer to micrometer (Li et al., 2008).The advantages of this technique are its ability to produce nanofibers that can provide high surface-area-to-volume ratio, its morphological similarities to the collagen nanofibers in natural bone, its degradability, its biocompatibility, and its high mechanical strength, which all promote cell adhesion, proliferation and differentiation (Chua et al., 2005; Schindler et al., 2005). These characteristics are favorable for better cellular growth and tissue organization for *in vitro* and *in vivo* bone tissue engineering applications (Kim, Lee, & Knowles, 2006; Yoshimoto, Shin, Terai, & Vacanti, 2003).

Self-Assembly

Self-assembly, one of the nanofibrous scaffold fabrication techniques, is the organization of the molecules into an ordered structure required for a specific function (Zhang, 2003). The structure of the self-assembled molecules are maintained by non-covalent bonds such as hydrogen bonds, ionic bonds, and hydrophobic interactions (Whitesides & Grzybowski, 2002). Peptides and proteins can be self-assembled to form nanofibrous, well-defined functional structures (Huang et al., 2003; (Silva et al., 2004). The advantage of the self-assembly technique over phase separation and electrospinning is the formation of much smaller nanofibers. Nanofibrous scaffolds developed by molecular self-assembly have been studied for bone tissue engineering applications. For instance, Hamada et al. encapsulated mesenchymal stem cells (MSCs) in self assembling peptide hydrogel (RADA16) and examined the osteogenic differentiation of MSCs. It was demonstrated that MSCs can extensively differentiate into osteoblasts and form the mineralized structure of bone tissue within the nanofibrous based hydrogel structures (Hamada, 2008). However, the drawbacks of low mechanical strength structure of self-assembled peptides prevent them from being ideal scaffolds for bone tissue engineering applications. In order to overcome this limitation, Gungormus et al. developed self-assembling peptide hydrogel, which can induce the mineralization of calcium phosphate by designed heptapeptide MDG1 (Mineral Directing Gelator) epitope. It was obtained that mineralization was induced when the self-assembled peptide sequence modified with MDG1 epitope (Gungormus, 2010). In another study, Li et al. fabricated self-assembly peptide/demineralized bone matrix composite (SAP/DBM) scaffolds and studied the osteogenic efficacy of these scaffolds *in vitro* and *in vivo*. It was shown that compared to the control group of DBM scaffolds in vitro and in vivo bone tissue formation was significantly higher on SAP/DBM composite scaffolds (Li et al, 2016). The current studies prove that mineralized structures of self-assembled nanofibrous scaffolds have great potential for further clinical studies.

Phase Separation

The phase separation technique for fibrous scaffold fabrication relies on temperature changes that separate the polymeric solution into two phases, one having low polymer concentration (polymer lean phase), and the other having the high polymer concentration (polymer rich phase) (Shin, Jo, & Mikos, 2003). The phase separation process for nanofiber formation consists of five steps: polymer dissolution, phase separation and gelation, solvent extraction, freezing, and freeze-drying under vacuum (P. X. Ma, 2008). The characteristics of the fiber network depend on the solvent of the polymer solution and the gelation temperature. The fibers formed in this manner have diameters ranging from 50 nm to 500 nm, and have porosity in excess of 98% (X. Liu, Smith, Hu, & Ma, 2009; (P. X. Ma, 2008). The advantage of the phase separation technique is that it can be combined with other processing techniques, such as particulate leaching, to fabricate biomimetic scaffolds with well-defined pore morphologies (J. Y. Zhang, Beckman, Piesco, & Agarwal, 2000). Limited control on fiber morphology and orientation, relatively longer fabrication time, and inadequate mechanical properties are the drawbacks of nanofibrous scaffolds fabricated by phase separation technique.

SURFACE FUNCTIONALIZATION OF NANOFIBERS

The common problem that is faced for all the fabrication techniques of nanofibrous scaffolds is their inadequate mechanical properties when they are used for bone tissue engineering applications. Biomineralization is a natural process, which occurs during bone tissue formation where calcium-phosphate (CaP) crystals grow on the collagen nanofibers to form stiff bone tissue. The function of calcium phosphate crystals (mainly hydroxyapatite) in a composite scaffold is to provide good osteoconductivity, while the fibrous structure of natural or synthetic polymer supports design flexibility to achieve the high porosity and high surface area for cell survival and differentiation (X. Liu et al., 2009; P. X. Ma, 2008). To enhance mechanical properties of nanofibers, biomineralization is one of the technique that is being used for developing composite nanofibrous scaffolds (Karaman et al., 2013). However, most of the fabricated nanofibers have inert surface where biomineralization cannot occur when immersed in the simulated body fluid (SBF). Therefore, research in the area of biomimetic mineralization has been mainly focused on the surface modification of the nanofibers. Several surface modification methods have been developed for enhancing CaP deposition (Alves, Leonor, Azevedo, Reis, & Mano, 2010; Cai et al., 2011).

Many biomimetic approaches to mineralization involve soaking a polymer construct in simulated body fluid (SBF), a type of medium with ion concentrations approximately similar to human plasma. Different compositions of SBF with different molar ratios of the components relative to body fluid, such as 1xSBF, 5xSBF, and 10xSBF, have been developed in order to deposit different structures of CaP crystals and reduce the duration of biomineralization (Kokubo & Takadama, 2006;

W. Y. Liu et al., 2011; Muller & Muller, 2006). For example, Cowan et al. fabricated mineralized poly(lactic-co-glycolic acid) (PLGA) nanofibers by soaking the scaffold in 5xSBF and seeding with BMS cells. They have shown that mineralized PLGA scaffolds produced new bone formation after 4 weeks; however, the non-mineralized scaffolds showed no bone formation at the same time point (Cowan et al., 2004).

The limitation of using synthetic biodegradable polymers as biomimetic scaffolds is that they cannot promote cell adhesion, proliferation, or differentiation to the extent that a natural polymer is able to because synthetic polymers often lack biological cues. Although growing hydroxyapatite on synthetic scaffold improves osteoconductivity and cellular interactions, further surface modifications have been done to completely overcome the problem of limited cell recognition points. For example, Langer et al. synthesized poly(L-lactic acid-co-L-lysine) and conjugated Arg-Gly-Asp (RGD) peptide to the lysine residue of the copolymer to enhance cell adhesion. The RGD peptides of ECM proteins interact with BMS cells through integrin surface receptors to facilitate cell spreading and adhesion. The results demonstrated that the mean area over which cells spread was significantly increased on scaffolds with RGD peptide (Cook et al., 1997; (R. Y. Zhang & Ma, 1999). Hu et al. fabricated RGD peptide attached to porous poly(lactic acid) (PLA), then seeded osteogenic precursor cells on it to investigate the RGD peptide attachment on cell proliferation and differentiation. The results of this study indicated that RGD peptide attachment to the surface of PLA polymer scaffolds enhances bone cell attachment and differentiation (Hu, Winn, Krajbich, & Hollinger, 2003). Our group recently produced RGD and bone morphogenic peptide-2 (BMP-2) grafted Poly (lactide-*co*-ethylene oxide-*co*-fumarate) (PLEOF) based hydrogels and investigated the effects of RGD and BMP peptide on osteogenic differentiation and mineralization of BMS cells. Our results demonstrated that RGD and BMP peptides, grafted to a hydrogel substrate, act synergistically to enhance osteogenic differentiation and mineralization of BMS cells (He, Ma, & Jabbari, 2008).

Wet Chemical Technique

Murphy et al. functionalized the surface of PLGA by hydrolysis, which increases the amount of surface carboxyl and hydroxyl groups due to scission of the polyester chains. The availability of carboxyl and hydroxyl groups enhanced the apatite growth (Murphy & Mooney, 2002).

Plasma Treatment

Yang et al. fabricated electrospun polycaprolactone (PCL) nanofibers and functionalized them with plasma surface treatment to replace the weak chemical bonding of

PCL with highly reactive carbonyl (–CO–), carboxyl (–COOH), and hydroxyl (–OH) groups. They demonstrated that significantly higher CaP nucleation on the surface functionalized PCL scaffolds (Cai et al., 2011; Ngiam et al., 2009).

Functional Peptide Conjugation

The functionalized surfaces of the biomimetic scaffolds act as an analogous template of osteonectin and bone sialoprotein, which have high glutamic acid and aspartic acid residues, and function on mineralization in natural bone. Therefore, in order to mimic the biomineralization that occurs in natural bone tissue, our group has done several studies. We synthesized osteonectin-derived peptide containing a 6-glutamic acid sequence by solid phase peptide synthesis. In order to covalently attach the peptide to in situ cross-linkable PLEOF hydrogel, one end of the synthesized peptide was functionalized with an acrylate group. We investigated the effect of osteonectin derived peptide on the viscoelastic properties of PLEOF/apatite composite. The results demonstrated that the osteonectin derived peptide significantly affects the shear modulus of the scaffold because of the ionic interactions between the peptide and HA crystals (Sarvestani et al., 2008; (Sarvestani et al., 2007). Following this study, we functionalized the surface of the PLGA nanofibers with two residues glutamic acid (GLU) containing peptide and investigated the effect of 2-residue GLU conjugation to the surface of aligned synthetic nanofibers on calcium phosphate (CaP) nucleation, growth, and osteogenic differentiation of marrow stromal cells (MSCs). The EEGGC peptide, with glycine residues to increase solubility in the electrospinning solvent, was conjugated to a low molecular weight acrylate-terminated polylactide (PLAA). The PLA-GLU conjugate was mixed with high molecular weight PLGA and electrospun to form surface functionalized nanofibers (NF). GLU-NF microsheets were incubated in a 10 times concentrated modified SBF to initiate nucleation and growth of CaP crystals on the fiber surface. Mineralization of GLU-NF is illustrated in Figure 1. The SEM images proved that the functionalized NF surfaces with 2-residue GLU significantly enhanced the apatite deposition of the NF (Figure 2). The results also showed that enhancing apatite deposition on NFs significantly increased the tensile modulus of the NFs. In addition, MSCs seeded on apatite deposited NF showed significantly higher alkaline phosphates, calcium secretion, and expression of osteogenic markers compared to non-mineralized nanofibers (Karaman et al., 2013).

Our recent study showed that there is a limitation of depositing apatite crystals on PLGA nanofibers functionalized with 2-mer GLU due to limited diffusion of calcium and phosphate ions to the interior parts of the nanofibers. In order to overcome this limitation, we have developed layer-by-layer deposition technique. In the layer-by-layer approach, the first GLU-NF microsheet, approximately 5 μm thick

Figure 1. The schematic diagram of apatite deposition on 2-mer glutamic acid peptide conjugated nanofibers in simulated body fluid

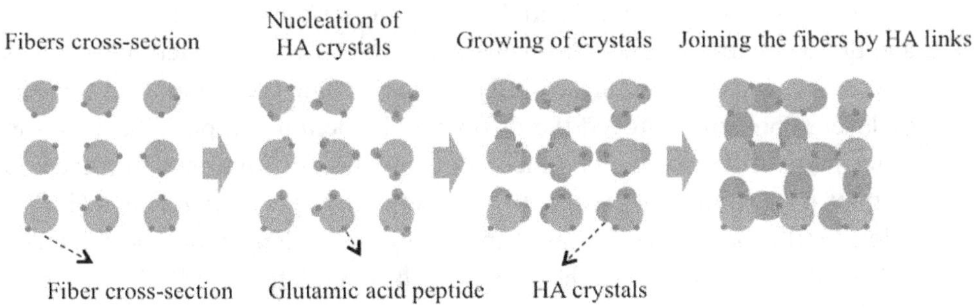

Figure 2. Schematic diagram of apatite deposited GLU-NF fabrication by layer-by-layer approach (a), low magnification image of 5 layers apatite/GLU-NF cross-section (b), high magnification image of 5 layers apatite/GLU-NF cross-section

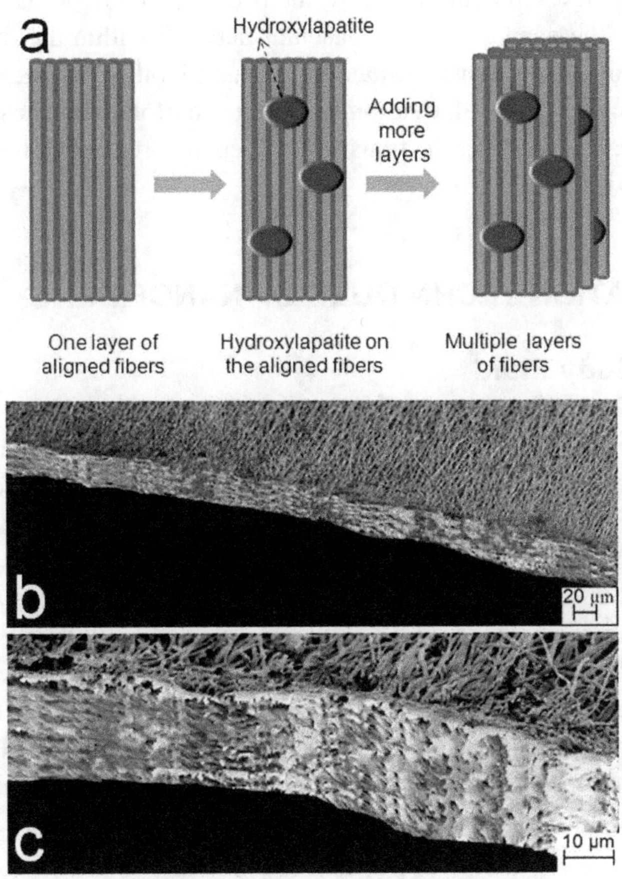

was incubated in m10SBF for the prescribed time, washed with DDI, and dried. After drying, the second GLU-NF layer was electrospun directly on the first layer. The bilayer sheet was incubated in m10SBF for a prescribed time, washed with DDI, and dried (Figure 3a). This layer-by-layer process was repeated many times to produce a mineralized multilayer GLU-NF sheet. Our results showed that the layer-by-layer approach improved the diffusion of calcium and phosphate ions to the interior parts of the sheet when incubated in m10SBF solution and increased the CaP/fiber ratio from 50 wt% to 200 wt%, between those of cancellous (160%) and cortical (310%) bone. The increased amount of apatite deposition also enhanced the tensile modulus of the NF and the osteogenic differentiation of MSCs (Karaman et al., 2013).

The cortical bone composed of osteons that have microtube like structures with a central vascular channel inside for ensuring suitable nutrient/waste transport to the bone tissue (van Oers, Ruimerman, van Rietbergen, Hilbers, & Huiskes, 2008). Laminated multilayers of CaP deposited collagen fibers form the osteon-mimicking microtubes. The development of three-dimensional (3-D) bone tissue in-vitro has been limited by the need to create a vascular network within the tissue to supply required oxygen and nutrients to maintain tissue viability and function (Garvin, Dalecki, & Hocking). Vascularization inside the scaffold enhances diffusion and provides optimal nutrient/waste transport for engineered scaffolds (Sun, Wang, Qian, & Hu, 2011).

MINERALIZATION TECHNIQUES OF NANOFIBERS

Simulated Body Fluid

Many biomimetic approaches to mineralization involve soaking a polymer construct in simulated body fluid (SBF), a type of medium with ion concentrations approximately similar to human plasma. Different compositions of SBF with different molar ratios of the components relative to body fluid, such as 1xSBF, 5xSBF, and 10xSBF, have been developed in order to deposit different structures of CaP crystals and reduce the duration of biomineralization (Kokubo & Takadama, 2006; (W. Y. Liu et al., 2011; (Muller & Muller, 2006). For example, Cowan et al. fabricated mineralized poly(lactic-co-glycolic acid) (PLGA) scaffolds by soaking the scaffold in 5xSBF and seeding with BMS cells. They have shown that mineralized PLGA scaffolds produced new bone formation after 4 weeks; however, the non-mineralized scaffolds showed no bone formation at the same time point (Cowan et al., 2004).

Most of the fabricated scaffolds have inert surfaces that do not allow for biomineralization to occur when immersed in the SBF. Therefore, research in the area of biomimetic mineralization has been mainly focused on the surface modification of the scaffolds. Several surface modification methods have been developed for enhancing CaP deposition (Alves et al., 2010; (Cai et al., 2011). For example, Kim et al. functionalized the surface of Titanium (Ti) by hydroxylation which forms Ti-OH groups on the surface, and then immersed into SBF to investigate the surface modification on bone-like apatite deposition. The results indicated that the Ti-OH groups enhanced the formation of apatite on the Ti alloys (H. M. Kim, Miyaji, Kokubo, & Nakamura, 1996). Yang et al. fabricated electrospun polycaprolactone (PCL) scaffolds and functionalized them with plasma surface treatment to replace the weak chemical bonding of PCL with highly reactive carbonyl (–CO–), carboxyl (–COOH), and hydroxyl (–OH) groups. They demonstrated that significantly higher CaP nucleation on the surface functionalized PCL scaffolds (Cai et al., 2011; (Ngiam et al., 2009). Murphy et al. functionalized the surface of PLGA by hydrolysis, which increases the amount of surface carboxyl and hydroxyl groups due to scission of the polyester chains. The availability of carboxyl and hydroxyl groups enhanced the apatite growth (Murphy & Mooney, 2002).

APPLICATIONS OF 3D MINERALIZED NANOFIBERS IN BONE TISSUE ENGINEERING

The limitation of using synthetic biodegradable polymers as biomimetic scaffolds is that they cannot promote cell adhesion, proliferation, or differentiation to the extent that a natural polymer is able to because synthetic polymers often lack biological cues. Although growing hydroxyapatite on synthetic scaffold improves osteoconductivity and cellular interactions, further surface modifications have been done to completely overcome the problem of limited cell recognition points. For example, Langer et al. synthesized poly(L-lactic acid-co-L-lysine) and conjugated Arg-Gly-Asp (RGD) peptide to the lysine residue of the copolymer to enhance cell adhesion. The RGD peptides of ECM proteins interact with BMS cells through integrin surface receptors to facilitate cell spreading and adhesion. The results demonstrated that the mean area over which cells spread was significantly increased on scaffolds with RGD peptide (Cook et al., 1997; (R. Y. Zhang & Ma, 1999). Hu et al. fabricated RGD peptide attached to porous poly(lactic acid) (PLA), then seeded osteogenic precursor cells on it to investigate the RGD peptide attachment on cell proliferation and differentiation. The results of this study indicated that RGD peptide attachment to the surface of PLA polymer scaffolds enhances bone cell attachment and differentiation (Hu et al., 2003). He et al. recently produced RGD and bone morphogenic peptide-2 (BMP-2) grafted Poly (lactide-*co*-ethylene oxide-*co*-fumarate) (PLEOF) based hydrogels and investigated the effects of RGD and BMP peptide on osteogenic differentiation and mineralization of BMS cells. Our results demonstrated that RGD and BMP peptides, grafted to a hydrogel substrate, act synergistically to enhance osteogenic differentiation and mineralization of BMS cells (He et al., 2008).

The function of calcium phosphate crystals (mainly hydroxyapatite) in a composite scaffold is to provide good osteoconductivity, while the fibrous structure of natural or synthetic polymer supports design flexibility to achieve the high porosity and high surface area for cell survival and differentiation (X. Liu et al., 2009; (P. X. Ma, 2008). Zhang et al. and Ma et al. developed highly porous poly(L-lactic acid) (PLLA) and PLLA/hydroxyapatite (HAP) composite scaffolds with a thermally induced phase separation technique and seeded them with osteoblastic cells. Phase separation techniques for scaffold design require temperature changes that separate the polymeric solution into two phases, one having low polymer concentration (polymer lean phase) and the other having the high polymer concentration (polymer rich phase). The proliferation of osteogenic cells and expression of bone sialoprotein and osteocalcin were significantly higher in the PLLA/HAP composite scaffold than in the PLLA scaffolds (P. X. Ma, Zhang, Xiao, & Franceschi, 2001; (R. Y. Zhang & Ma, 1999). Marra et al. fabricated hydroxyapatite incorporated biodegradable poly(caprolactone) and poly(D,L-lactic-co-glycolic acid) scaffolds by solvent-casting, followed by a particulate-leaching technique. The then seeded the scaffolds with rabbit bone marrow stromal cells. They have shown bone extracellular matrix formation (approximately 20% tissue ingrowth) after 8 weeks of in vitro culture (Marra, Szem, Kumta, DiMilla, & Weiss, 1999). In a particulate leaching technique, defined shape salt particles are poured into a mold with hydroxyapatite crystals and polymer solution. Polymer solution is then cast into the salt-filled mold. After the evaporation of the solvent, the salt crystals are leached away using water to form the pores of the scaffold (Thomson, Yaszemski, Powers, & Mikos, 1998).

Most of the fabricated scaffolds have inert surfaces that do not allow for biomineralization to occur when immersed in the SBF. Therefore, research in the area of biomimetic mineralization has been mainly focused on the surface modification of the scaffolds. Several surface modification methods have been developed for enhancing CaP deposition (Alves et al., 2010; (Cai et al., 2011). For example, Kim et al. functionalized the surface of Titanium (Ti) by hydroxylation which forms Ti-OH groups on the surface, and then immersed into SBF to investigate the surface modification on bone-like apatite deposition. The results indicated that the Ti-OH groups enhanced the formation of apatite on the Ti alloys (H. M. Kim et al., 1996). Yang et al. fabricated electrospun polycaprolactone (PCL) scaffolds and functionalized them with plasma surface treatment to replace the weak chemical bonding of PCL with highly reactive carbonyl (–CO–), carboxyl (–COOH), and hydroxyl (–OH) groups. They demonstrated that significantly higher CaP nucleation on the surface functionalized PCL scaffolds (Cai et al., 2011; (Ngiam et al., 2009). Murphy et al. functionalized the surface of PLGA by hydrolysis, which increases the amount of surface carboxyl and hydroxyl groups due to scission of the polyester chains. The availability of carboxyl and hydroxyl groups enhanced the apatite growth (Murphy & Mooney, 2002)

CONCLUSION AND FUTURE PERSPECTIVES

This chapter reviewed the development and applications of mineralized nanofibers in bone tissue engineering. The enhanced knowledge on single cell to cell and cell to extracellular matrix interactions will guide the researcher to develop better biomimetic scaffolds that will function on regeneration of defected bone tissue. Most of the recent studies focus on overcoming the limitations of mineralized nanofibers such as inadequate mechanical properties and relatively 2D structure for their bone tissue engineering applications.

Mineralized nanofibers match with the requirements of general bone mimetic scaffolds criteria. They provide sites for cells attachment, proliferation, differentiation, migrationas well as mechanical support during bone regeneration. The mimicking the extracellular matrix of natural bone tissue enhances the cell scaffold interaction, as well as cell adhesion, proliferation, differentiation, and *in vivo* bone tissue formation. Incorporation and biomineralization of apatite crystals on nanofibers mimic the apatite deposited collagen structure of bone, and enhance the mechanical strength and osteoconductivity of the scaffolds. The fabrication technique used to form the biomimetic scaffold should give the scaffold mechanical and morphological properties similar to natural bone. Surface modification of fibrous scaffolds increases cell scaffold interaction and *in vitro* apatite deposition, especially for synthetic scaffolds. Recent in vivo studies of 3D mineralized nanobiers used as bone mimetic scaffolds showed promising results. Overall, improvements on designing 3D mineralized nanofibers scaffolds will further allow designing bone mimetic scaffolds which have potential to use in clinic.

REFERENCES

Alves, N. M., Leonor, I. B., Azevedo, H. S., Reis, R. L., & Mano, J. F. (2010). Designing biomaterials based on biomineralization of bone. *Journal of Materials Chemistry*, *20*(15), 2911–2921. doi:10.1039/b910960a

Beachley, V., & Wen, X. J. (2010). Polymer nanofibrous structures: Fabrication, biofunctionalization, and cell interactions. *Progress in Polymer Science*, *35*(7), 868–892. doi:10.1016/j.progpolymsci.2010.03.003 PMID:20582161

Burg, K. J. L., Porter, S., & Kellam, J. F. (2000). Biomaterial developments for bone tissue engineering. *Biomaterials*, *21*(23), 2347–2359. doi:10.1016/S0142-9612(00)00102-2 PMID:11055282

Cai, Q., Xu, Q. Q., Feng, Q. F., Cao, X. Y., Yang, X. P., & Deng, X. L. (2011). Biomineralization of electrospun poly(L-lactic acid)/gelatin composite fibrous scaffold by using a supersaturated simulated body fluid with continuous CO2 bubbling. *Applied Surface Science*, *257*(23), 10109–10118. doi:10.1016/j.apsusc.2011.06.157

Chua, K. N., Lim, W. S., Zhang, P. C., Lu, H. F., Wen, J., Ramakrishna, S., & Mao, H. Q. et al. (2005). Stable immobilization of rat hepatocyte spheroids on galactosylated nanofiber scaffold. *Biomaterials*, *26*(15), 2537–2547. doi:10.1016/j.biomaterials.2004.07.040 PMID:15585256

Cook, A. D., Hrkach, J. S., Gao, N. N., Johnson, I. M., Pajvani, U. B., Cannizzaro, S. M., & Langer, R. (1997). Characterization and development of RGD-peptide-modified poly(lactic acid-co-lysine) as an interactive, resorbable biomaterial. *Journal of Biomedical Materials Research*, *35*(4), 513–523. doi:10.1002/(SICI)1097-4636(19970615)35:4<513::AID-JBM11>3.0.CO;2-C PMID:9189829

Cowan, C. M., Shi, Y. Y., Aalami, O. O., Chou, Y. F., Mari, C., Thomas, R., & Longaker, M. T. et al. (2004). Adipose-derived adult stromal cells heal critical-size mouse calvarial defects. *Nature Biotechnology*, *22*(5), 560–567. doi:10.1038/nbt958 PMID:15077117

Damien, C. J., & Parsons, J. R. (1991). Bone-Graft and Bone-Graft Substitutes - a Review of Current Technology and Applications. *Journal of Applied Biomaterials*, *2*(3), 187–208. doi:10.1002/jab.770020307 PMID:10149083

Elsdale, T., & Bard, J. (1972). Collagen Substrata for Studies on Cell Behavior. *Journal of Cell Biology, 54*(3), 626. Doi:10.1083/Jcb.54.3.626

Fang, J. A., Wang, H. X., Niu, H. T., Lin, T., & Wang, X. G. (2010). Evolution of Fiber Morphology During Electrospinning. *Journal of Applied Polymer Science*, *118*(5), 2553–2561. doi:10.1002/app.32569

Fantner, G. E., Birkedal, H., Kindt, J. H., Hassenkam, T., Weaver, J. C., Cutroni, J. A., & Hansma, P. K. et al. (2004). Influence of the degradation of the organic matrix on the microscopic fracture behavior of trabecular bone. *Bone*, *35*(5), 1013–1022. doi:10.1016/j.bone.2004.05.027 PMID:15542025

Farhadi, J., Jaquiery, C., Haug, M., Pierer, G., Zeilhofer, H. F., & Martin, I. (2006). Bone and cartilage tissue engineering for facial reconstructive surgery. *IEEE Engineering in Medicine and Biology Magazine*, *25*(1), 106–109. doi:10.1109/MEMB.2006.1578673 PMID:16485400

Frohlich, M., Grayson, W. L., Wan, L. Q., Marolt, D., Drobnic, M., & Vunjak-Novakovic, G. (2008). Tissue Engineered Bone Grafts: Biological Requirements, Tissue Culture and Clinical Relevance. *Current Stem Cell Research & Therapy*, *3*(4), 254–264. doi:10.2174/157488808786733962 PMID:19075755

Fujisawa, R., Wada, Y., Nodasaka, Y., & Kuboki, Y. (1996). Acidic amino acid-rich sequences as binding sites of osteonectin to hydroxyapatite crystals. *Biochimica Et Biophysica Acta-Protein Structure and Molecular Enzymology*, *1292*(1), 53–60. doi:10.1016/0167-4838(95)00190-5 PMID:8547349

Garvin, K. A., Dalecki, D., & Hocking, D. C. Vascularization of Three-Dimensional Collagen Hydrogels Using Ultrasound Standing Wave Fields. *Ultrasound in Medicine & Biology, 37*(11), 1853-1864. doi:10.1016/j.ultrasmedbio.2011.07.008

Greenwald, S., & Boden, S.D., & Barrack, R.L. (2010). The Evolving Role of Bone-Graft Substitutes. *Paper presented at the American Academy of Orthopaedic Surgeons*, New Orleans, Louisiana.

He, X. Z., Ma, J. Y., & Jabbari, E. (2008). Effect of Grafting RGD and BMP-2 Protein-Derived Peptides to a Hydrogel Substrate on Osteogenic Differentiation of Marrow Stromal Cells. *Langmuir*, *24*(21), 12508–12516. doi:10.1021/la802447v PMID:18837524

Hong, Y. S., Legge, R. L., Zhang, S., & Chen, P. (2003). Effect of amino acid sequence and pH on nanofiber formation of self-assembling peptides EAK16-II and EAK16-IV. *Biomacromolecules*, *4*(5), 1433–1442. doi:10.1021/bm0341374 PMID:12959616

Hu, Y. H., Winn, S. R., Krajbich, I., & Hollinger, J. O. (2003). Porous polymer scaffolds surface-modified with arginine-glycine-aspartic acid enhance bone cell attachment and differentiation in vitro. *Journal of Biomedical Materials Research. Part A*, *64A*(3), 583–590. doi:10.1002/jbm.a.10438 PMID:12579573

Huang, Z.-M., Zhang, Y. Z., Kotaki, M., & Ramakrishna, S. (2003). A review on polymer nanofibers by electrospinning and their applications in nanocomposites. *Composites Science and Technology*, *63*(15), 2223–2253. doi:10.1016/S0266-3538(03)00178-7

Karageorgiou, V., & Kaplan, D. (2005). Porosity of 3D biornaterial scaffolds and osteogenesis. *Biomaterials*, *26*(27), 5474–5491. doi:10.1016/j.biomaterials.2005.02.002 PMID:15860204

Karaman, O., Kumar, A., Moeinzadeh, S., He, X., Cui, T., & Jabbari, E. (2013). Effect of surface modification of nanofibres with glutamic acid peptide on calcium phosphate nucleation and osteogenic differentiation of marrow stromal cells. *Journal of Tissue Engineering and Regenerative Medicine*. doi:10.1002/term.1775 PMID:23897753

Kim, H. M., Miyaji, F., Kokubo, T., & Nakamura, T. (1996). Preparation of bioactive Ti and its alloys via simple chemical surface treatment. *Journal of Biomedical Materials Research*, *32*(3), 409–417. doi:10.1002/(SICI)1097-4636(199611)32:3<409::AID-JBM14>3.0.CO;2-B PMID:8897146

Kim, H. W., Lee, H. H., & Knowles, J. C. (2006). Electrospinning biomedical nanocomposite fibers of hydroxyapaite/poly(lactic acid) for bone regeneration. *Journal of Biomedical Materials Research. Part A*, *79A*(3), 643–649. doi:10.1002/jbm.a.30866 PMID:16826596

Kokubo, T., & Takadama, H. (2006). How useful is SBF in predicting in vivo bone bioactivity? *Biomaterials*, *27*(15), 2907–2915. doi:10.1016/j.biomaterials.2006.01.017 PMID:16448693

Kretlow, J. D., Young, S., Klouda, L., Wong, M., & Mikos, A. G. (2009). Injectable Biomaterials for Regenerating Complex Craniofacial Tissues. *Advanced Materials*, *21*(32-33), 3368–3393. doi:10.1002/adma.200802009 PMID:19750143

Li, X. R., Xie, J. W., Yuan, X. Y., & Xia, Y. N. (2008). Coating Electrospun Poly(ε-caprolactone) Fibers with Gelatin and Calcium Phosphate and Their Use as Biomimetic Scaffolds for Bone Tissue Engineering. *Langmuir*, *24*(24), 14145–14150. doi:10.1021/la802984a PMID:19053657

Liu, W. Y., Yeh, Y. C., Lipner, J., Xie, J. W., Sung, H. W., Thomopoulos, S., & Xia, Y. N. (2011). Enhancing the Stiffness of Electrospun Nanofiber Scaffolds with a Controlled Surface Coating and Mineralization. *Langmuir*, *27*(15), 9088–9093. doi:10.1021/la2018105 PMID:21710996

Liu, X., Smith, L. A., Hu, J., & Ma, P. X. (2009). Biomimetic nanofibrous gelatin/apatite composite scaffolds for bone tissue engineering. *Biomaterials*, *30*(12), 2252–2258. doi:10.1016/j.biomaterials.2008.12.068 PMID:19152974

Logeart-Avramoglou, D., Anagnostou, F., Bizios, R., & Petite, H. (2005). Engineering bone: Challenges and obstacles. *Journal of Cellular and Molecular Medicine*, *9*(1), 72–84. doi:10.1111/j.1582-4934.2005.tb00338.x PMID:15784166

Ma, P. X. (2008). Biomimetic materials for tissue engineering. *Advanced Drug Delivery Reviews*, *60*(2), 184–198. doi:10.1016/j.addr.2007.08.041 PMID:18045729

Ma, P. X., & Zhang, R. Y. (1999). Synthetic nano-scale fibrous extracellular matrix. *Journal of Biomedical Materials Research*, *46*(1), 60–72. doi:10.1002/(SICI)1097-4636(199907)46:1<60::AID-JBM7>3.0.CO;2-H PMID:10357136

Ma, P. X., Zhang, R. Y., Xiao, G. Z., & Franceschi, R. (2001). Engineering new bone tissue in vitro on highly porous poly(alpha-hydroxyl acids)/hydroxyapatite composite scaffolds. *Journal of Biomedical Materials Research*, *54*(2), 284–293. doi:10.1002/1097-4636(200102)54:2

Ma, Z. W., Kotaki, M., Inai, R., & Ramakrishna, S. (2005). Potential of nanofiber matrix as tissue-engineering scaffolds. *Tissue Engineering*, *11*(1-2), 101–109. doi:10.1089/ten.2005.11.101

Marra, K. G., Szem, J. W., Kumta, P. N., DiMilla, P. A., & Weiss, L. E. (1999). In vitro analysis of biodegradable polymer blend/hydroxyapatite composites for bone tissue engineering. *Journal of Biomedical Materials Research*, *47*(3), 324–335. doi:10.1002/(SICI)1097-4636(19991205)47:3<324::AID-JBM6>3.0.CO;2-Y PMID:10487883

Muller, L., & Muller, F. A. (2006). Preparation of SBF with different HCO3- content and its influence on the composition of biomimetic apatites. *Acta Biomaterialia*, *2*(2), 181–189. doi:10.1016/j.actbio.2005.11.001 PMID:16701876

Murphy, W. L., & Mooney, D. J. (2002). Bioinspired growth of crystalline carbonate apatite on biodegradable polymer substrata. *Journal of the American Chemical Society*, *124*(9), 1910–1917. doi:10.1021/ja012433n PMID:11866603

Ngiam, M., Liao, S. S., Patil, A. J., Cheng, Z. Y., Chan, C. K., & Ramakrishna, S. (2009). The fabrication of nano-hydroxyapatite on PLGA and PLGA/collagen nanofibrous composite scaffolds and their effects in osteoblastic behavior for bone tissue engineering. *Bone*, *45*(1), 4–16. doi:10.1016/j.bone.2009.03.674 PMID:19358900

Rho, J. Y., Kuhn-Spearing, L., & Zioupos, P. (1998). Mechanical properties and the hierarchical structure of bone. *Medical Engineering & Physics*, *20*(2), 92–102. doi:10.1016/S1350-4533(98)00007-1 PMID:9679227

Sarvestani, A. S., He, X., & Jabbari, E. (2008). Osteonectin-derived peptide increases the modulus of a bone-mimetic nanocomposite. *European Biophysics Journal with Biophysics Letters*, *37*(2), 229–234. doi:10.1007/s00249-007-0198-3 PMID:17609937

Sarvestani, A. S., He, X. Z., & Jabbari, E. (2007). Effect of osteonectin-derived peptide on the viscoelasticity of hydrogel/apatite nanocomposite scaffolds. *Biopolymers*, *85*(4), 370–378. doi:10.1002/bip.20659 PMID:17183515

Scheller, E. L., Krebsbach, P. H., & Kohn, D. H. (2009). Tissue engineering: State of the art in oral rehabilitation. *Journal of Oral Rehabilitation*, *36*(5), 368–389. doi:10.1111/j.1365-2842.2009.01939.x PMID:19228277

Schindler, M., Ahmed, I., Kamal, J., Nur-E-Kamal, A., Grafe, T. H., Chung, H. Y., & Meiners, S. (2005). A synthetic nanofibrillar matrix promotes in vivo-like organization and morphogenesis for cells in culture. *Biomaterials*, *26*(28), 5624–5631. doi:10.1016/j.biomaterials.2005.02.014 PMID:15878367

Shin, H., Jo, S., & Mikos, A. G. (2003). Biomimetic materials for tissue engineering. *Biomaterials*, *24*(24), 4353–4364. doi:10.1016/S0142-9612(03)00339-9 PMID:12922148

Silva, G. A., Czeisler, C., Niece, K. L., Beniash, E., Harrington, D. A., Kessler, J. A., & Stupp, S. I. (2004). Selective differentiation of neural progenitor cells by high-epitope density nanofibers. *Science*, *303*(5662), 1352–1355. doi:10.1126/science.1093783 PMID:14739465

Stevens, B., Yang, Y. Z., Mohanda, S. A., Stucker, B., & Nguyen, K. T. (2008). A review of materials, fabrication to enhance bone regeneration in methods, and strategies used engineered bone tissues. *Journal of Biomedical Materials Research. Part B, Applied Biomaterials*, *85B*(2), 573–582. doi:10.1002/jbm.b.30962 PMID:17937408

Sun, J. X., Wang, Y. L., Qian, Z. Y., & Hu, C. B. (2011). An approach to architecture 3D scaffold with interconnective microchannel networks inducing angiogenesis for tissue engineering. *Journal of Materials Science. Materials in Medicine*, *22*(11), 2565–2571. doi:10.1007/s10856-011-4426-0 PMID:21861076

Thomson, R. C., Yaszemski, M. J., Powers, J. M., & Mikos, A. G. (1998). Hydroxyapatite fiber reinforced poly(alpha-hydroxy ester) foams for bone regeneration. *Biomaterials*, *19*(21), 1935–1943. doi:10.1016/S0142-9612(98)00097-0 PMID:9863527

van Oers, R. F. M., Ruimerman, R., van Rietbergen, B., Hilbers, P. A. J., & Huiskes, R. (2008). Relating osteon diameter to strain. *Bone*, *43*(3), 476–482. doi:10.1016/j.bone.2008.05.015 PMID:18619937

Vasita, R., & Katti, D. S. (2006). Nanofibers and their applications in tissue engineering. *International Journal of Nanomedicine*, *1*(1), 15–30. doi:10.2147/nano.2006.1.1.15 PMID:17722259

Venugopal, J., Low, S., Choon, A. T., & Ramakrishna, S. (2008). Interaction of cells and nanofiber scaffolds in tissue engineering. *Journal of Biomedical Materials Research. Part B, Applied Biomaterials*, *84B*(1), 34–48. doi:10.1002/jbm.b.30841 PMID:17477388

Whited, B. M., Whitney, J. R., Hofmann, M. C., Xu, Y., & Rylander, M. N. (2011). Pre-osteoblast infiltration and differentiation in highly porous apatite-coated PLLA electrospun scaffolds. *Biomaterials*, *32*(9), 2294–2304. doi:10.1016/j.biomaterials.2010.12.003 PMID:21195474

Whitesides, G. M., & Grzybowski, B. (2002). Self-assembly at all scales. *Science*, *295*(5564), 2418–2421. doi:10.1126/science.1070821 PMID:11923529

Xu, C. Y., Inai, R., Kotaki, M., & Ramakrishna, S. (2004). Aligned biodegradable nanotibrous structure: A potential scaffold for blood vessel engineering. *Biomaterials*, *25*(5), 877–886. doi:10.1016/S0142-9612(03)00593-3 PMID:14609676

Xue, Y. H., Wang, H. X., Yu, D. S., Feng, L. F., Dai, L. M., Wang, X. G., & Lin, T. (2009). Superhydrophobic electrospun POSS-PMMA copolymer fibres with highly ordered nanofibrillar and surface structures. *Chemical Communications*, (42): 6418–6420. doi:10.1039/b911509a PMID:19841795

Yoshimoto, H., Shin, Y. M., Terai, H., & Vacanti, J. P. (2003). A biodegradable nanofiber scaffold by electrospinning and its potential for bone tissue engineering. *Biomaterials*, *24*(12), 2077–2082. doi:10.1016/S0142-9612(02)00635-X PMID:12628828

Zhang, J. Y., Beckman, E. J., Piesco, N. P., & Agarwal, S. (2000). A new peptide-based urethane polymer: Synthesis, biodegradation, and potential to support cell growth in vitro. *Biomaterials*, *21*(12), 1247–1258. doi:10.1016/S0142-9612(00)00005-3 PMID:10811306

Zhang, R. Y., & Ma, P. X. (1999). Poly(alpha-hydroxyl acids) hydroxyapatite porous composites for bone-tissue engineering. I. Preparation and morphology. *Journal of Biomedical Materials Research*, *44*(4), 446–455. doi:10.1002/(SICI)1097-4636(19990315)44:4<446::AID-JBM11>3.0.CO;2-F PMID:10397949

Zhang, Y. Z., Reddy, V. J., Wong, S. Y., Li, X., Su, B., Ramakrishna, S., & Lim, C. T. (2010). Enhanced Biomineralization in Osteoblasts on a Novel Electrospun Biocomposite Nanofibrous Substrate of Hydroxyapatite/Collagen/Chitosan. *Tissue Engineering Part A*, *16*(6), 1949–1960. doi:10.1089/ten.tea.2009.0221 PMID:20088700

Chapter 10
Recent Advances in Synthesis and Biomedical Applications of Magnetic Nanoparticles:
Magnetic Nanoparticles for Biomedical Applications

Irshad Ahmad Wani
Department of School Education, J&K, India

ABSTRACT

Magnetic nanoparticles due to their unique magnetic phenomenon, are gaining immense interest due to the utilization of these properties for a wide variety of applications in various arena especially in biomedical field. This book chapter, therefore, summarizes the synthesis of various types of magnetic nanoparticles using different approaches depending of their ability to generate either single core of multcore magnetic nanoparticles. The various biomedical applications of magnetic nanoparticles like Magnetic Resonance Imaging (MRI), drug delivery etc. along with possible limitations and challenges for their extended applications in medicine are also discussed.

DOI: 10.4018/978-1-5225-1043-7.ch010

INTRODUCTION

Nanotechnology is now a days working at the cellular and molecular levels to produce major progress in the life sciences and healthcare. Although real applications of nanomaterials in life sciences are still uncommon but the excellent properties of these materials provide a very promising future for their use in this field (Grainger & Okano, 2003; Hood et al., 2002; Niemeyer, 2001). Nanoclusters located at the transition region between molecules and micron size particles are ultrafine particles of nanometer dimensions. When viewed from the molecular perspective, their particle size is so large that they provide access to realms of quantum behaviour which are not otherwise accessible and from the materials perspective, their particle size is so small that they exhibit characteristics not observed in the larger structures (>100 nm). Recent advances in biology, physics and chemistry has been mainly made in this size regime (Roco, Williams, & Alivisatos, 2000). For example, the band gap of semiconductor material increases when the particle dimension of semiconductor materials become comparable to, or smaller than the Bohr radius (El-Sayed, 2001; Heath, 1995). In noble metals, intense absorption is observed in the visible to near UV region as the particle size decreases below the electron mean free path (Bohren & Huffman, 1983). Other fascinating mechanical behaviors such as superplasticity is also observed in metal nanoparticles in this size range (Kung & Foecke, 1999). Powder ceramic materials with particle size in nanorange are also receiving attention due to significant enhancement in the sintering rates or lowering of sintering temperatures (Hahn, Logas, & Averback, 1990; J. Tartaj, Zärate, Tartaj, & Lachowski, 2002; P Tartaj & Tartaj, 2002). Ceramic matrix composites with dispersed nanoparticles have better mechanical properties (Kung & Foecke, 1999; Niihara, 1991; a G. Roca et al., 2009).

Nanotechnology has made major contribution in medicine and pharmaceutical industries, like controlled drug delivery, as biosensors, in disease detection, in tissue engineering etc. However, nanoparticles designed for drug delivery should be biocompatible and biodegradable (Gupta & Gupta, 2005a; Xie et al., 2006). Among many other nanosystems, special interest is focused on magnetic nanoparticles (MNPs), which have a potential to revolutionize current clinical diagnostic and therapeutic techniques (Mailander & Landfester, 2009; Mishra, Patel, & Tiwari, 2010; Suh, Suslick, Stucky, & Suh, 2009). Many applications of MNPs have been widely studied including magnetically induced hyperthermia, magnetically enhanced transfection, magnetically assisted gene therapy and magnetic-force-based tissue engineering (Chomoucka et al., 2010; Corchero & Villaverde, 2009).

However, the development of highly effective medicine requires continuous and timely monitoring of the medical treatment process. This monitoring in combination with therapeutics commonly referred to as "theranostics" allows a large degree of control on the treatment efficacy in different individuals (Sambasivarao, 2013). Magnetic nanoparticles (MNPs) containing ferromagnetic iron (Fe) and cobalt (Co) as well as their alloys and oxides has proved to be the most promising probes for theranostics with therapeutic efficacy and desired imaging sensitivity (J. Kim, Piao, & Hyeon, 2009; Lacroix, Ho, & Sun, 2010; Sambasivarao, 2013).

Magnetic nanoparticles show a wide range of applications such as magnetic fluids (Chikazumi et al., 1987), catalysis, (A. Lu et al., 2004; S. C. Tsang, Caps, Paraskevas, Chadwick, & Thompsett, 2004), data storage (Hyeon, 2003) and environmental remediation (Elliott & Zhang, 2001; Takafuji, Ide, Ihara, & Xu, 2004). However, Magnetic nanoparticles (MNPs) owing to their unique magnetic properties and size comparable to biologically important objects are very useful for biomedical applications. Magnetic nanoparticles allow some new and exiciting approaches to bioseperation and biodetection along with targeted drug delivery due to unique response to external magnetic fields which allow biomolecules to be tagged and detected magnetically (Dobson, 2006; Gupta & Gupta, 2005b; Sunderland, Steiert, Talmadge, Derfus, & Barry, 2006). Due to their resonating response to the alternating magnetic field they MNPs function as heaters, thereby, offering a promising therapeutic solution by magnetic fluid hyperthermia (Jeun et al., 2009; Rui et al., 2010; Thomas et al., 2009).

In most of the applications mentioned above, the particle gives best performance below a certain critical value depending upon the material but a range is 10-20 nm. In this size regime each magnetic nanoparticles shows a supermagnetic behaviour above a certain temperature called as blocking temperature. Such individual nanoparticles acting as single domains has a large constant magnetic moment and behave like a giant paramagnetic atom with negligible remanence (residual magnetism) and coercivity (the field required to bring the magnetization to zero) due to the fast response to applied magnetic fields, thus making them very attractive for a broad range of biomedical applications due to the negligible risk of forming agglomerates (Takafuji et al., 2004).

The most commonly used magnetic materials for biomedical applications are iron oxides such as magnetite (Fe_3O_4) and maghemite (γ-Fe_2O_3) due to the reason that human body has the organs and metabolic mechanisms to transfer iron into safe storage for later use (L Gutiérrez et al., 2011; Lucía Gutiérrez et al., 2006). It should be noted MNPs intended for in vivo applications are not used directly after synthesis due to the less intolerability but often several post synthesis steps are employed so as to completely remove the potentially toxic compounds remaining from the synthesis steps (L Gutiérrez et al., 2015).

Basic Physical Properties

The orbital and spin motion of the electrons is responsible for the microscopic origin of magnetic properties in materials (Martin & Martin, 1967), whose spin and angular momentum are associated with a magnetic moment. Below a certain critical temperature, the interaction between the magnetic moments of atoms from the same material causes magnetic order. Thus bulk materials can be classified on the basis of these interactions and material behaves differently in response to magnetic field at different temperature (e.g. ferromagnetism, ferrimagnetism, etc.) (Jiles, 1998). Bulk magnetic materials contain magnetic domains within which there is an alignment of the magnetic moments. If the volume of the material is reduced to nanodimensions as in MNPs, single magnetic domains are formed in the material and the magnetic properties are no longer similar to bulk materials but a supermagnetic behaviour is achieved meaning that the thermal energy may be enough to change spontaneously the magnetization within each MNP. Due to temperature influence the magnetic moment of each MNP will be able to rotate randomly in reference to the orientation of the MNP. For this reason, in the absence of electromagnetic field the net magnetic moment of a system containing MNPs is zero at high enough temperatures (Figure 1). However, in the presence of a field, net stastical alignment of magnetic moments occurs analogous to what happens to paramagnetic materials, but the magnetic moment is that of the MNP containing various atoms instead of a single atom and this magnetic mimetic moment can be up to 104 times larger than for a paramagnetic material. This property, marked by the lack of remanent magnetisation after removal of external fields, enables the MNPs to maintain their colloidal stability and avoid agglomeration, which is important for biomedical applications. Many new interesting phenomena have been observed in MNPs that are not shared with their bulk counterparts (Jun, Seo, & Cheon, 2008). Shape and size play an important role in determining the magnetic behaviour and magnetic properties such as coercivity (Hc) and susceptibility (w) of MNPs apart from the main parameters which are used to characterize the magnetic properties of bulk materials (discussed in section below). In general, smaller size means lower transition temperature from ferromagnetic to superparamagnetic behaviour. Likewise, reduction in size from 12 nm to 5 nm in maghemite (g-Fe_2O_3) may lead to reduction in saturation magnetization to half the theoretical value due to relative contribution of surface effects and surface disorders (M. Morales et al., 1999). Besides size effects, shape of the MNP strongly influences the magnetic anisotropic barrier and K thereby influencing the whole magnetic properties of the nanomaterial. For most of the applications a larger magnetic moments are preferred thereby reducing the amount of MNP needed. However, biocompatibility is given due priority as the toxic effects of MNPs needs to be reduced as much as possible, thereby, necessitating the need for maintain the

Figure 1.

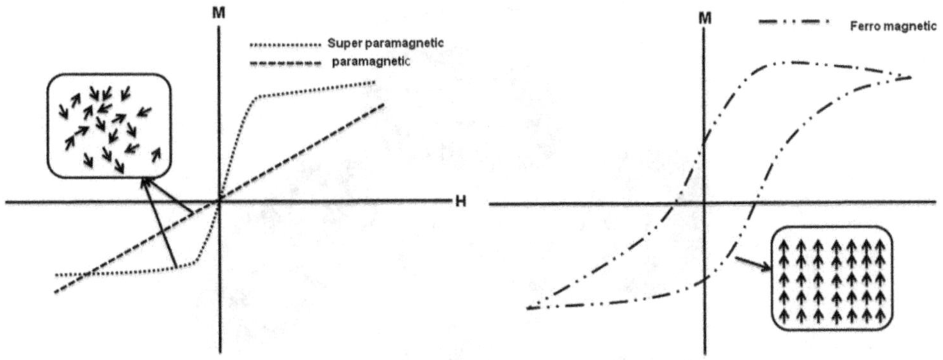

balance between biocompatibility and larger magnetic moments. This is the reason why iron-based MNPs are often preferred compared to others based on more toxic transition metals (Sandhu, Handa, & Abe, 2010).

Nanomagnetism Scaling Laws: Size and Surface Effects in Magnetic Nanoparticles

When the size of the magnetic material is reduced to the nanoscale regime, at least two additional parameters are strongly involved in determining the magnetic properties of the material. These are size and shape effects (Cornell & Schwertmann,

Figure 2.

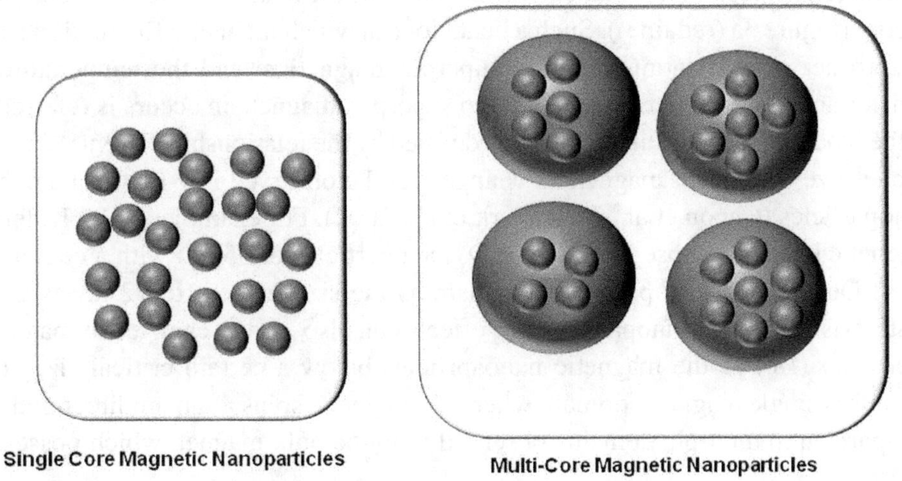

Figure 3.

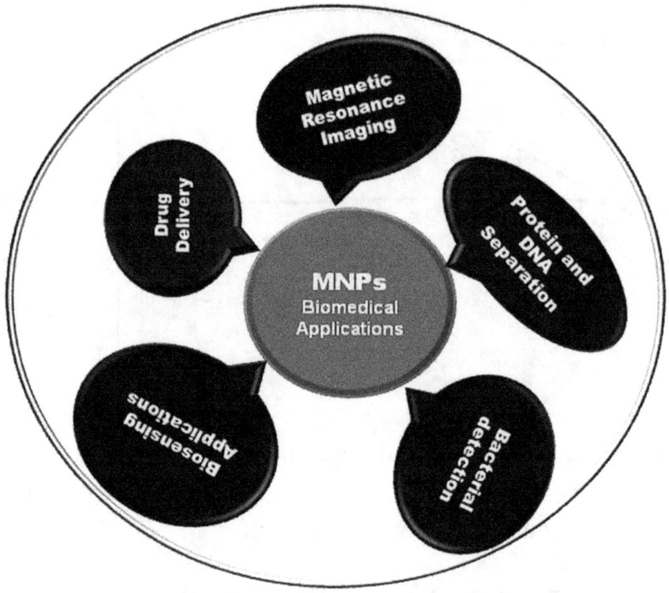

2003; Jiles, 1998; McCurrie, 1994). Superparamagnetism is one of the interesting size-dependent phenomena of nanoparticles. We know that bulk materials have magnetic anisotropic energy barrier from spin up state to spin down sate much larger than the thermal energy (kT). The magnetic anisotropic energy barrier of the magnet is proportional to the product of the magnetic anisotropic constant (Ku) and the volume (V) of the magnet (Jiles, 1998).36. However, in case of nanoparticles, the thermal energy of the nanoparticles is sufficient to readily invert the magnetic spin direction, though it is inadequate to overcome the spin-spin exchange coupling energy (Figure 3a (red line)). Such a behaviour in which magnetic fluctuation leads to zero net magnetization is called superparamagnetism and the temperature at which transition from ferromagnetism to superparamagnetism occurs is referred to as the blocking temperature (Tb) and is defined by the relationship Tb) KuV/25k.36. Such a size-dependent magnetism changes can be observed in γ-Fe_2O_3 and cobalt nanoparticles (Cheon et al., 2004; Park et al., 2002). For example at 300 K, ferromagnetic behaviour observed in γ-Fe_2O_3 nanoparticles of 55 nm with a coercivity of 52 Oe changes to superparamagnetism as the size reduces to 12 nm with no hysteresis behavior. Nanoparticle size effects can also lead to changes in magnetic coercivity (Hc) as the magnetic nanoaprticles below a certain critical size (Dc) possesses single magnetic domain where all magnetic spins align unidirectinally in comparison to multiple domains observed in to the bulk magnet, which possesses multiple magnetic domain structures. In this single-domain regime, the magnetic coercivity increases as the size of the nanoparticle increases with the relationship $Hc = 2K_u/m_s[1\text{-}5(kT/KuV)^{1/2}]$ where ms is the saturation magnetization (Jiles, 1998).

Since the multidomain magnetism begins above the critical size ($D > Dc$), thereby requiring a small reversal magnetic field to make the net magnetization zero. It has been observed that magnetic coercivity of the Co nanoparticles increases from 370 to 1680 Oe in a single-domain regime as the size increases from 4 to 8 nm (Park et al., 2002). Conversely, further increase in size to 10, 12, and 13 nm leads to decrease in the magnetic coercivity values to 1600, 1100, and 250 Oe by forming multimagnetic domains. Therefore, the critical single-domain of Co nanoparticles is expected to be around 8–10 nm. Saturation magnetization of nanoparticles is also strongly dependent on their size. Inherently magnetic materials due to the reduced spin-spin exchange coupling energy at the surface possess magnetically disordered spin glass like layers (M. P. Morales, Serna, Bødker, & Mørup, 1997; M. Morales et al., 1999). In bulk cases, such as surface spin canting effects are negligible since the disordered surface layer is minimal compared with the total volume of the magnet. Decrease in size of the material to nanoregime however leads to the dramatic pronunciation such surface canting in the saturation magnetization value (m_s), described as

$$m_s = \mathrm{Ms}[(r - d)/r]^3$$

where r is the size, Ms is the saturation magnetization of bulk materials, and d is the thickness of disordered surface layer. Such size effects are more noticeable for very small nanoparticles less than 5 nm, in size since internal spins of the nanoparticle also start to be canted as well as the surface spins due to increased interactions between the surface and internal spins (M. Morales et al., 1999).

Likewise, in order to understand the surface effects, we know that as the particles size decreases, a large percentage of all the atoms in a nanoparticle are surface atoms implying thereby that surface and interface effects become more important. For example, in case of face-centered cubic (fcc) cobalt nanoparticles with a diameter of around 1.6nm, about 60% of the total number of spins are surface spins (Batlle & Labarta, 2002). Therefore, surface spins make an important contribution to the magnetization due to this large surface atoms/bulk atoms ratio. Due to this local breaking of the symmetry changes in the band structure, lattice constant or/and atom coordination might be observed. Also under these conditions, some surface or interface related effects such as surface anisotropy core-surface exchange anisotropy can occur (A.-H. Lu, Salabas, & Schüth, 2007).

SYNTHESIS OF MAGNETIC NANOPARTICLES

Several chemical routes have been developed to synthesize magnetic nanoparticles such as sol-gel synthesis, hydrothermal synthesis, microemulsion and hydrolysis and thermolysis of precursors etc (Laurent et al., 2008).30However during synthesis aggregation of nanoparticles may take place owing to the large surface area. In order to prevent aggregation, use of surfactant is necessitated in order to attain single core nanoparticles. But problem with the surfactant arises as many surfactants are toxic (e.g. Triton X 31 or didodecyldimethylammonium bromide) and needs to be removed or replaced by biocompatible species prior to in vivo applications (Dayeh, Chow, Schirmer, Lynn, & Bols, 2004).

Single-Core MNPs

Most of the single-core magnetic nanoparticles synthesis routes in aqueous phases has been produced by the alkaline precipitation of $FeCl_3$ and $FeCl_2$ as developed by Massart(Massart, 1981). In this method polydispersity of the particles has been reduced by precipitating larger ones from colloidal solutions by the addition of an electrolyte solution or antisolvant leaving nearly monodisperse smaller particles in the supernatant. These monodisperse particles are then coated to prevent aggregation. Non agglomerated single core nanoparticles having hydrodynamic particle diameter smaller than 20 nm have been obtained with sodium citrate or sodium polyacrylate or silica coatings.

Using citrate coating, Iron oxide nanoparticles with a core diameter of around 5 nm have been obtained with a hydrodynamic diameter of 8 nm (Taupitz et al., 2000). Briefly, a mixture of $FeCl_3$ and $FeCl_2$ is heated to 100 C in the absence of oxygen to which a mixture of ammonia water and citric acid are added. The resulting solution is boiled for 10 min and then cooled down to room temperature (Pilgrimm, 2003). Relaxivities in water, R_1 and R_2 of these particles were 20.1 mM^{-1} s^{-1} and 37.1 mM^{-1} s^{-1} respectively, (Wagner, SCHNORR, Pilgrimm, Hamm, & Taupitz, 2002)36 suggesting their use as dynamic T1-weighted MRI contrast agents.(Schnorr et al., 2006)

Using polyacrylate, magnetic nanoparticles with hydrodynamic sizes 5 nm greater than uncoated particles have been produced. (Chanteau, Fresnais, & Berret, 2009).40,41. This method involves the addition of Poly (sodium acrylate) (Mw = 2000 g mol^{-1} or 5000 g mol^{-1}) to the iron oxide suspension at pH 2 resulting in the formation of a precipitate which is separated by centrifugation. Afterwards, upon increasing the pH up to 7-8 results in the spontaneous redispersion of the particles. These magnetic nanoparticles samples produced have been used to generate a model to predict the efficiency of magnetic nanoparticles as T_2 MRI contrast agents (Vuong, Berret, Fresnais, Gossuin, & Sandre, 2012).

Silica coating of MNPs is performed by modifications of the Stöber method, originally described to prepare monodisperse silica particles in 1968. (Stöber, Fink, & Bohn, n.d.). The Stöber method involves the hydrolysis of TEOS (Tetraethyl orthosilicate) in alcoholic solutions. Though fast and simple, it has several drawbacks, such as secondary nucleation of silica particles, difficulty in controlling the silica shell thickness and the problems of coating particles individually. However, a better control of the silica coating can be attained by hydrolysis of TEOS in a microemulsion, where size of the particles is controlled by the size of the water droplets (Santra et al., 2001; Pedro Tartaj & Serna, 2002). Briefly, two separate inverse microemulsions have to be prepared using a surfactant such as Brij-97 or Igepal CO-520. The first microemulsion contains iron salts such as $FeSO_4$ and $FeCl_3$, while the second microemulsion contains a base, either NaOH or NH_4OH, and neat TEOS. After stirring, the second microemulsion is added dropwise to the first one, placed in an ultrasonic bath. Collision of the water droplets resulting the mixing of the reactants, thereby producing nanoparticles. The base catalyzed hydrolysis of TEOS produces silicic acid, which polymerizes to form the silica coating (Santra et al., 2001). The significance of this coating for biomedical applications is due to the strong effect of the silica coating thickness on the relaxivity properties of these particles in aqueous suspensions (Pinho et al., 2010)

Single-core nanoparticles can also be produced from organic phase by suing two different organic phase approaches such as the thermal decomposition of organometallic compounds, and the polyol method.

Thermal decomposition of organometallic compounds produces highly monodispersed magnetic nanoparticles with good crystallinity usually with hydrophobic properties (Hyeon, Lee, Park, Chung, & Na, 2001). This approach offers very good control of the nucleation and growth processes either by the injection of organometallic compounds into a hot surfactant solution, resulting in the formation of nuclei almost instantaneously or by the controlled heating of organometallic compounds in a surfactant solution to generate the nuclei. After the nucleation has occurred, particles grow at high temperature. After obtaining the optimum size, further growth can be checked by the quick decrease of the reaction temperature. Single-core iron oxide nanocubes with sizes between 20 nm and 160 nm have been synthesized using iron acetylacetonate in oleic acid and benzyl ether (D. Kim et al., 2008) other iron precursors have also been used such as iron oleate(A G Roca, Morales, O'Grady, & Serna, 2006)49 and iron pentacarbonyl (Alejandro G Roca, Marco, Morales, & Serna, 2007). Particle size however can be controlled either by varying the precursor concentration, the Fe/oleic acid ratio51 or the solvents used (Baaziz et al., 2014; A G Roca et al., 2006). Particle shape may be tuned from from spheres to cubes by adding sodium oleate to the synthesis (Wetterskog et al., 2014).

However, the organic phase synthesis has the major disadvantage that in order for their use in biomedical applications additional steps are needed after synthesis to stabilize the particles in aqueous medium which can lead to irreversible aggregation of the particles.

Multi-Core Nanoparticles

Like single core nanoparticles synthesis of multicore nanoparticles may also be classified depending upon whether the processing phase is aqueous or although a third phase is also possible, viz. the gaseous phase, as used in laser pyrolysis. Most of the aqueous phase synthesis of multicore magnetic nanoparticles is based upon the Massart procedure, (Massart, 1981). For instance, iron oxide nanoparticles have been prepared by basic precipitation followed by acidification with nitric acid. (Fauconnier, Bee, Roger, & Pons, 1999) and the same procedure could also be used for further modification, e.g. silanisation and PEGylation to obtain multi-core particles. Some alternative synthesis routes have been developed by precipitating of $FeSO_4$ in presence of NaOH and a mild oxidant (KNO_3 or $NaNO_3$), with subsequent aging. the order of adding $NaNO_3$ to NaOH and Fe^{2+} was found to have a significant influence on the nanoparticle properties (Z. Li et al., 2010). Coatings play a pivotal role in aqueous-phase multi-core syntheses in terms of establishing the final structure and magnetic character of the nanoparticles as well as their biocompatibility and biodistribution in the human body. Polysccharide coating can be used to encapsulate multicore nanoparticles containing several magnetic cores (Figure 4) such as magnetic resonance imaging (MRI) contrast agents Endorem® (no longer commercially available) and Resovist® (Fujifilm RI Pharma, Japan), both of which comprise 4 to 8 nm cores coated with dextran or carboxydextran, respectively, to make particles with mean hydrodynamic sizes of 150 and 60 nm, respectively (Jung & Jacobs, 1995). Starch coated nanoparticles to be used as MRI contrast agents have also been described (D. K. Kim et al., 2003). For hyperthermia cancer treatment multi-core particles have been prepared by applying a high pressure homogenization process during the iron oxide precipitation, resulting in Bionized Nano Ferrite (BNF) particles (Grüttner et al., 2007) useful for hyperthermia cancer treatment as they provide high heating rates at magnetic field strengths >30 kA m^{-1}(Bordelon et al., 2011; Hoopes et al., 2009; Krycka et al., 2011). In order to maintain the colloidal stability via electrostatic repulsion, Iron oxide nanoparticle clusters can also be coated with various molecules such as citric acid (L. Li et al., 2013), tartaric acid (Yan et al., 2009), gluconic acid or dimercaptosuccinic acid (Luengo, Nardecchia, Morales, & Serrano, 2013) with the colloidal stability of these coated particles strongly depends on the pH, i.e., protonation/deprotonation of the coating molecules, as well as on the ionic strength of the suspension medium(L Gutiérrez et al., 2015).

Figure 4.

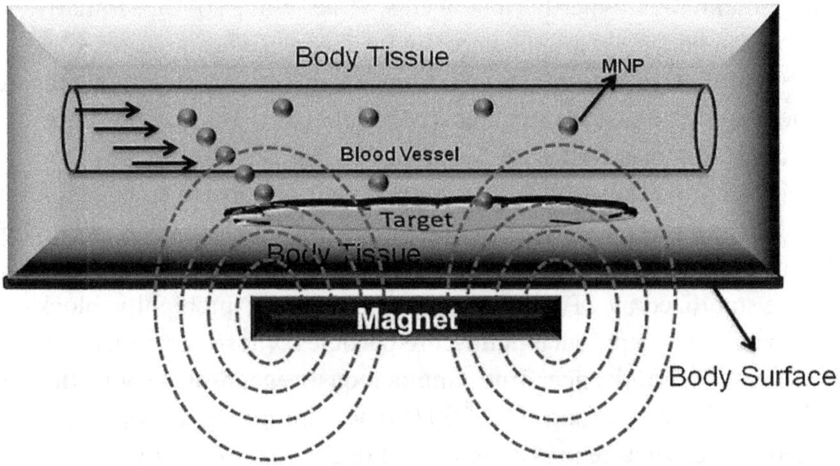

Likewise hydrophilic polymers such as poly(trimethylammonium ethylacrylate methyl sulfate)-b-poly(acrylamide) (PTEA-b-PAM) has been used to coat Iron oxide nanoparticles of around 6 nm in size.This procedure involves the mixing of a solution containing the particles with another solution with another solution containing the polymer at the same concentration and pH leading to the formation of multi-core particles with hydrodynamic diameter between 70–150 nm containing between tens and hundreds of cores (Berret et al., 2006). The relaxivities of these materials indicate a better performance as MRI contrast agents than their single-core counterparts (Berret et al., 2006).

Magnetic nanoparticles have been encapsulated in liposomes of around 200 nm formed by egg phosphatidylcholine (EPC) and distearoyl-SN-glycero-3-phosphoethanolamine-N-[methoxy(poly(ethyleneglycol)) 2000] (Martina et al., 2005) and also within the membrane of poly(trimethylene carbonate)-b-poly(L-glutamic acid) (PTMC-b-PGA) block copolymer vesicles with sizes in the range of 100–400 nm (Sanson et al., 2011).

Multi-core nanoparticles have also been obtained by organic phase synthesis by clustering of single core nanoparticles obtained by thermal decomposition of organometallic compounds (D. Kim et al., 2008). As prepared single core nanoparticles have been clustered by several methods to improve their physicochemical properties especially for their use on hyperthermia treatments or as MRI contrast agents. For instance, magnetic nanoparticles have been encapsulated in a shell of DOPA (3,4-dihydroxyphenylalanine)-conjugated chitosan oligosaccharide (Bae et al., 2012) that has led to the formation of mulicore nanoparticles with a hydrodynamic diameter of around 150 nm containing several 30 nm iron oxide cores inside.

This approach resulted in the formation of very stable iron oxide nanopartilces with interesting potential as hyperthermia agents. (Bae et al., 2012). Similarly, sol-gel chemistry and superficial fluid technology has been used to produce silica-coated multi-core particles of around 150–300 nm using cores that where prepared by thermal decomposition of iron pentacarbonyl ($Fe(CO)_5$) in the presence of oleic acid. (Taboada, Solanas, Rodriguez, Weissleder, & Roig, 2009).

In order to improve the stability in water magnetic nanoparticles have been encapsulated inside the hydrophobic core of a polymeric micelle formed by an amphiphilic diblock copolymer of PCL-b-PEG (poly (ε-caprolactone)-b-poly(ethyleneglycol)) with surface stabilized by a PEG shell (Ai et al., 2005). Amphiphilic block copolymers have been used to produce multi-core particles whose cores were prepared in organic solvents with anticancer drug simultaneously encapsulated with the cores, to be used for drug delivery (Yang et al., 2007). These multi-core nanoparticles display a controlled microstructure providing a good opportunity to study the effect of the geometric distribution on the magnetic properties of these systems. Citrate coated multi-core flower-shaped maghemite structures, consisting of smaller grains between 11 and 30 nm have been prepared by the polyol route, using a DEG/ NMDA mixture (Hugounenq et al., 2012; Lartigue et al., 2012). The close contact of the cores in these multi-core flower-shaped structures enhances their magnetic properties and allows exchange coupling among the cores comparison to single-core counterparts.

MAGNETIC NANOPARTICLES FOR BIOMEDICAL APPLICATIONS

The surface of the magnetic nanoparticles can be tailored with appropriate ligands to attain the specific targeting ability, providing them high colloidal and biostability, low nonspecific binding affinity, and facile bioconjugation with target-specific chemical and biological molecules, thus making them useful for various biomedical applications. The various biomedical applications of magnetic nanoparticles are briefly discussed below.

Magnetic Resonance Imaging of Biological Targets

Paramagnetic materials like those of gadolinium (Gd), europium (Eu), neodymium (Nd), and manganese (Mn) and also some superparamagnetic materials such as iron oxide in the form of maghemite (g-Fe_2O_3) and/or magnetite (Fe_3O_4)) compounds can be used as MRI contrast materials. However, some of these materials such as Gd-contrast materials may induce severe adverse effects on animal health as is observed in patients with compromised renal function and subsequent deposition in different organs/tissues and release of Gd^{3+} which is called nephrogenic systemic fibrosis (NSF). With respect to molecular imaging, iron oxide based MNPs have been favored because iron oxide induces a stronger contrast effect on surrounding tissues under MRI scans due to their effect on spin-spin relaxation time of surrounding water protons and therefore have served as important probes for MR imaging. However due to the poor magnetic contrast effects of conventional iron oxide based contrast agents such as superparamagnetic iron oxide (SPIO) and related nanoparticle probes (e.g., cross-linked iron oxide (CLIO)) have limited their uses for advanced MR imaging (Pankhurst, Connolly, Jones, & Dobson, 2003)(Weissleder et al., 1991) (Tanimoto & Kuribayashi, 2006)25,40,41 Therefore, enhancement of the magnetic contrast by significant shortening of the spin-spin relaxation time (*T*2) of the proton is possible by using a new type of magnetic nanoparticle such as MEIO with high and tunable mass magnetization values(Tanimoto & Kuribayashi, 2006)42 In particular, significant shortening of the spin-spin relaxation time (*T*2) of the proton is possible by using a new type of magnetic nanoparticles. Higher magnetization values of larger MEIO nanoparticles produces stronger MR contrast effects, which are observed as darker MR contrast and blue in color MRI maps. (Jun et al., 2005)13 The relaxivity coefficient (r_2), which is a direct indication of contrast enhancement effects gradually increases from 78 mM^{-1} s^{-1} for 4 nm MEIO to 106, 130, and to 218 mM^{-1} s^{-1} for 6, 9, and 12 nm size MEIO nanoparticles. In addition to the size effects, magnetic dopant effects of MEIO nanoparticles are also significant. For example, Mn-MEIO nanoparticles with the highest magnetization values of 110 emu/g(Mn+Fe) exhibit the best MR signal enhancement effects with an r_2 of 358 mM^{-1} s^{-1} (J.-H. Lee et al., 2007). Other metal-doped MEIO nanoparticles including MEIO, Co-MEIO, and Ni-MEIO with diminished mass magnetization values of 101 emu/g(Fe), 99 emu/ g(Co+Fe), and 85 emu/g(Ni+Fe), respectively, show less MR contrast effects with r_2 of 218, 172, and 152 mM^{-1} s^{-1}, respectively, which is consistent with the trend of mass magnetization of nanoparticles. It is noteworthy that the *r*2 of 12 nm Mn-MEIO nanoparticles is nearly 5.8 times higher than that of conventional molecular MR imaging contrast agents such as crosslinked iron oxide (CLIO) nanoparticles. Such MR signal enhancement effects of Mn-MEIO nanoparticles enable successful ultrasensitive *in vivo* detection of biological targets (J.-H. Lee et al., 2007).

MNPs for Drug Delivering

To overcome the nonspecific nature of the chertherpautic drugs, focus has been made on MNP. These nanoparticles can be used to treat tumors either by (i) conjugating MNPs to antibodies to selectively bind to related receptors and inhibit tumor growth or (ii) targeted MNPs can be used for hyperthermia for tumor therapy or (iii) drugs can also be loaded onto the MNPs for targeted therapy (Peng, Qian, Mao, & Wang, 2008; Subramani, Hosseinkhani, Khraisat, Hosseinkhani, & Pathak, 2009). The drug loaded MNPs are concentrated at the target site with the aid of an external magnet and the drugs are then released on the desired area (Fernández-Pacheco et al., 2007). Magnetic particles smaller than 4 nm are eliminated by cells of the RES, mainly in the liver (60–90%) and spleen (3–10%). Particles larger than 200 nm are usually filtered to the spleen, whose cut-off point extends up to 250 nm, while particles up to 100 nm are mainly phagocytosed through liver cells. In general, the larger the particles are, the shorter their plasma half-life-period (Arruebo, Fernández-Pacheco, Ibarra, & Santamaría, 2007).

However, there are several factors related to the size and magnetism of the biocompatible nanoparticles which control the effective use of MNPs for biomedical applications. Similarly, other parameters such as the physicochemical properties of the drug-loaded MNPs, field strength and geometry, depth of the target tissue, rate of blood flow, and vascular supply, all play a role in determining the effectiveness of this method of drug delivery (Chomoucka et al., 2010). The nanparticles must be small enough to obtain superparamagnetic character so as to avoid agglomeration after stopping magnetic field and to remain in circulation without being removed by the body's natural filters such as the liver or the immune system (Pankhurst et al., 2003). Superparamagnetic nanosystem is preferred due to its ability to become magnetized upon exposure to a magnetic field but have no permanent magnetization (remanence) once the field is turned off.

Protein and DNA Separation

Magnetic naoparticles, MNPs via a simple magnetic separation method represents an attractive alternative method for the selective and reliable capture of specific proteins, DNA and entire cells, by making use of the cheap materials and less time for sample preparation (Gu, Xu, Xu, & Xu, 2006; Obata, Tajima, Yohda, & Matsunaga, 2002). MNPs functionalized with ion-exchange groups or an immobilized affinity tag based ligands or hydrophobic ligands, are mixed with the samples containing the desired molecules such as crude cell lysates, whole blood, plasma, urine, or any biological fluid or fermentation broth. The speceies having affinity tightly bind to the ligands anchored on the MNPs and after a suitable incubation time the complexes are isolated by magnetic decantation and the contaminants washed out. Finally, by employing suitable elution procedures the purified target molecules are recovered by displacement from the MNPs. At present MNPs functionalized with ligands bearing Ni^{2+}-chelating species, such as nitrilotriacetic acid (NTA) are the most thoroughly investigated affinity tag for magnetic separation of proteins which allows for the selective sequestration of (6-His)-tagged proteins with highly conserved

Figure 5.

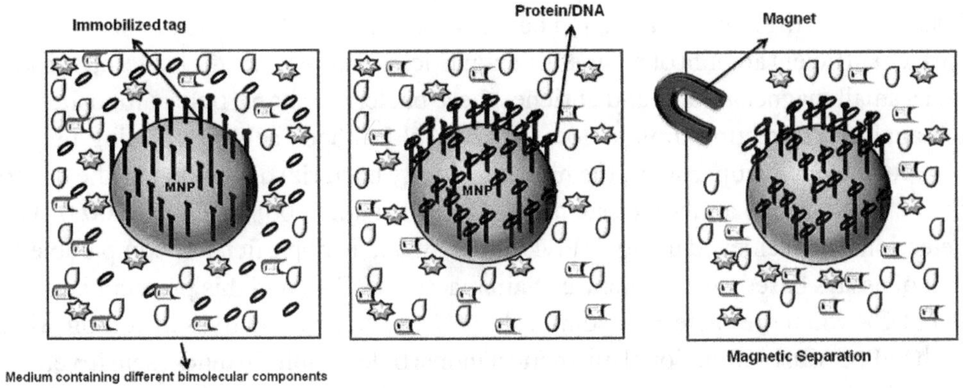

folding down to picomolar concentrations (Kim, Valencia, Liu, & Lin, 2007; Xu et al., 2004). Multivalent action of the chelating agent plays an important role in enhancing the binding selectivity of His-tagged proteins at low concentration, the choice of the anchoring strategy as well as the chelating ligand might significantly affect the binding capacity and reusability of biofunctional MNPs without losing efficiency (Woo, Ponvel, Ahn, & Lee, 2010). New advances made in the separation/extraction of biomolecules by MNPs suggest that this technology could be general and versatile.

Due to the high surface to volume ratio of nanosized sequestrants only small amounts of materials are necessary for the separation process. Moreover, the delicate equipments like centrifuges, filters or liquid chromatography systems are not required during the magnetic separation. However, one of the problems often encountered by researchers when using cationic extractors, including amine-functionalized MNPs, to purify nucleic acids resides in the low release of captured genetic materials. Indeed, the efficiency is generally modest using phosphate buffer as a competitor and even much lower with other ions. It has been found that desorption efficiency can be significantly enhanced using deoxynucleotide triphosphates in place of phosphate buffer, thus improving the chances of successful PCR analyses. (Tanaka, Sakai, Kobayashi, Hatakeyama, & Matsunaga, 2009).

Bacterial Detection

Bacteria at low concentrations are hard to detect and usually require long induction times before further analysis. A simple strategy has been developed using vancomycin-conjugated FePt nanoparticles (FePt@Van conjugates) at exceptionally low concentrations to capture and detect pathogens such as vancomycin-resistant enterococci (VRE) and other Gram-positive bacteria. (Gu, Ho, Tsang, Wang, & Xu, 2003). Due to the strong interaction between Van and D-Ala-D-Ala on the bacterial surface sufficient amount of magnetic nanoparticles binding onto the bacterial surface and a small magnet attracts and enriches these bacteria-nanoparticle composites for the analysis. Detection limit as low as 4 cfu/ml has been achieved. Using FePt@ Van polyvalent nanoparticles, the high-sensitivity bacterial detection may be due to their similar size to antibodies such as IgG (Kell et al., 2008; K. Tsang, 2003). The detection limit achieved using FePt@Van magnetic nanoparticles is comparable to that of assays based on polymerase chain reaction (PCR), and this protocol is faster than PCR when the bacterium count is low. Using fluorescent dyes in combination with FePt@Van biofunctional magnetic nanoparticles results in quick, sensitive, and low-cost detection of bacteria in blood.9(Gao et al., 2006). The fluorescent vancomycin and fluorescein amine (Van-FLA) stains the enriched bacteria for the quick detection using fluorescence microscopy. Figure 3 illustrates the straightforward steps for detecting bacteria in a blood sample:

1. Mixing,
2. Separation,
3. Staining, and
4. Washing.

However, the main disadvantage of this protocol is limited accuracy and render high cost (Gao, Gu, & Xu, 2009).

MAGNETIC NANOSWITCHES IN BIOSENSING APPLICATIONS

By exploiting peculiar magnetic properties of MNPs have led to a novel class of nanosensors. Magnetic relaxation nanoswitches consisting of 3–5 nm iron oxide MNPs coated with a 10 nm thick dextran layer stabilized by crosslinking (CLIOs) and functionalized with amino groups have been demonstrated to be the efficient biosensors for the accurate and sensitive detection of a broad range of biological species, including proteins, DNA, pathogens, and processes such as enzymatic function (Josephson, Tung, Moore, & Weissleder, 1999; Koh, Hong, Weissleder, & Josephson, 2008; H. Lee, Yoon, & Weissleder, 2009; Perez, O'Loughin, Simeone, Weissleder, & Josephson, 2002; Taktak, Sosnovik, Cima, Weissleder, & Josephson, 2007). Such nanoswitches undergo reversible assembly in the presence of a specific molecule that is selectively recognized by the affinity ligands immobilized on the MNPs, resulting in a change in transverse magnetic relaxivity (R2 = 1/T2) of water protons adjacent to the floating nanodipole. As per the outer-sphere diffusion theory, when clusters of MNPs are sufficiently small of the order of few hundred nanometres, the effect of assembly is to reduce the average T2 value. In contrast, T2 increases when large agglomerates are formed with size range of several micrometres compared with that of individual MNPs dispersed in the same fluid or matrix. (Hong, Cima, Weissleder, & Josephson, 2008). Magnetic relaxation nanosensor assays being designed to form reversible nanoassemblies upon MNP interaction with specific analytes in solution both in a forward (clustering) or reverse (declustering) setup. The concentration of analytes, such as glucose and calcium in solution has been quantitatively determined by employing this approach (Atanasijevic, Shusteff, Fam, & Jasanoff, 2006; Sun, Weissleder, & Josephson, 2006). In principle, the same concept could be applied to simultaneous analysis of multiple metabolites in a continuous and noninvasive way by MRI in vivo (Sun, Josephson, Kelly, & Weissleder, 2006). Several other pathological biomarkers have been detected by using functional MNPs endowed with higher inherent magnetic susceptibilities including autoantibodies, toxins and human plasma glycoproteins (Cai et al., 2011; Colombo et al., 2009). By using a library of magnetic glycol-Nps, sets of primary and metastatic cancer cells could be detected (El-Boubbou et al., 2010)192 Recently, a chip-based diagnostic magnetic resonance (DMR) system for multiplexed, quantitative and rapid analysis of unprocessed biological samples has been developed. (H. Lee, Sun, Ham, & Weissleder, 2008). Miniaturized DMR has several advantages over the reported conventional bulk experiments such as only small volume, i.e,. microlitre volumes of the sample are required, with a remarkable increase in detection sensitivity. Also the magnetic field is generated using a small, portable magnet with more homogeneous radiofrequency magnetic fields and less electrical resistance compared with conventional relaxometers (Colombo et al., 2012).

CONCLUSION AND FUTURE OUTLOOK

Past few years have witnessed a substantial progress in the synthesis of MNPs covering a wide range of compositions and tunable sizes. Moreover, complex chemical, physical and physicochemical multiple interactions are involved in the fabrication and surface engineering of iron oxide MNPs. MNPs can be easily synthesized, they can be made colloidally stable, they are inexpensive, and they can be conjugated with biological molecules in a straightforward way. However, in the areas of synthesis, heterogeneity of nanomaterials remains a major problem, and it is hard to precisely control the number of functional molecules on the surface of nanoparticles. Researchers need to develop better strategies for producing nanoparticles that have precise composition, uniform surface modification, and reproducible functionalization. The potential of MNPs stems from the intrinsic properties of their magnetic core combined with their biochemical properties that can be bestowed on them by means of a suitable coating. MNPs based on superparamagnetic iron oxide offer the privileged status of being accepted for clinical purposes such as in humans for MRI diagnosis. But in a near future they are expected to be also used for therapeutic issues and thus becoming theranostic agents. The use of nonradiative and non-toxic detection techniques, and the possibility of analyte magnetic separation and collection are among the peculiar advantages of the use of MNPs for diagnosis. MNPs are specially interesting in delivery applications because apart of the possibility of tagging their surface, they can be driven inside the organism by the application of an external magnetic field gradient to the target area of the body where the therapy has specifically to act. However, the long-term outcome of the MNPs in the body will still need to be determined if their use in medicine wants to be extended regardless of the good tolerance that some MNPs have shown. Due to factors related with the epigenetic phenomena and long term cardiovascular effects in depth analysis of the potential risks associated with the intensive use of inorganic MNPs cannot be further delayed. MNPs represent an ideal model to attempt to set up a comprehensive and acceptable nanotechnology platform for the accurate classification of such risks, for the identification of general protocols for the evaluation of nanomaterial safety toward human health and environmental protection, and for the certification of nanoparticle- based drugs and contrast agents for extensive medical application. For *in vivo* biomedical applications, the purity, dispersity, and stability of the multifunctional magnetic nanoparticles in a physiological environment are highly important (Gao et al., 2009; Gao & Xu, 2009).45 Therefore, it is necessary to further study and explore multifunctional magnetic nanoparticles for creating successful nanobiotechnology of biomedical applications.

REFERENCES

Ai, H., Flask, C., Weinberg, B., Shuai, X., Pagel, M. D., & Farrell, D. (2005). Magnetite-loaded polymeric micelles as ultrasensitive magnetic-resonance probes. *Advanced Materials*, *17*(16).

Arruebo, M., Fernández-Pacheco, R., Ibarra, M. R., & Santamaría, J. (2007). Magnetic nanoparticles for drug delivery. *Nano Today*, *2*(3), 22–32. doi:10.1016/S1748-0132(07)70084-1

Atanasijevic, T., Shusteff, M., Fam, P., & Jasanoff, A. (2006). Calcium-sensitive MRI contrast agents based on superparamagnetic iron oxide nanoparticles and calmodulin. *Proceedings of the National Academy of Sciences of the United States of America, 103*(40), 14707–14712. doi:10.1073/pnas.0606749103 PMID:17003117

Baaziz, W., Pichon, B. P., Fleutot, S., Liu, Y., Lefevre, C., Greneche, J.-M., & Begin-Colin, S. et al. (2014). Magnetic iron oxide nanoparticles: Reproducible tuning of the size and nanosized-dependent composition, defects, and spin canting. *The Journal of Physical Chemistry C, 118*(7), 3795–3810. doi:10.1021/jp411481p

Bae, K. H., Park, M., Do, M. J., Lee, N., Ryu, J. H., Kim, G. W., & Hyeon, T. et al. (2012). Chitosan oligosaccharide-stabilized ferrimagnetic iron oxide nanocubes for magnetically modulated cancer hyperthermia. *ACS Nano, 6*(6), 5266–5273. doi:10.1021/nn301046w PMID:22588093

Batlle, X., & Labarta, A. (2002). Finite-size effects in fine particles: Magnetic and transport properties. *Journal of Physics. D, Applied Physics, 35*(6), R15–R42. doi:10.1088/0022-3727/35/6/201

Berret, J.-F., Schonbeck, N., Gazeau, F., El Kharrat, D., Sandre, O., Vacher, A., & Airiau, M. (2006). Controlled clustering of superparamagnetic nanoparticles using block copolymers: Design of new contrast agents for magnetic resonance imaging. *Journal of the American Chemical Society, 128*(5), 1755–1761. doi:10.1021/ja0562999 PMID:16448152

Bohren, C. F., & Huffman, D. R. (1983). Absorption and scattering by a sphere. In *Absorption and Scattering of Light by Small Particles* (pp. 82–129).

Bordelon, D. E., Cornejo, C., Grüttner, C., Westphal, F., DeWeese, T. L., & Ivkov, R. (2011). Magnetic nanoparticle heating efficiency reveals magneto-structural differences when characterized with wide ranging and high amplitude alternating magnetic fields. *Journal of Applied Physics, 109*(12), 124904. doi:10.1063/1.3597820

Cai, S., Liang, G., Zhang, P., Chen, H., Zhang, S., Liu, B., & Kong, J. (2011). Rational strategy of magnetic relaxation switches for glycoprotein sensing. *Analyst (London), 136*(1), 201–204. doi:10.1039/C0AN00362J PMID:20949140

Chanteau, B., Fresnais, J., & Berret, J.-F. (2009). Electrosteric enhanced stability of functional sub-10 nm cerium and iron oxide particles in cell culture medium. *Langmuir, 25*(16), 9064–9070. doi:10.1021/la900833v PMID:19572532

Cheon, J., Kang, N.-J., Lee, S.-M., Lee, J.-H., Yoon, J.-H., & Oh, S. J. (2004). Shape evolution of single-crystalline iron oxide nanocrystals. *Journal of the American Chemical Society, 126*(7), 1950–1951. doi:10.1021/ja038722o PMID:14971924

Chikazumi, S., Taketomi, S., Ukita, M., Mizukami, M., Miyajima, H., Setogawa, M., & Kurihara, Y. (1987). Physics of magnetic fluids. *Journal of Magnetism and Magnetic Materials*, *65*(2), 245–251. doi:10.1016/0304-8853(87)90043-6

Chomoucka, J., Drbohlavova, J., Huska, D., Adam, V., Kizek, R., & Hubalek, J. (2010). Magnetic nanoparticles and targeted drug delivering. *Pharmacological Research*, *62*(2), 144–149. doi:10.1016/j.phrs.2010.01.014 PMID:20149874

Colombo, M., Carregal-Romero, S., Casula, M. F., Gutiérrez, L., Morales, M. P., Bohm, I. B., & Parak, W. J. et al. (2012). Biological applications of magnetic nanoparticles. *Chemical Society Reviews*, *41*(11), 4306–4334. doi:10.1039/c2cs15337h PMID:22481569

Colombo, M., Ronchi, S., Monti, D., Corsi, F., Trabucchi, E., & Prosperi, D. (2009). Femtomolar detection of autoantibodies by magnetic relaxation nanosensors. *Analytical Biochemistry*, *392*(1), 96–102. doi:10.1016/j.ab.2009.05.034 PMID:19481052

Corchero, J. L., & Villaverde, A. (2009). Biomedical applications of distally controlled magnetic nanoparticles. *Trends in Biotechnology*, *27*(8), 468–476. doi:10.1016/j.tibtech.2009.04.003 PMID:19564057

Cornell, R. M., & Schwertmann, U. (2003). The Iron Oxides. Wiley.

Dayeh, V. R., Chow, S. L., Schirmer, K., Lynn, D. H., & Bols, N. C. (2004). Evaluating the toxicity of Triton X-100 to protozoan, fish, and mammalian cells using fluorescent dyes as indicators of cell viability. *Ecotoxicology and Environmental Safety*, *57*(3), 375–382. doi:10.1016/S0147-6513(03)00083-6 PMID:15041260

Dobson, J. (2006). Magnetic nanoparticles for drug delivery. *Drug Development Research*, *67*(1), 55–60. doi:10.1002/ddr.20067

El-Boubbou, K., Zhu, D. C., Vasileiou, C., Borhan, B., Prosperi, D., Li, W., & Huang, X. (2010). Magnetic glyco-nanoparticles: A tool to detect, differentiate, and unlock the glyco-codes of cancer via magnetic resonance imaging. *Journal of the American Chemical Society*, *132*(12), 4490–4499. doi:10.1021/ja100455c PMID:20201530

El-Sayed, M. A. (2001). Some interesting properties of metals confined in time and nanometer space of different shapes. *Accounts of Chemical Research*, *34*(4), 257–264. doi:10.1021/ar960016n PMID:11308299

Elliott, D. W., & Zhang, W.-X. (2001). Field assessment of nanoscale bimetallic particles for groundwater treatment. *Environmental Science & Technology*, *35*(24), 4922–4926. doi:10.1021/es0108584 PMID:11775172

Fauconnier, N., Bee, A., Roger, J., & Pons, J. N. (1999). Synthesis of aqueous magnetic liquids by surface complexation of maghemite nanoparticles. *Journal of Molecular Liquids*, *83*(1), 233–242. doi:10.1016/S0167-7322(99)00088-4

Fernández-Pacheco, R., Marquina, C., Valdivia, J. G., Gutiérrez, M., Romero, M. S., Cornudella, R., & García, A. et al. (2007). Magnetic nanoparticles for local drug delivery using magnetic implants. *Journal of Magnetism and Magnetic Materials*, *311*(1), 318–322. doi:10.1016/j.jmmm.2006.11.192

Gao, J., Gu, H., & Xu, B. (2009). Multifunctional Magnetic Nanoparticles: Design, Synthesis, and Biomedical Applications. *Accounts of Chemical Research*, *42*(8), 1097–1107. doi:10.1021/ar9000026 PMID:19476332

Gao, J., Li, L., Ho, P., Mak, G. C., Gu, H., & Xu, B. (2006). Combining fluorescent probes and biofunctional magnetic nanoparticles for rapid detection of bacteria in human blood. *Advanced Materials*, *18*(23), 3145–3148. doi:10.1002/adma.200601058

Gao, J., & Xu, B. (2009). Applications of nanomaterials inside cells. *Nano Today*, *4*(1), 37–51. doi:10.1016/j.nantod.2008.10.009

Grainger, D. W., & Okano, T. (2003). Biomedical micro and nano-technology. *Advanced Drug Delivery Reviews*, *55*(3), 311–313. doi:10.1016/S0169-409X(02)00222-3

Grüttner, C., Müller, K., Teller, J., Westphal, F., Foreman, A., & Ivkov, R. (2007). Synthesis and antibody conjugation of magnetic nanoparticles with improved specific power absorption rates for alternating magnetic field cancer therapy. *Journal of Magnetism and Magnetic Materials*, *311*(1), 181–186. doi:10.1016/j.jmmm.2006.10.1151

Gu, H., Ho, P.-L., Tsang, K. W. T., Wang, L., & Xu, B. (2003). Using biofunctional magnetic nanoparticles to capture vancomycin-resistant enterococci and other gram-positive bacteria at ultralow concentration. *Journal of the American Chemical Society*, *125*(51), 15702–15703. doi:10.1021/ja0359310 PMID:14677934

Gu, H., Xu, K., Xu, C., & Xu, B. (2006). Biofunctional magnetic nanoparticles for protein separation and pathogen detection. *Chemical Communications*, (9): 941–949. doi:10.1039/b514130c PMID:16491171

Gupta, A. K., & Gupta, M. (2005a). Synthesis and surface engineering of iron oxide nanoparticles for biomedical applications. *Biomaterials*, *26*(18), 3995–4021. doi:10.1016/j.biomaterials.2004.10.012 PMID:15626447

Gupta, A. K., & Gupta, M. (2005b). Synthesis and surface engineering of iron oxide nanoparticles for biomedical applications. *Biomaterials*, *26*(18), 3995–4021. doi:10.1016/j.biomaterials.2004.10.012 PMID:15626447

Gutiérrez, L., Costo, R., Grüttner, C., Westphal, F., Gehrke, N., Heinke, D., & Morales, M. P. et al. (2015). Synthesis methods to prepare single- and multi-core iron oxide nanoparticles for biomedical applications. *Dalton Transactions (Cambridge, England)*, *44*(7), 2943–2952. doi:10.1039/C4DT03013C PMID:25564784

Gutiérrez, L., Lázaro, F. J., Abadía, A. R., Romero, M. S., Quintana, C., Morales, M. P., & Arranz, R. et al. (2006). Bioinorganic transformations of liver iron deposits observed by tissue magnetic characterisation in a rat model. *Journal of Inorganic Biochemistry*, *100*(11), 1790–1799. doi:10.1016/j.jinorgbio.2006.06.010 PMID:16899298

Gutiérrez, L., Mejías, R., Barber, D. F., Veintemillas-Verdaguer, S., Serna, C. J., Lázaro, F. J., & Morales, M. P. (2011). Ac magnetic susceptibility study of in vivo nanoparticle biodistribution. *Journal of Physics. D, Applied Physics*, *44*(25), 255002. doi:10.1088/0022-3727/44/25/255002

Hahn, H., Logas, J., & Averback, R. S. (1990). Sintering characteristics of nanocrystalline TiO 2. *Journal of Materials Research*, *5*(03), 609–614. doi:10.1557/JMR.1990.0609

Hao, R., Xing, R., Xu, Z., Hou, Y., Gao, S., & Sun, S. (2010). Synthesis, functionalization, and biomedical applications of multifunctional magnetic nanoparticles. *Advanced Materials*, *22*(25), 2729–2742. doi:10.1002/adma.201000260 PMID:20473985

Heath, J. R. (1995). The chemistry of size and order on a nanometer scale. *Science*, *270*(5240), 1315–1316. doi:10.1126/science.270.5240.1315

Hong, R., Cima, M. J., Weissleder, R., & Josephson, L. (2008). Magnetic microparticle aggregation for viscosity determination by MR. *Magnetic Resonance in Medicine*, *59*(3), 515–520. doi:10.1002/mrm.21526 PMID:18306403

Hood, J. D., Bednarski, M., Frausto, R., Guccione, S., Reisfeld, R. A., Xiang, R., & Cheresh, D. A. (2002). Tumor regression by targeted gene delivery to the neovasculature. *Science*, *296*(5577), 2404–2407. doi:10.1126/science.1070200 PMID:12089446

Hoopes, P. J., Tate, J. A., Ogden, J. A., Strawbridge, R. R., Fiering, S. N., & Petryk, A. A. … Ivkov, R. (2009). Assessment of intratumor non-antibody directed iron oxide nanoparticle hyperthermia cancer therapy and antibody directed IONP uptake in murine and human cells. In SPIE BiOS: Biomedical Optics (p. 71810P).

Hugounenq, P., Levy, M., Alloyeau, D., Lartigue, L., Dubois, E., Cabuil, V., & Gazeau, F. et al. (2012). Iron oxide monocrystalline nanoflowers for highly efficient magnetic hyperthermia. *The Journal of Physical Chemistry C*, *116*(29), 15702–15712. doi:10.1021/jp3025478

Hyeon, T. (2003). Chemical synthesis of magnetic nanoparticles. *Chemical Communications*, (8): 927–934. doi:10.1039/b207789b PMID:12744306

Hyeon, T., Lee, S. S., Park, J., Chung, Y., & Na, H. (2001). Synthesis of highly crystalline and monodisperse maghemite nanocrystallites without a size-selection process. *Journal of the American Chemical Society*, *123*(51), 12798–12801. doi:10.1021/ja016812s PMID:11749537

Jeun, M., Bae, S., Tomitaka, A., Takemura, Y., Park, K. H., Paek, S. H., & Chung, K.-W. (2009). Effects of particle dipole interaction on the ac magnetically induced heating characteristics of ferrite nanoparticles for hyperthermia. *Applied Physics Letters*, *95*(8), 82501. doi:10.1063/1.3211120

Jiles, D. C. (1998). *Introduction to magnetism and magnetic materials*. CRC press.

Josephson, L., Tung, C.-H., Moore, A., & Weissleder, R. (1999). High-efficiency intracellular magnetic labeling with novel superparamagnetic-Tat peptide conjugates. *Bioconjugate Chemistry*, *10*(2), 186–191. doi:10.1021/bc980125h PMID:10077466

Jun, Y., Huh, Y.-M., Choi, J., Lee, J.-H., Song, H.-T., Kim, S., & Shin, J.-S. et al. (2005). Nanoscale size effect of magnetic nanocrystals and their utilization for cancer diagnosis via magnetic resonance imaging. *Journal of the American Chemical Society*, *127*(16), 5732–5733. doi:10.1021/ja0422155 PMID:15839639

Jun, Y., Seo, J., & Cheon, J. (2008). Nanoscaling laws of magnetic nanoparticles and their applicabilities in biomedical sciences. *Accounts of Chemical Research*, *41*(2), 179–189. doi:10.1021/ar700121f PMID:18281944

Jung, C. W., & Jacobs, P. (1995). Physical and chemical properties of superparamagnetic iron oxide MR contrast agents: Ferumoxides, ferumoxtran, ferumoxsil. *Magnetic Resonance Imaging*, *13*(5), 661–674. doi:10.1016/0730-725X(95)00024-B PMID:8569441

Kell, A. J., Stewart, G., Ryan, S., Peytavi, R., Boissinot, M., Huletsky, A., & Simard, B. et al. (2008). Vancomycin-modified nanoparticles for efficient targeting and preconcentration of Gram-positive and Gram-negative bacteria. *ACS Nano*, *2*(9), 1777–1788. doi:10.1021/nn700183g PMID:19206416

Kim, D., Lee, N., Park, M., Kim, B. H., An, K., & Hyeon, T. (2008). Synthesis of uniform ferrimagnetic magnetite nanocubes. *Journal of the American Chemical Society*, *131*(2), 454–455. doi:10.1021/ja8086906 PMID:19099480

Kim, D. K., Mikhaylova, M., Wang, F. H., Kehr, J., Bjelke, B., Zhang, Y., & Muhammed, M. et al. (2003). Starch-coated superparamagnetic nanoparticles as MR contrast agents. *Chemistry of Materials*, *15*(23), 4343–4351. doi:10.1021/cm031104m

Kim, J., Piao, Y., & Hyeon, T. (2009). Multifunctional nanostructured materials for multimodal imaging, and simultaneous imaging and therapy. *Chemical Society Reviews*, *38*(2), 372–390. doi:10.1039/B709883A PMID:19169455

Kim, J. S., Valencia, C. A., Liu, R., & Lin, W. (2007). Highly-efficient purification of native polyhistidine-tagged proteins by multivalent NTA-modified magnetic nanoparticles. *Bioconjugate Chemistry*, *18*(2), 333–341. doi:10.1021/bc0601951 PMID:17311440

Koh, I., Hong, R., Weissleder, R., & Josephson, L. (2008). Sensitive NMR sensors detect antibodies to influenza. *Angewandte Chemie*, *120*(22), 4187–4189. doi:10.1002/ange.200800069 PMID:18428168

Krycka, K. L., Jackson, A. J., Borchers, J. A., Shih, J., Briber, R., Ivkov, R., … Dennis, C. L. (2011). Internal magnetic structure of dextran coated magnetite nanoparticles in solution using small angle neutron scattering with polarization analysis. *Journal of Applied Physics, 109*(7).

Kung, H., & Foecke, T. (1999). Mechanical behavior of nanostructured materials. *MRS Bulletin*, *24*(2), 14–15.

Lacroix, L.-M., Ho, D., & Sun, S. (2010). Magnetic nanoparticles as both imaging probes and therapeutic agents. *Current Topics in Medicinal Chemistry*, *10*(12), 1184–1197. doi:10.2174/156802610791384207 PMID:20388109

Lartigue, L., Hugounenq, P., Alloyeau, D., Clarke, S. P., Lévy, M., Bacri, J.-C., & Gazeau, F. et al. (2012). Cooperative organization in iron oxide multi-core nanoparticles potentiates their efficiency as heating mediators and MRI contrast agents. *ACS Nano*, *6*(12), 10935–10949. PMID:23167525

Laurent, S., Forge, D., Port, M., Roch, A., Robic, C., Vander Elst, L., & Muller, R. N. (2008). Magnetic iron oxide nanoparticles: Synthesis, stabilization, vectorization, physicochemical characterizations, and biological applications. *Chemical Reviews*, *108*(6), 2064–2110. doi:10.1021/cr068445e PMID:18543879

Lee, H., Sun, E., Ham, D., & Weissleder, R. (2008). Chip–NMR biosensor for detection and molecular analysis of cells. *Nature Medicine*, *14*(8), 869–874. doi:10.1038/nm.1711 PMID:18607350

Lee, H., Yoon, T., & Weissleder, R. (2009). Ultrasensitive Detection of Bacteria Using Core–Shell Nanoparticles and an NMR-Filter System. *Angewandte Chemie International Edition*, *48*(31), 5657–5660. doi:10.1002/anie.200901791 PMID:19554581

Lee, J.-H., Huh, Y.-M., Jun, Y., Seo, J., Jang, J., Song, H.-T., & Suh, J.-S. et al. (2007). Artificially engineered magnetic nanoparticles for ultra-sensitive molecular imaging. *Nature Medicine*, *13*(1), 95–99. doi:10.1038/nm1467 PMID:17187073

Li, L., Mak, K. Y., Leung, C. W., Chan, K. Y., Chan, W. K., Zhong, W., & Pong, P. W. T. (2013). Effect of synthesis conditions on the properties of citric-acid coated iron oxide nanoparticles. *Microelectronic Engineering*, *110*, 329–334. doi:10.1016/j.mee.2013.02.045

Li, Z., Kawashita, M., Araki, N., Mitsumori, M., Hiraoka, M., & Doi, M. (2010). Preparation of magnetic iron oxide nanoparticles for hyperthermia of cancer in a FeCl2-NaNO3-NaOH aqueous system. *Journal of Biomaterials Applications*. PMID:20207773

Lu, A., Schmidt, W., Matoussevitch, N., Bönnemann, H., Spliethoff, B., Tesche, B., & Schüth, F. et al. (2004). Nanoengineering of a magnetically separable hydrogenation catalyst. *Angewandte Chemie*, *116*(33), 4403–4406. doi:10.1002/ange.200454222 PMID:15368378

Lu, A.-H., Salabas, E. L., & Schüth, F. (2007). Magnetic Nanoparticles: Synthesis, Protection, Functionalization, and Application. *Angewandte Chemie International Edition*, *46*(8), 1222–1244. doi:10.1002/anie.200602866 PMID:17278160

Luengo, Y., Nardecchia, S., Morales, M. P., & Serrano, M. C. (2013). Different cell responses induced by exposure to maghemite nanoparticles. *Nanoscale*, *5*(23), 11428–11437. doi:10.1039/c3nr02148c PMID:23963338

Mailander, V., & Landfester, K. (2009). Interaction of nanoparticles with cells. *Biomacromolecules*, *10*(9), 2379–2400. doi:10.1021/bm900266r PMID:19637907

Martin, D. H., & Martin, D. H. (1967). *Magnetism in solids*. MA: MIT Press Cambridge.

Martina, M.-S., Fortin, J.-P., Ménager, C., Clément, O., Barratt, G., Grabielle-Madelmont, C., & Lesieur, S. et al. (2005). Generation of superparamagnetic liposomes revealed as highly efficient MRI contrast agents for in vivo imaging. *Journal of the American Chemical Society*, *127*(30), 10676–10685. doi:10.1021/ja0516460 PMID:16045355

Massart, R. (1981). Preparation of aqueous magnetic liquids in alkaline and acidic media. *IEEE Transactions on Magnetics*, *17*(2), 1247–1248. doi:10.1109/TMAG.1981.1061188

McCurrie, R. A. (1994). *Ferromagnetic Materials: Structure and Properties. New York, Academic.*

Mishra, B., Patel, B. B., & Tiwari, S. (2010). Colloidal nanocarriers: A review on formulation technology, types and applications toward targeted drug delivery. *Nanomedicine; Nanotechnology, Biology, and Medicine*, *6*(1), 9–24. doi:10.1016/j.nano.2009.04.008 PMID:19447208

Morales, M., Veintemillas-Verdaguer, S., Montero, M. I., Serna, C. J., Roig, A., Casas, L., & Sandiumenge, F. et al. (1999). Surface and internal spin canting in γ-Fe2O3 nanoparticles. *Chemistry of Materials*, *11*(11), 3058–3064. doi:10.1021/cm991018f

Morales, M. P., Serna, C. J., Bødker, F., & Mørup, S. (1997). Spin canting due to structural disorder in maghemite. *Journal of Physics Condensed Matter*, *9*(25), 5461–5467. doi:10.1088/0953-8984/9/25/013

Niemeyer, C. M. (2001). Nanoparticles, proteins, and nucleic acids: Biotechnology meets materials science. *Angewandte Chemie International Edition*, *40*(22), 4128–4158. doi:10.1002/1521-3773(20011119)40:22<4128::AID-ANIE4128>3.0.CO;2-S

Niihara, K. (1991). New design concept of structural ceramics—ceramic nanocomposites. *Nippon Seramikkusu Kyokai Gakujutsu Ronbunshi*, *99*(10), 974–982. doi:10.2109/jcersj.99.974

Obata, K., Tajima, H., Yohda, M., & Matsunaga, T. (2002). Recent developments in laboratory automation using magnetic particles for genome analysis. *Pharmacogenomics*, *3*(5), 697–708. doi:10.1517/14622416.3.5.697 PMID:12223054

Pankhurst, Q. A., Connolly, J., Jones, S. K., & Dobson, J. J. (2003). Applications of magnetic nanoparticles in biomedicine. *Journal of Physics. D, Applied Physics*, *36*(13), R167–R181. doi:10.1088/0022-3727/36/13/201

Park, J., Kang, N., Jun, Y., Oh, S. J., Ri, H., & Cheon, J. (2002). Superlattice and magnetism directed by the size and shape of nanocrystals. *ChemPhysChem*, *3*(6), 543–547. doi:10.1002/1439-7641(20020617)3:6<543::AID-CPHC543>3.0.CO;2-E PMID:12465496

Peng, X.-H., Qian, X., Mao, H., & Wang, A. Y. (2008). Targeted magnetic iron oxide nanoparticles for tumor imaging and therapy. *International Journal of Nanomedicine*, *3*(3), 311. PMID:18990940

Perez, J. M., O'Loughin, T., Simeone, F. J., Weissleder, R., & Josephson, L. (2002). DNA-based magnetic nanoparticle assembly acts as a magnetic relaxation nanoswitch allowing screening of DNA-cleaving agents. *Journal of the American Chemical Society*, *124*(12), 2856–2857. doi:10.1021/ja017773n PMID:11902860

Pilgrimm, H. (2003, October 28). Super-paramagnetic particles with increased R1 relaxivity, process for producing said particles and use thereof. Google Patents.

Pinho, S. L. C., Pereira, G. A., Voisin, P., Kassem, J., Bouchaud, V., Etienne, L., & Geraldes, C. F. G. C. et al. (2010). Fine tuning of the relaxometry of γ-Fe2O3@ SiO2 nanoparticles by tweaking the silica coating thickness. *ACS Nano*, *4*(9), 5339–5349. doi:10.1021/nn101129r PMID:20795638

Roca, A. G., Marco, J. F., Morales, M. del P., & Serna, C. J. (2007). Effect of nature and particle size on properties of uniform magnetite and maghemite nanoparticles. *The Journal of Physical Chemistry C*, *111*(50), 18577–18584. doi:10.1021/jp075133m

Roca, A. G., Morales, M. P., O'Grady, K., & Serna, C. J. (2006). Structural and magnetic properties of uniform magnetite nanoparticles prepared by high temperature decomposition of organic precursors. *Nanotechnology*, *17*(11), 2783–2788. doi:10.1088/0957-4484/17/11/010

Roca, a G., Costo, R., Rebolledo, a F., Veintemillas-Verdaguer, S., Tartaj, P., González-Carreño, T., … Serna, C. J. (2009). Progress in the preparation of magnetic nanoparticles for applications in biomedicine. *Journal of Physics D: Applied Physics, 42*(22), 224002. doi:10.1088/0022-3727/42/22/224002

Roco, M. C., Williams, R. S., & Alivisatos, P. (2000). Nanotechnology Research Directions: *IWGN Workshop Report: Vision for Nanotechnology in the Next Decade*. Springer Science & Business Media.

Sandhu, A., Handa, H., & Abe, M. (2010). Synthesis and applications of magnetic nanoparticles for biorecognition and point of care medical diagnostics. *Nanotechnology*, *21*(44), 442001. doi:10.1088/0957-4484/21/44/442001 PMID:20935358

Sanson, C., Diou, O., Thevenot, J., Ibarboure, E., Soum, A., Brûlet, A., & Brisson, A. et al. (2011). Doxorubicin loaded magnetic polymersomes: Theranostic nanocarriers for MR imaging and magneto-chemotherapy. *ACS Nano*, *5*(2), 1122–1140. doi:10.1021/nn102762f PMID:21218795

Santra, S., Tapec, R., Theodoropoulou, N., Dobson, J., Hebard, A., & Tan, W. (2001). Synthesis and characterization of silica-coated iron oxide nanoparticles in microemulsion: The effect of nonionic surfactants. *Langmuir*, *17*(10), 2900–2906. doi:10.1021/la0008636

Schnorr, J., Wagner, S., Abramjuk, C., Drees, R., Schink, T., Schellenberger, E. A., & Taupitz, M. et al. (2006). Focal Liver Lesions: SPIO-, Gadolinium-, and Ferucarbotran-enhanced Dynamic T1-weighted and Delayed T2-weighted MR Imaging in Rabbits 1. *Radiology*, *240*(1), 90–100. doi:10.1148/radiol.2393040884 PMID:16684917

Subramani, K., Hosseinkhani, H., Khraisat, A., Hosseinkhani, M., & Pathak, Y. (2009). Targeting nanoparticles as drug delivery systems for cancer treatment. *Current Nanoscience*, *5*(2), 135–140. doi:10.2174/157341309788185406

Suh, W. H., Suslick, K. S., Stucky, G. D., & Suh, Y.-H. (2009). Nanotechnology, nanotoxicology, and neuroscience. *Progress in Neurobiology*, *87*(3), 133–170. doi:10.1016/j.pneurobio.2008.09.009 PMID:18926873

Sun, E. Y., Josephson, L., Kelly, K. A., & Weissleder, R. (2006). Development of nanoparticle libraries for biosensing. *Bioconjugate Chemistry*, *17*(1), 109–113. doi:10.1021/bc050290e PMID:16417258

Sun, E. Y., Weissleder, R., & Josephson, L. (2006). Continuous analyte sensing with magnetic nanoswitches. *Small*, *2*(10), 1144–1147. doi:10.1002/smll.200600204 PMID:17193579

Sunderland, C. J., Steiert, M., Talmadge, J. E., Derfus, A. M., & Barry, S. E. (2006). Targeted nanoparticles for detecting and treating cancer. *Drug Development Research*, *67*(1), 70–93. doi:10.1002/ddr.20069

Taboada, E., Solanas, R., Rodriguez, E., Weissleder, R., & Roig, A. (2009). Supercritical-Fluid-Assisted One-Pot Synthesis of Biocompatible Core (γ-Fe2O3)/Shell (SiO2) Nanoparticles as High Relaxivity T2-Contrast Agents for Magnetic Resonance Imaging. *Advanced Functional Materials*, *19*(14), 2319–2324. doi:10.1002/adfm.200801681

Takafuji, M., Ide, S., Ihara, H., & Xu, Z. (2004). Preparation of poly (1-vinylimidazole)-grafted magnetic nanoparticles and their application for removal of metal ions. *Chemistry of Materials*, *16*(10), 1977–1983. doi:10.1021/cm030334y

Taktak, S., Sosnovik, D., Cima, M. J., Weissleder, R., & Josephson, L. (2007). Multiparameter magnetic relaxation switch assays. *Analytical Chemistry*, *79*(23), 8863–8869. doi:10.1021/ac701976p PMID:17983206

Tanaka, T., Sakai, R., Kobayashi, R., Hatakeyama, K., & Matsunaga, T. (2009). Contributions of phosphate to DNA adsorption/desorption behaviors on aminosilane-modified magnetic nanoparticles. *Langmuir*, *25*(5), 2956–2961. doi:10.1021/la8032397 PMID:19437706

Tanimoto, A., & Kuribayashi, S. (2006). Application of superparamagnetic iron oxide to imaging of hepatocellular carcinoma. *European Journal of Radiology*, *58*(2), 200–216. doi:10.1016/j.ejrad.2005.11.040 PMID:16414230

Tartaj, J., Zärate, J., Tartaj, P., & Lachowski, E. E. (2002). Sol-gel Cyclic Self-Production of α-Al2O3 Nanoseeds as a Convenient Route for the Low Cost Preparation of Dense Submicronic Alumina Sintered Monoliths. *Advanced Engineering Materials*, *4*(1-2), 17–21. doi:10.1002/1527-2648(20020212)4:1/2<17::AID-ADEM17>3.0.CO;2-R

Tartaj, P., & Serna, C. J. (2002). Microemulsion-assisted synthesis of tunable superparamagnetic composites. *Chemistry of Materials*, *14*(10), 4396–4402. doi:10.1021/cm021214d

Tartaj, P., & Tartaj, J. (2002). Preparation, characterization and sintering behavior of spherical iron oxide doped alumina particles. *Acta Materialia*, *50*(1), 5–12. doi:10.1016/S1359-6454(01)00336-6

Taupitz, M., Schnorr, J., Abramjuk, C., Wagner, S., Pilgrimm, H., Hünigen, H., & Hamm, B. (2000). New generation of monomer-stabilized very small superparamagnetic iron oxide particles (VSOP) as contrast medium for MR angiography: Preclinical results in rats and rabbits. *Journal of Magnetic Resonance Imaging*, *12*(6), 905–911. doi:10.1002/1522-2586(200012)12:6<905::AID-JMRI14>3.0.CO;2-5 PMID:11105029

Thomas, L. A., Dekker, L., Kallumadil, M., Southern, P., Wilson, M., Nair, S. P., & Parkin, I. P. (2009). Carboxylic acid-stabilised iron oxide nanoparticles for use in magnetic hyperthermia. *Journal of Materials Chemistry*, *19*(36), 6529–6535. doi:10.1039/b908187a

Tsang, K. (2003). Using biofunctional magnetic nanoparticles to capture Gram-negative bacteria at an ultra-low concentration. *Chemical Communications*, (15): 1966–1967. PMID:12932056

Tsang, S. C., Caps, V., Paraskevas, I., Chadwick, D., & Thompsett, D. (2004). Magnetically Separable, Carbon-Supported Nanocatalysts for the Manufacture of Fine Chemicals. *Angewandte Chemie*, *116*(42), 5763–5767. doi:10.1002/ange.200460552 PMID:15495206

Vuong, Q. L., Berret, J., Fresnais, J., Gossuin, Y., & Sandre, O. (2012). A Universal Scaling Law to Predict the Efficiency of Magnetic Nanoparticles as MRI T2-Contrast Agents. *Advanced Healthcare Materials*, *1*(4), 502–512. doi:10.1002/adhm.201200078 PMID:23184784

Wagner, S., Schnorr, J., Pilgrimm, H., Hamm, B., & Taupitz, M. (2002). Monomer-coated very small superparamagnetic iron oxide particles as contrast medium for magnetic resonance imaging: Preclinical in vivo characterization. *Investigative Radiology*, *37*(4), 167–177. doi:10.1097/00004424-200204000-00002 PMID:11923639

Weissleder, R., Lee, A. S., Fischman, A. J., Reimer, P., Shen, T., Wilkinson, R., & Brady, T. J. (1991). Polyclonal human immunoglobulin G labeled with polymeric iron oxide: Antibody MR imaging. *Radiology*, *181*(1), 245–249. doi:10.1148/radiology.181.1.1887040 PMID:1887040

Wetterskog, E., Agthe, M., Mayence, A., Grins, J., Wang, D., Rana, S., & Bergström, L. (2014). Precise control over shape and size of iron oxide nanocrystals suitable for assembly into ordered particle arrays. *Science and Technology of Advanced Materials*, *15*(5), 55010. doi:10.1088/1468-6996/15/5/055010

Woo, E., Ponvel, K. M., Ahn, I.-S., & Lee, C.-H. (2010). Synthesis of magnetic/silica nanoparticles with a core of magnetic clusters and their application for the immobilization of His-tagged enzymes. *Journal of Materials Chemistry*, *20*(8), 1511–1515. doi:10.1039/B918682D

Xie, J., Xu, C., Xu, Z., Hou, Y., Young, K. L., Wang, S. X., & Sun, S. (2006). Linking hydrophilic macromolecules to monodisperse magnetite (Fe3O4) nanoparticles via trichloro-s-triazine. *Chemistry of Materials*, *18*(23), 5401–5403. doi:10.1021/cm061793c PMID:18176627

Xu, C., Xu, K., Gu, H., Zhong, X., Guo, Z., Zheng, R., & Xu, B. (2004). Nitrilotriacetic acid-modified magnetic nanoparticles as a general agent to bind histidine-tagged proteins. *Journal of the American Chemical Society*, *126*(11), 3392–3393. doi:10.1021/ja031776d PMID:15025444

Yan, J., Mo, S., Nie, J., Chen, W., Shen, X., Hu, J., & Tong, H. (2009). Hydrothermal synthesis of monodisperse Fe 3 O 4 nanoparticles based on modulation of tartaric acid. *Colloids and Surfaces. A, Physicochemical and Engineering Aspects*, *340*(1), 109–114. doi:10.1016/j.colsurfa.2009.03.016

Yang, J., Lee, C., Ko, H., Suh, J., Yoon, H., & Lee, K. Haam, S. (2007). Multifunctional magneto-polymeric nanohybrids for targeted detection and synergistic therapeutic effects on breast cancer. Angewandte Chemie, 46(46), 8836–8839.

Chapter 11
Stratagems of Nanotechnology Augmenting the Bioavailability and Therapeutic Efficacy of Traditional Medicine to Formulate Smart Herbal Drugs Combating

Anita Margret
Bishop Heber College, India

ABSTRACT

Mental illness is one of the most fundamental emotional states of conscious being which becomes unbalanced and leads to neurological disorders. It is a significant contributor to the global burden of disease and there is a strong desire to devise a remedy. Ayurveda represents a traditional medicine system of India that endorses antiquity than western medicine and relies on formulations rather than their active components. It has categorised a group of herbal medicines to improve mental abilities. Conversely, the mechanistic details of the therapy are not available in ayurvedic literature and there is a need to fortify this system with modern scientific analysis. The design of nanosystems encompasses promising characteristics in the field of drug delivery with a limited dosage thereby decreasing adverse effects. This chapter confers stratagems of devising polymeric nano herbal formulations as smart nano brain drugs to espouse mental health.

DOI: 10.4018/978-1-5225-1043-7.ch011

INTRODUCTION

The venture to maintain an optimal state of health includes both physical and mental wellbeing and a stable mind symbolizes an ideal physiology which is an imperative cause of existence. Mental health is an essential parameter that reflects the emotional, psychological and social wellbeing of an individual. Presently the populace stumble upon various emotional issues pertaining to their professional and personal life that evokes stress in their day today life. The unbalanced state of conscious being leads to mental disorders that erode subsistence. Trapped with pessimistic thoughts they feel hopeless and helpless thereby desire suicide as their only redeemer. Mental illnesses are a sober condition which can affect a person's thinking, mood, and behaviour. There are various types of mental illness and mental health problems, such as:

- Anxiety disorders,
- Mood disorders,
- Post-traumatic stress disorder (PTSD),
- Obsessive compulsive disorder(OCD),
- Impulse control and addiction disorders,
- Personality disorders.

Though medical field has advanced technologies, the threat of mental ailments intrudes the normal survival rate. There is a sturdy necessitate to treat these distressing conditions and fetch a remedy to stir up the universal welfare. There are several, medications available to combat mental illness but are accompanied with severe adverse effects that would lead to other physiological ailments. The advancement in modern science has to resolve the instinct of studying the context, nonclinical elements and alternative therapies required for enhancing mental health. Therefore, there is a strong desire to elevate research in the field of neuroscience with the resourceful mechanisms revealed, by manoeuvring the innovative technologies. Drug delivery is the most essential feature in drug administration facility and the design of nanosystems makes the action more consistent and appropriate. Polymeric nano particles bind efficiently to the drug and disperse the drug proficiently into the system. This mechanism significantly reduces the side effects of the synthetic drugs. Nanoscience focuses on the approaches of drug delivery in a nano scale, which needs a minimum dosage of drug than the conventional medicinal use. Nanomedicine is the medical application of nanotechnology which is considered as the future medicine. Biomedical nanotechnology, nanobiotechnology, and nanomedicine are the areas used to evolve this hybrid field of research (Figure 1).

Figure 1. Venn diagram indicating the integration of various research fields as a hybrid

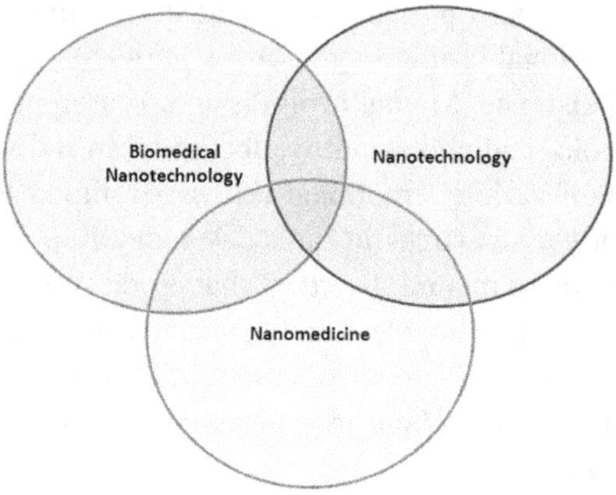

The size of nanomaterials is similar to that of most biological molecules and structures. Therefore, nanomaterials can be useful for both *in vivo* and *in vitro* biomedical research and applications. Nanotechnology has provided the possibility of delivering drugs to specific cells using nanoparticles.

It has the following possibilities of facilitating:

- Lowered overall drug consumption and side-effects,
- Reduces expenses,
- Targeted action and develops personalized medicine,
- To improve drug bioavailability.

Drug delivery is the most essential feature in drug administration facility and the design of Nanosystems makes the action more consistent and appropriate. Drug delivery is the most essential feature in drug administration facility and the design of nanosystems makes the action more consistent and appropriate. Polymer nano particles bind efficiently to the drug and disperse the drug proficiently into the system. The efficacy of many drugs is often limited by their potential to reach the site of therapeutic action. In most cases (conventional dosage forms), only a small amount of administered dose reaches the target site, while the majority of the drug distributes throughout the rest of the body in accordance with its physicochemical and biochemical properties. Therefore, developing a drug delivery system that optimizes the pharmaceutical action of a drug while reducing its toxic side effects *in vivo* is a challenging task.

Biocompatibility is an essential feature for drug administration and polymeric nanoparticles represent a significant efficiency and effectiveness, with unique physiochemical characteristics. There are various advantage of using polymers as encapsulating agents in drug delivery where, the foremost is the augmentation of stability and exact dispersion of drugs. The polymeric nanoparticles have been devised with specificity, which channelizes them to deliver a higher concentration of pharmaceutical agent to a desired location. Polymeric nanocomposites represent a new alternative to conventional polymers which exhibits markedly improved properties such as:

- Increased modulus and strength,
- Outstanding barrier properties,
- Improved solvent and heat resistance,
- Decreased flammability.

Thus polymers can abet imminent drugs with smart formulations to facilitate drug delivery and improve the pharmaceutical index. Alternative medicine is the use of a nonconventional approach instead of standard synthetic drugs. Traditional system of medicine has a deep rooted contextual derived from ancient communities that endorses antiquity than western medicine. Ayurveda, the Indian traditional system of medicine, has categorized a group of herbal medicines to improve mental abilities and restore metal health effectively. The treaties mention a number of single and compound drug formulations of plant origin that are used in the treatment of psychiatric disorders which have a less adverse effect profile, and exhibit efficacy in treating physical ailments. However, an ambiguity is reflected in the interpretation and mode of action pursued by the drug formulations. The current scenario has realized the significance of Ayurveda, where biomedical and ayurvedic scientists are making attempts to combine the best of different healing traditions to meet the challenge faced in healthcare requirements of modern era. Therefore, there is a robust requirement to promote ayurvedic medicine to reinstate mental illness with ingenious scientific technologies.

The considerable focus of this chapter is to reflect the advance strategies of nanotechnology on herbal medicine that can reduce adverse effects by improving the efficacy against mental illness.

The efficacy of many drugs is often limited by their potential to reach the site of therapeutic action. In conventional dosage system only a small amount of administered dose reaches the target site, while the majority of drug distributes throughout the body in accordance with its physicochemical and biochemical properties. Polymers synchronized with the drug facilitates its delivery precisely across the targeted sites precisely. India is sanctified with a profound herbal resources whose attributes needs to be fortified with novel technologies. The deliverables discussed provides insights to hoist the implication of ayurvedic medicine on par with the western medicine.

BACKGROUND

Mental ailment is a serious physiological condition that affects the wellbeing of individuals. Globally, there is a periodic increase in psychological disorders which is a threat to the society, and depression ascertains a menace among mental illness. The World Health Organization (WHO) has ranked depression the 4th leading cause of disability worldwide and projects that by 2020, it will be the second leading cause (Murray et al., 1996). It is one of the major concerns in current scenario where, people are stumbled and is considered as a threat that intrudes the normal subsistence. Further it has reported that only 12.3% of the global population with this infirmity accept this condition and go for treatment. (Who Fact sheet N0.369, 2015). It is essential to investigate this existing issue and need to pay attention over the mental health worldwide. Fortunately, there is a wide range of synthetic drugs to treat these ailments but are accompanied with side effects which makes the treatment appalling. Commonly, alternative medicines incorporate the practice of herbal remedies which are effective with a low potential of toxic effects. Indian traditional medicine has formulated rich medical expertise knowledge of treatise which was handwritten in Sanskrit language as pre-European information. But after the establishment of British rule western medicine was looked upon as the leading system of medicine. This might be one of the reasons due to which ayurveda could not stand parallel to the western medicine. An ambiguity is reflected in the interpretation and mode of action in the formulations and drugs (Perinchery et al., 2013). The ancient Ayurvedic literature available today does not elaborate the effects of any therapy in terms of our current understanding of biology or physiology. The mode of action and the mechanism involved is not mentioned comprehensibly in the written treatise (Valiathan 2006). Although, there has been an increased interest in traditional and herbal medicine systems in recent times, most studies have used specific extracts or ''active principles'' derived from herbal or other traditional drugs and formulations. However current scenario has realised the significance of ayurveda, where biomedical and ayurvedic scientists are making attempts to combine the best of different healing traditions to meet the challenge faced in healthcare requirements of modern era and nanotechnology is one such advanced field.

Drug delivery is the most essential feature in drug administration facility and the design of nanosystems makes the action more consistent and appropriate. Delivering drugs across the physiological barrier is one of the most promising applications of nanotechnology in clinical neuroscience. Nanoparticles could potentially carry out multiple tasks in a predefined sequence, which is very important in the delivery of drugs across these barriers. Nanoscience is the study that involves the control of matter on an atomic and molecular scale which range usually below 100 nm. Initially nanomaterials were analyzed for its physical and electronic applications, but pharmaceutical appliance has been geared towards the area of drug delivery. Nanoparticles have the ability to defend the drugs from the degradation in the gastrointestinal tract, which channelizes the target delivery of drugs to various areas of the body. The oral bioavailability of drugs is profoundly increased due to absorptive endocytosis which is a cellular mechanism that sustains the drug for an extended time. This controls the release of drugs that minimizes the side-effects with increased effectiveness and safety that ultimately reducing health care costs. This technology also has the ability to enhance the performance of drugs that are unable to pass clinical trial phases. There is definite promise of nanodrugs to serve as drug delivery carrier of choice for the more challenging conventional drugs used for the treatment and management of various human ailments. Polymeric nano particles bind efficiently to the drug and disperse the drug proficiently into the system. The design of polymer-drug conjugates was propagated by Ringsdorf who used to deliver hydrophobic small molecules (Ringsdorf, 1975). The interpretation of his research described that, small molecule drugs, especially hydrophobic compounds, have a low aqueous solubility and a broad tissue distribution profile that may result in serious side effects. Consequently, conjugation of these compounds to hydrophilic, biocompatible polymers would significantly increase their aqueous solubility, modify their tissue distribution profile and enhance their plasma circulation half-life. The structural organization of a nanosystem is based on its composition and the presence of compartments within nanocapsules (Dong et al., 2013) leads to oily or aqueous cores surrounded by thin polymer membranes,(Arora et al.,2011) whereas, nanospheres provide a matrix-based organization of the polymeric chains .Nanoparticles have been prepared using several methods, including polymer polymerization(Chu, et al., 2013;Hu et al., 2011) ionic gelation or coacervation (emulsion solvent evaporation (Alshamsan, 2014), spontaneous emulsification or solvent diffusion, (Mainardes et al., 2012) nanoprecipitation, spray drying(Ali et al., 2014), supercritical fluid technology (Mazzarino et al., 2014), and (PRINT) particle replication in non-wetting templates(Galloway et al., 2013).Size is definitely a main advantage of these polymeric nanocomposites which enables them to penetrate capillaries (Brigger et al., 2002). Polymeric nanoparticles have been used brain drug targeting for neurodegenerative disorders such as Alzheimer's disease, to enhance penetration and distribution in and across the blood brain barrier (Medina et al., 2007). Nanoparticle based delivery methods are proved to be the best methods to transfer drugs across central nervous system (Jain, 2012). These strategies are proved to be the best methods to facilitate the delivery of drugs and biological therapeutics across the blood brain barrier (Gaillard et al., 2010). Due to their smaller size, nanoparticles penetrate into minute capillaries that distribute the drug after its delivery. Therefore, an efficient drug accumulation takes place at targeted sites in the body. However, to enhance the therapeutic action of drug its maximum absorption is required and polymers facilitate this aspect (Kreuter, 2014). The application of nanoscience in particular to drug delivery has extended immense accomplishments and has a hopeful elevation in the field of neuroscience. Subsequently, the chapter discusses the description and distinct features of nanotechnology in engendering smart herbal drugs to enhance mental health.

APPROACHES TO CONTEST THE INTIMIDATION OF MENTAL ILLNESS

Synthetic Drugs as a Redeemer of Mental Infirmity: A Blessing Transformed as Desecrate

The preparatory phase for providing effective treatment against mental illness is to recognize the sort of problem and offer a proper diagnosis (NICE, 2010). There are fragmentary discussions regarding the effectiveness of the management of common mental health problems in general practice settings. Although evidence based clinical guidelines are available, initiation and adherence to effective treatment are usually poor (Richards, 2012). Untreated, depression can lead to serious complications that can lay life at risk. Fortunately, most clinical depression is treatable with the wide usage of antidepressant medications whose mode of action in the body occurs at the level of the synapse -the site in the nervous system where neurons communicate with each other or with other types of cells (e.g. muscle cell) by means of neurotransmitters. Some neurotransmitters (e.g. norepinephrine, serotonin and dopamine) are taken back into the nerve ending after release. This process is called uptake, reuptake or transport. This transport is a mechanism that prevents overstimulation of receptors in the receiving neuron. By blocking the (re-)uptake of these neurotransmitters, or inhibiting the mitochondrial enzyme monoamine oxidase, antidepressants alter the magnitude of the effects of neurotransmitter at these synapses (Richelson, 2001). The loss of sensitivity of the cell to the neurotransmitter is called desensitization, while the loss of the receptor protein from the cellular surface is called down-regulation. Though the specific mechanism of the therapeutic action of drugs remains unclear, it is known, that they adapt different mechanisms which can desensitize or down-regulate certain receptors for neurotransmitters by causing a local abundance of neurotransmitters. There are numerous synthetic drugs available to treat these psychological disturbances but are often prone to cause adverse effects, and the difficulty in tolerating these is the most common reason for discontinuing an effective medication.

The Major types of inhibitors are listed below:

- **Monoamine Oxidase Inhibitors:** MAOIs act by inhibiting the activity of the enzyme monoamine oxidase, thus preventing the breakdown of monoamine neurotransmitters and thereby increasing their availability. Well known examples of medicines in this class are phenelzine, isocarboxazid and trylcypromine. There are two isoforms of monoamine oxidase, MAO-A and MAO-B. MAO-A preferentially deaminates serotonin, melatonin, epinephrine, and norepinephrine. MAO-B preferentially deaminates phenylethylamine and trace amines. Dopamine is equally deaminated by both types. Due to the lethal dietary and drug interactions, monoamine oxidase inhibitors are usually reserved as a last line of treatment and used only when other classes of antidepressant drugs (for example selective serotonin reuptake inhibitors and tricyclic antidepressants) fail (Grady,2012).
- **Tricyclic Antidepressants:** TCAs were first discovered in the early 1950s and were subsequently registered later in the decade. The medicine is named after their chemical structure, which contains three rings of atoms. Examples of medicines from the TCA class are amitriptyline, imipramine and desipramine. The tetracyclic antidepressants, which contain four rings of atoms, are a closely related group of antidepressant compounds. The majority of the TCAs act primarily as serotonin-norepinephrine reuptake inhibitors (SNRIs) by blocking the serotonin transporter (SERT) and the norepinephrine transporter (NET), respectively, which results in an elevation of the synaptic concentrations of these neurotransmitters. In addition to their reuptake inhibition, many TCAs also have high affinity as antagonists at several receptors, which may contribute to their therapeutic efficacy, as well as their side effects (Gillman, 2007).
- **Selective Serotonin Reuptake Inhibitors:** SSRIs were first developed in the 1970s with the aim of targeting a specific site of the neuron, the reuptake protein, while not influencing other parts of the neuron such as the receptors. SSRIs are believed to increase the extracellular level of the neurotransmitter serotonin by inhibiting its reuptake into the presynaptic cell, increasing the level of serotonin in the synaptic cleft available to bind to the postsynaptic receptor. They have varying degrees of selectivity for the other monoamine transporters, with the more specific SSRIs having only weak affinity for the noradrenaline and dopamine transporter. Examples of SSRIs are escitalopram, fluoxetine, sertraline, paroxetine and fluvoxamine (Royal Pharmaceutical Society of Great Britain 2010).

- **Serotonin and Noradrenaline Reuptake Inhibitors:** SNRIs were first introduced in 1994. This type of antidepressant acts by inhibiting the reuptake of the neurotransmitters serotonin and norepinephrine. In relation to major depressive disorder, norepinephrine is thought to be related to alertness and energy as well as anxiety, attention, and interest in life; while (a lack of) serotonin is thought to be related to anxiety, obsessions, and compulsions. The inhibition of the reuptake of these neurotransmitters results in an increase in their extracellular concentrations and, therefore, an increase in neurotransmission. Venlafaxine and milnacipran are examples of SNRIs (Royal Pharmaceutical Society of Great Britain 2010).
- **Noradrenergic and Specific Serotonergic Antidepressants:** NaSSAs are a class of psychiatric drugs used that act by antagonizing the α2-adrenergic receptor and certain serotonin receptors. Antidepressants mianserin and its successor mirtazapine are examples of medicines in this class. By blocking α2-adrenergic autoreceptors and heteroreceptors, NaSSAs enhance adrenergic and serotonergic neurotransmission in the brain involved in mood regulation. In addition, due to their blockade of certain serotonin receptors, serotonin transmission is not facilitated in unwanted areas. This specificity is what is referred to in the name of this class of medicines (Julien, 2004).

In spite of the novel development for pharmacotherapy against mental illness, it is unfortunate that this disorder goes undiagnosed and untreated in many patients. It has been shown that less than 50% of patients experience a clinical response to treatment with the foremost drug prescribed. Yet, the presently prescribed molecules provide some improvement in the clinical condition of patients, it is at a cost of having to bear the burden of their adverse effects (Tripathi 2008, Hardman et.al, 2007). Many patients are not able to tolerate available (Figure 2) medications due to its deleterious side effects.

Studies show that as many as 50% of subjects may discontinue antidepressant treatments within a six month period, reporting adverse effects as a main reason for discontinuation (Cipriani et al., 2009). Along with the classical theory of decrease in the neurotransmitter levels in the brain leading to the pathogenesis of clinical disorders, recent studies have also shown the involvement of oxidative stress in the phenomenon (Sarandolet.al, 2007, Ibrahim, et.al. 2007).

Figure 2. Signifies the common adverse effects caused by synthetic drugs

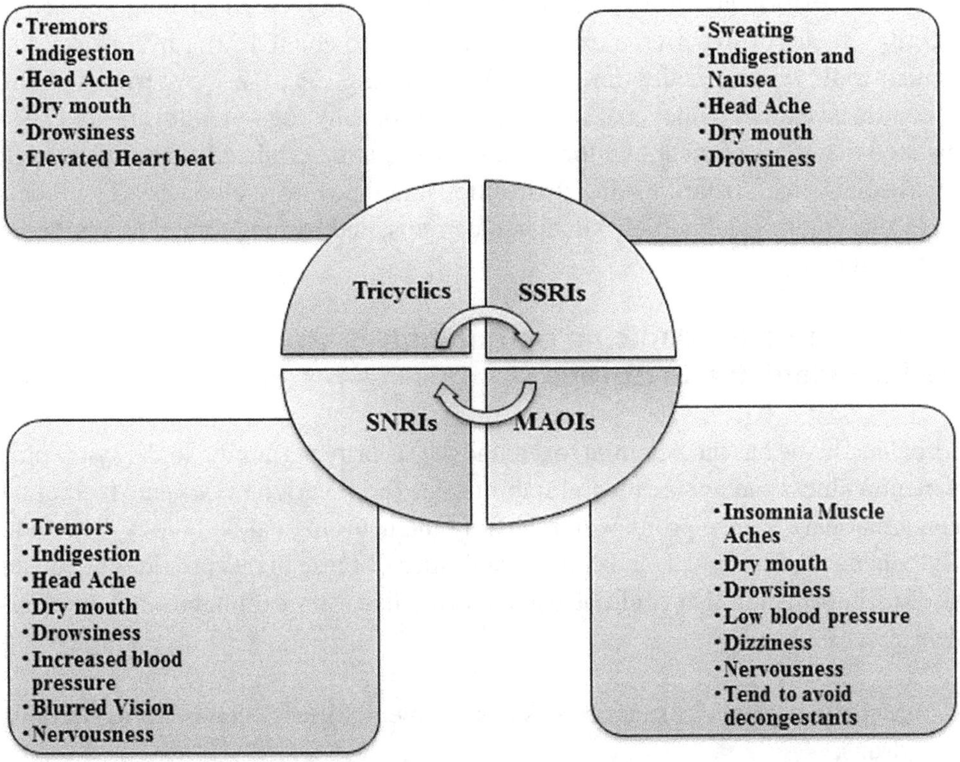

Alternative Herbal Remedies Formulated as Traditional Medicine to Combat Mental Illness: Plant Formulations as Potential Antidepressants

Although a number of synthetic drugs have been used as antidepressants they have adverse effects that compromise the treatment. These conditions create an opportunity for alternative treatment of depression by the use of medicinl plants or plant based antidepressant formulations. Several plants prove to contest depression but *Hypericum Perforatum* L. (St. John's Wort) is an established European medicine which is commercialized in large scale. The plant has profound medicinal properties and primarily, an extensive amount of clinical research shows that St. John's Wort extracts are effective in treating major depressive disorder. Botanical formulations are commercially available as herbal preparations over-the-counter in the forms of tea, tincture, tablet, and capsule usually containing its biologically active compounds such as hypericin (0.19 – 0.3%), hyperforin (1.3 – 3.9%) and flavonoids (calculated as rutin 4.8 – 11.4%.).The vital phytocompound Hypericin is a major psychoactive component of the plant is segregated with a phloroglucinol derivative compounds such as Hyperforin and Adhyperforin (Von Eggelkraut – Gottanka et al., 2002). A 300 mg - tablet of hypericum extract counterpart to about 15 mg hyperforin and clinical trials report a daily dose of 900 – 1500 mg (i.e., 0.3% hypericin) of the standardized extract (Biber et al., 1998). The European drug - monitoring study has reported that these commercial formulation associates certain adverse effects such as gastrointestinal irritation, allergic reactions, fatigue and restlessness. The blemish of this commercial herbal extract can be reinstated by the prominent ayurvedic medicinal plants.

Indian Ayurvedic Medicine on Par with the Folk Medicine in Europe

Herbal medicine has the potential to decline the severity pertaining to the symptoms of mental illness and restore mental stability. There are various *ayurvedic* medicinal plants that have shown promise in reversing the neurological pathology. The vital phytochemical, biological, and cellular activities of these plants provides adequate baseline information that could be used in drug discovery campaigns and development processes.

- Medicinal plants as functional leads for mental illness classified as Medhya drugs:
    - **Ashwagandha (*Withania somnifera*):** Ashwagandha (*Withania somnifera*, family: *Solanaceae*) (Figure 3) is commonly known as "Indian Winter cherry" or "Indian Ginseng" which is widely spread in North western regions of India. It is one of the most important herb of *ayurveda* that has been used for ages as a Rasayana which is reputed to promote physical and mental health. The plant consists of vital phytochemical constituents that comprises of alkaloids (isopelletierine, anaferine, cuseohygrine, anahygrine), steroidal lactones (withanolides, withaferins) and saponins (Mishra, et al.,2000). Sitoindosides and acylsterylglucosides are considered as active ingredient of the herb that is established as anti-stress agents against acute models of experimental stress (Bhattacharya et al., 1987). As mentioned earlier in the text the plant belongs to a sub-group of Rasayanas known as Medhya rasayanas. Medhya is derived from Sanskrit language that refers to the mind and mental/intellectual capacity. The cognition intellectual effect of this ayurvedic drug enhances the learning skills in children, restores memory deficits during several circumstances such as head injury, protracted illness and old age (Singh and Udupa, 1993). Neuritic atrophy is

a pathological condition that leads to synaptic loss which is the foremost cause of neurodegenerative diseases like Alzheimer's, Parkinson's, Huntington's and several other mental disorders. Numerous studies exhibit the realistic medication that can cease and reverse neuritic atrophy. Glyco withanolides withaferin- A and sitoindosides are the dynamic phyto compounds that have induced cognitive defects in Alzheimer's disease model (Bhattacharya et al., 1995).The furnished scientific details sustain the finale that Ashwagandha is a real potent herb that ingredients in regenerative tonic (Rasayana of *ayurveda*), due to its manifold pharmacological accomplishments like anti-stress and neuroprotective activity.

- **Brahmi (*Bacopa monnieri*)**: *Bacopa monnieri* (Brahmi) (Figure 4) belonging to the family *Schropulariaceae* is an ayurvedic herb which has a long primordial history of almost 3000 years and catalogued as an ingredient in medhya rasayana. The plant is habitat all over India particularly in wetlands and marshy tropical regions. The name of the plant is derived from Sanskrit language that reflects 'Brahma' who is considered as the god of creation according to Hindu religion. Brahmi has an alternative meaning "bringing knowledge of the supreme reality' that relates to brain which is the central organ of intellectual. In India this plant is utilized immensely as a revitalizing herb that fortifies the brain function and memory. The medicinal plant comprises of imperative phytocompounds among which saponins termed as "Bacosides" are considered to be a major bioactive marker that has therapeutic index (Deepak and Amit 2004). The whole plant extract of Brahmi has undergone various clinical trials that substantiate the reality to improve cognitive function in children and adults. A research conducted among school children by Sharma et al., 1987 improved their learning, memory and perception skills reaction without inducing significant side-effects. Brahmi has been commenced in the market as single or associated with other phytocompounds in treating attention deficient disorders. Pravina et al., 2007 devised a commercial formulation labeled as BacoMindR, which highlighted the non-noxious nature of the herbal extract reinforces ayurvedic medicine at the summit of healing mental illness. Hence there is a firm requisite to elevate this herbal medicine with advent technologies that would stimulate it to be on par with the western medicines.
- **Shankhpushpi *Convolvulus pluricaulis***: Shankhpushpi is a reputed drug constituted in medhya rasayana the name of the plant is derived from Sanskrit language that illustrates a plant with flowers shaped like a conch or shankha which is a sacred instrument used in ritual wor-

ships. According to the pharmacopoeia of India, *Convolvulus pluricaulis* (Convulvulaceae) (Figure 5a) as a whole plant rightfully claim the name of Shankapushpi but, Ayurvedic practitioners have used three other medicinal herbs such as *Canscora decussate* Schult. (Gentianaceae) (Figure 5b), *Evolvulus alsinoids* Linn. (Convulvulaceae) (Figure 5c), and (Figure 5d) *Clitoria ternatea* Linn. (Papilionaceae) (Nahata et al., 2008,2010) Indian Council of Medical Research(ICMR) has approved *C. pluricaulis as a herbal* drug with systematic potential as a brain drug, functioning as anxiolytic, tranquillizing, antidepressant, antistress, neurodegenerative, antiamensic, antioxidant agent. Phytochemical assays of the plant report to contain significant constituents like alkaloids, flavanoids, and coumarins (Sethiya et al., 2010).Though all the four plants are catalogued under the same class *Magnoliopsida*, *Clitoria ternatea* separates itself at the level of sub-class belonging to *Rosidae*, while the other three belong to *Asteridae*.

- ***Clitoria ternatea*:** *Clitoria ternatea,* is distinct among other herbs and has the property of being a good brain drug (Sivaranjan 1994). The plant has a rich source of vital phytoconstituents such as tri terpenoids, flavonol, glycosides, anthocyanins and steroids which are therapeutically significant. Phytochemical screening has yielded tannins, resins, taraxerol and ternatins.

Figure 3. Pictorial representation of the medicinal plant Ashwagandha (Withania somnifera)

Ashwagandha (*Withania somnifera*)

Figure 4. Portrays the pictorial representation of the medicinal plant Bacopa monnieri (Brahmi)

Brahmi (*Bacopa monnieri*)

Figure 5. (a, b, c & d): Pictorial representation of the medicinal plants Convolvulus pluricaulis, Canscora decussate, Evolvulus alsinoides & Clitoria ternatea

Interestingly, the floral petals of the plant yielded three flavonol glycosides - kaempferol 3-O-(2″-O-α-rhamnosyl-6″-O-malonyl)-β-glucoside, quercetin 3-O-(2″-O-α-rhamnosyl-6″-O-malonyl)-β-glucoside, and myricetin 3-O-(2″,6″-di-O-α-rhamnosyl)-β-glucoside (Ramaswami et al., 2011). Banerjee and Chakravarthi (1963) reported the isolation and identification of pentacyclic triterpenoids, taraxerol and taraxerone from the roots. Yadava and Verma (2003) isolated antimicrobial flavonol glycoside3,5,4_-trihydroxy-7-methoxyflavonol-3-O-_-d-xylopyranosyl-(1,3)-O-d-galactopyranosyl (1,6)-O-d-glucopyranoside from the ethyl acetate soluble fraction of the defatted seeds of the plants.

The GC/MS profile of root extract extricated a major compound 9,17-Octadeca-dienal, (Z)- o with a elevated retention time of 32.99 .The in silico docking analysis furnished a minimum binding affinity energy values of -6.5 / -7.71 kcal/mol against both the isoforms of Monoamine oxidase(MAO) A and B there by establishing the plant as MAO-inhibitors (Anita Margret et al.,2015)The roots, seeds and leaves of plant have been widely used as a brain tonic and is believed to promote memory and intelligence (Mukherjee et al., 2007). The whole plant of *Clitoria ternatea* has been widely screened for its various pharmacological activities. The aqueous root extract of the plant affects the brain structures concernedwith learning and memory regions of brain namely hippocampus and amygdala that can restore permanent changes in the brain. It may also affect the biosynthesis of neurotransmitter viz. acetylcholine, which has been implicated in the process of learning and memory. On the other hand, *Clitoria ternatea* may also induce long-term potentiation (Rai et al., 2002) that not only promotes neurogenesis in the amygdala, but also stimulates the release of hormones or neuromodulators that modulate the activity of neurotransmitters and neuromodulators involved in learning and memory, thereby contributing to enhanced learning and memory. After detailed clinical trials in adults, aqueous root extract treatment of the plant may be envisaged as a memory enhancer and also might be useful in the treatment of neuronal degenerative disorders involving the amygdala (Rai et al., 2005). *Clitoria ternatea*, a traditional ayurvedic medicinal plant elevates itself as an efficient botanical medicine combating neurological ailments. It can be used as a lead drug for developing new phytoceuticals for the treatment of neuro-logical disorders and may be used effectively as a memory enhancer.

Though pharmacotherapy of modern medicine is well developed with chemical units it faces several challenges among which, adverse effects, lethal dosage levels and ability to surpass physiological barriers have been highlighted. Consequently, research features a novel facet of shifting from mono therapy (single active com-pound) to multi therapy that involves a collection of synchronized active compounds. Traditional herbs head this revolution in embracing phytochemicals which possess highly beneficial medicinal properties. Interestingly this novel attempt contests the western medicine and remain a challenge to the therapeutics. The advent of newer technologies like nano science can elevate herbal formulations as smart drugs that can fetch natural remedy against depression and mental illness.

THE CONDUCT OF NANOTECHNOLOGY IN REFORMING THE MENACE OF MENTAL ILLNESS TO WELL-BEING

Nano particulate system is well devised to facilitate drug delivery that increases the therapeutic index of the drug. Nanoparticles are nanoscale sized polymeric particles which are made up of natural or artificial polymers that range in size between 10 and 1000nm (1 mm). Polymeric Nanoparticles have been shown to be promising carriers for central nervous system drug delivery due to their potential both in encapsulating drugs, and in delivering them across the blood-brain barrier without inflicting any damage to the barrier. The biodistribution of pharmaceuticals in a controlled manner and consequently protect them from excretion and metabolism. The capacity of the surfactants in the polymeric nanoparticles offer stability to the encased drugs and facilitates smart release and delivery. Studies have revealed ample number of mechanistic details pertaining to these particles such as site-targeted, controlled, and enhanced bioavailability of hydrophobic drugs (Panyam, et al., 2003).

Polymers Facilitating Bioavailability and Efficacy of Drugs

The efficacy of many drugs is often limited by their potential to reach the site of therapeutic action. In most cases only a small amount of administered dose reaches the target site, while the majority of the drug distributes throughout the rest of the body in accordance with its physicochemical and biological properties. Therefore, developing a drug delivery system that optimizes the pharmaceutical action of drug while reducing its toxic side effects *in vivo* is a challenging risk. One of the approaches is the use of colloidal drug carriers that can provide site specific or targeted drug delivery combined with optimal drug release profiles. There are several drug delivery systems among which three efficient organizations are listed below:

- **Liposomes:** They are spherically shaped vesicles composed of natural phospholipids. The properties of liposomes can be varied with composition, size, and surface charge of lipids and also the method used for preparation. Liposomes can carry both hydrophobic and hydrophillic molecules inside them. The surface of liposomes can also be modified by attaching polyethylene glycol (PEG) chain to the lipid bilayer (stealth liposomes) to enhance their circulation time in the blood. Drugs can be actively targeted by conjugating liposomes with antibodies or ligands.

Table 1. Criteria for ideal polymeric carriers for nanoparticles and delivery systems

Polymeric Carriers	Nano Particle Delivery System
Easy to synthesize and characterize	Simple and inexpensive to manufacture and scale –up
Biocompatible	No heat high shear forces or organic solvents involved in the preparation process
Biodegradable	Reproducible and stable
Inexpensive	Applicable to a broad category of drugs, small molecules, proteins and polynucleotides
Non immunogenic	Ability to lyophilize
Non toxic	Non toxic
Water soluble	Stability after administration

- **Dendrimers:** They are hyper branched structures that comprise of an inner core, a series of branches and outer surface with functional groups. Due to their nanometer size range, ease of preparation and functionalization they are attractive drug delivery systems. Due to the presence of internal cavities it is possible to encapsulate therapeutic agents in the inner core. The properties of dendrimers can be controlled by the functional groups on outer surface. Drug can either be encapsulated in the interior of the dendrimers or attached to surface functional groups.
- **Polymeric Micelles:** They consist of amphiphillic block copolymers, which can self-assemble to form micelles in aqueous solution. They have a narrow size distribution in the nanometer range and core-shell structure, in which hydrophobic segments are separated from the hydrophillic exterior (Table 1). Drugs can be partitioned in the hydrophobic core of micelles and the outer hydrophillic layer forms a stable dispersion in aqueous media.

Like liposome they can also be functionalized with PEG for stealth properties and with targeting ligands including antibodies to the micelle surface (Labhasetwar, 1993). Among these carriers, liposomes and nanoparticles have been the most extensively investigated. Liposomes present some technological limitations including poor reproducibility and stability, low drug entrapment efficiency. Polymeric nanoparticles which possesses a better reproducibility and stability profiles than liposomes have been proposed as alternative drug carriers that overcome many of these problems.

Chitosan: A Dextrous Biopolymer

Biopolymers are polymers produced by living organisms; in other words, they are polymeric biomolecules. Since they are polymers, biopolymers contain monomeric units that are covalently bonded to form larger structures. Chitosan is a natural polymer obtained by deacetylation of chitin. After cellulose chitin is the second most abundant polysaccharide in nature. It is biologically safe, non-toxic, biocompatible and biodegradable polysaccharide. Chitosan nanoparticles have gained more attention as drug delivery carriers because of their better stability, low toxicity, simple and mild preparation method and providing versatile routes of administration. Their sub-micron size is also suitable for mucosal routes of administration i.e. oral, nasal and ocular mucosa which is non-invasive route. Chitosan is a linear polysaccharide, composed of glucosamine and N-acetylglucosamine (Figure 6), produced by partial deacetylation of chitin by alkaline or enzymatic hydrolysis. Chitin is the major component of the exoskeleton of crustaceans, insects, cell wall of fungi and yeast.

The most remarkable difference between chitin and chitosan is their solubility characteristics. There are few solvents for chitin and most of them are toxic and therefore inappropriate for pharmaceutical applications. Chitosan is insoluble at neutral and alkaline pH, but forms positively charged salts with inorganic and organic acids (e.g., glutaminic acid, hyaluronic acid, lactic acid, and acetic acid), whereas the positive charge of the protonated amine group promotes solubility in aqueous solutions. Chitosan is available in a wide range of different molecular weights (50 kDa-2000 kDa) and degrees of deacetylation (40%-98%). The most

Figure 6. The chemical structures of Chitin and Chitosan

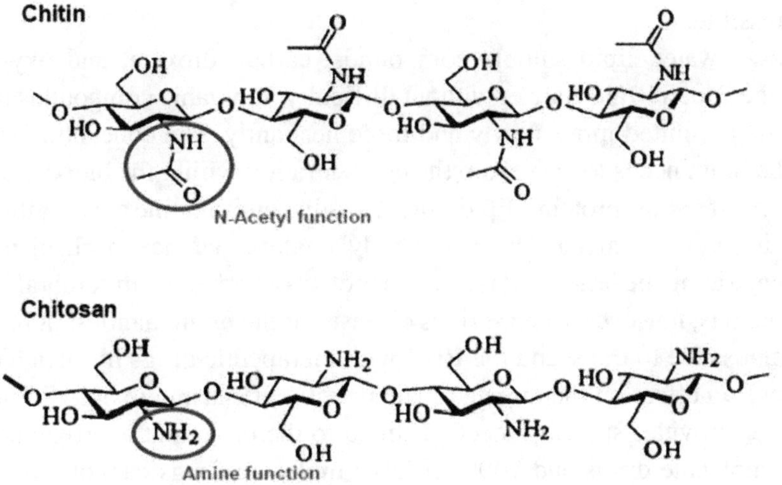

commonly used forms of chitosan salts are chitosan glutamate and chitosan hydrochloride. Chitosan features different conformations of the molecule for high and low deacetylated chitosan. At a high degree of deacetylation (highly charged) chitosan has extended conformation with a more flexible chain, whereas at low degree of deacetylation the molecule adopts a more coiled shape due to lower charge. The amino group in chitosan has a pKa value of ~6.5, thus, chitosan is positively charged and soluble in acidic to neutral solutions. Chitosan with a relative low degree of deacetylation (40%) are soluble only at pH < 6.5 (Illum 2001).

Among water-soluble polymers available, chitosan is one of the most extensively studied. This is because chitosan possesses some ideal properties of polymeric carriers for nanoparticles such as biocompatible, biodegradable, nontoxic, and inexpensive. Furthermore, it possesses positively charge and exhibits absorption enhancing effect. These properties render chitosan a very attractive material as a drug delivery carrier. In the last two decades, chitosan nanoparticles (chitosan NP) have been extensively developed and explored for pharmaceutical applications.

Physiological Barriers Hindering Therapeutic Molecules to the Targeted Sites and the Polymers Facilitating the Brain Drug

The Blood Brain Barrier (BBB) is the protective barrier that separates the brain from the circulatory system, which blocks harmful compounds from entering the brain tissue. It is a capillary network of tightly woven endothelial cells lining the blood vessels in the brain, regulating transport of essential molecules and maintaining a stable environment. It is a highly sophisticated structural design of the human body which determines what fluids are permitted in and near the brain and which would prove harmful. Capillaries (Figure 7) which surround the connective tissue and neuroglia known as astrocytes "stand guard" and select the substances which are permissible.

Glucose, water, lipid soluble compounds, carbon dioxide, and oxygen pass through the Blood Brain Barrier without difficulty. Inorganic compounds may pass but will be permitted more slowly and more hesitantly. The concentration of ions within the brain needs to vary from the concentration within the bloodstream. Antibiotics, most toxins, proteins, lipids, urea, insulin, and creatine are prohibited from passing through the barrier. This is the body's natural defence mechanism against toxins penetrating the brain, and is how we protect ourselves from cerebral invasion. The problem is, however, when serious diseases of the brain manifest, it has proven nearly impossible to transcend the BBB with therapeutic drugs in sufficient quantity to have an effect on the targeted malady without causing toxicity in the brain. The barrier provides such a protective shield to the brain that approximately 98% of small molecule drugs and 100% of large molecule drugs cannot cross it. This

Figure 7. Morphological representation of the blood brain barrier

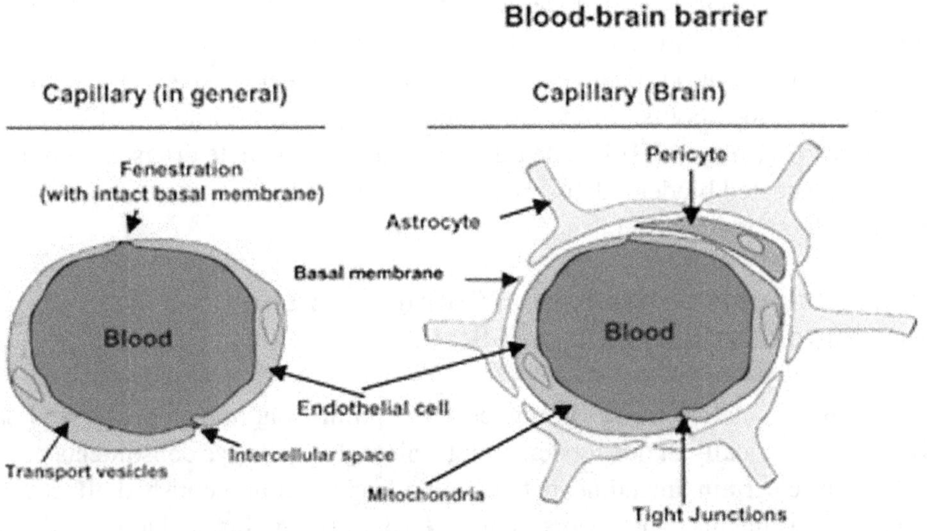

very tight regulation of the brain homeostasis results in the inability of some small and large therapeutic compounds to cross the blood–brain barrier (BBB). Therefore, various strategies are being developed to enhance the amount

Polymeric Nanoparticles as Drug Delivery Vehicles Encasing Brain Drug

Biodegradable polymers such as polylactic acid, polycaprolactone, poly(lactic-co-glycolic acid), the poly(fumaric-cosebacic) anhydride chitosan, and modified chitosan, as well as solid lipids, have revealed immense prospective in the delivery of proteins and peptidal drugs in nanao scale. Theses polymers channelize the drug *in vivo* is reported to mimic molecules that are easily surpassed through brain (example) polysorbate-coated nanoparticles are thought to mimic low density lipoprotein (LDL), allowing them to be transported across the capillary wall and into the brain by loading onthe LDL receptor (Kreuter, 2014) This mechanism adapted by the polymer improve the efficacy of the drug and allows it to pass through the Blood brain barrier. (Liu et al., 2012). It offers a sustained drug release at the targeted site after injection over a period of several days and weeks (. Nanoparticle based delivery methods are proved to be the best methods to transfer drugs across CNS (Jain, 2012). These strategies require the multifaceted attributes of nanoparticles which are combined and controlled to pass across the BBB. These are proved to be the best methods to facilitate the delivery of drugs and biological therapeutics for brain across the BBB. HAS (human serum albumin) is an illustrious example of a polymer that is used as nanoparticle and resists to the system by exhibiting no serious side effect. More exceptionally albumin functional groups can be utilized for surface modification of barrier that allows specific cell uptake (Vinogradov et al., 2002). It also acts like as a transforming growth factor in microbubble based drug delivery. (Kreuter, 2014). The efficacy of nanoparticulate drugs is, improved by these are coated biodegradable materials which make them more permeable to cross the blood brain barrier.

Smart Nanoparticulate Herbal Formulations to Contribute Mental Health and Stability

The herbal formulation throws in immense of optimism in human health, and has advanced periodically in this modern era. Deplorably the active constituents of the plants require certain amendments that would lead to an unblemished effect. The vital phyoconstituents constituents such as flavonoids, tannins, and terpenoids, are highly soluble in water, but have low absorption, due to which they are unable to cross the lipid membranes of the cells. An additional snag pertaining to the phytocompounds that limits its usage is the intense high molecular size, which poorly absorbed, resulting in loss of bioavailability and efficacy. Nanotechnology provides solution to these deficits and hence has been widely proposed to synchronize with herbal medicine. Some prospective features of nanostructures in elevating herbal medicine are:

- Reduced dosage,
- Lessen adverse effects,
- Increase bioavailability and absorption,
- Control drug release and metabolism,
- Increased endurance of the drug.

Size is a blessing in nanotechnology where the therapeutic agent is distributed into the miniature blood capillaries (smaller than 5 mm) without forming aggregates, and ensure that the particles are not agglomerated. Polymers as nanocarriers provide more surface area and have the potential to increase solubility, enhance bioavailability, improve controlled release and enable precision targeting of the entrapped compounds to a greater extent. As a consequence, the expenditure intensity decreases widely, since isolated phytocompounds are expensive. Further, polymers are efficient for both hydrophilic and hydrophobic herbal drugs and can be channelized through the lymphatic system, blood brain barrier and arterial walls. Several clinical trials and experimentation has been performed by encasing herbal formulation into polymers to combat human infirmity. A wide number of plants are being used to as nanoformulations using various suspension modes against human infirmity Table 2 shows the details of the nano system incorporated.

Table 2. List of potential Nano formulated medicinal plants targeting several human ailments

Name and Family of the Medicinal Plant	Nano System Employed	Purpose/Application	Reference
Panax ginseng(Araliaceae)	lipid-based systems	increase the absorption of the active components	Kesarwani and Gupta 2013
Artemisia arborescens L. (Asteraceae)	Liposomes	Facillitate the active compounds to penetrate the cytoplasmic viral barrier	Sinico C et al., 2005
Phytolacca decandra(Phytolaccaceae)	PLGA-encapsulate	lung cancer	Das et al 2012
Ocimum sanctum (Lamiaceae)	Sodium alginate chitosan nanoparticles(Cotton fabrics)	antimicrobial activity	Rajendran et al., 2013
Curcuma longa (Zingiberaceae)	copolymers of N-isopropylacrylamide, N-vinyl-2-pyrrolidone, and poly(ethylene glycol) monoacrylate.	pancreatic Cancer	Bisht et al 2007
Curcuma longa (Zingiberaceae)	PLGA	Prostate cancer	Mukerjee and Vishwanatha 2009
Magnolia officinalis (Magnoliaceae)	PEG/PLGA	anxiolytic, central nervous system depressant, and muscle relaxant activities	Zheng et al.,2010
Cuscuta chinensis Lam. (Convolvulaceae) nanoparticles	Eudragit (E) and PVA	Oxidation stress AND hepatoprotective effects	Yen et al., 2008
Camptotheca acuminata Decne (Cornaceae)	hydrophobically modified glycol chitosan (HGC)	Breast Cancer	Min et al., 2008
Ginkgo biloba (Ginkgoaceae)	polyvinyl alcohol (PVA)	anti-inflammatory, antitumor, antiviral, inhibitory of cataracts in diabetics, antihistamine (anti-allergic), cardiovascular, antioxidant, and hepatoprotective effects	Wu et al., 2008

Various research works has substantiated smart herbal formulations as nano drugs extensively. The proportion of research related to neuro drugs is comparatively low. Yet, few studies establish the proficiency of herbal drugs as nano formulations. Ginkgo biloba L. (Ginkgoaceae) is prominent medicinal plant that has been widely marketed for its brain cell activation properties

techniques) It was demonstrated (Shinji et al., 2011, Williams et al., 2004) that extract of G. biloba containing nanoparticles increase the release of acetylcholine from cerebral cortical synapses there by stimulating the response of hippocampal pyramidal cell. Thus, the nanosized G. biloba extract is expected to activate the brain cell and restore the habitual symptoms of Alzheimer's dementia (like loss of memory, thinking, language, judgement and behavior).

Several conditions such as oxidative stress, mitochondrial dysfunction, infections, hypoxia, apoptosis, and others, play important roles in the development and prevalence of these diseases (Cacciatore et al., 2012). It has been reported that oxidative stress (OS) is implicated in the development of neurodegeneration, mainly by the presence and attack of reactive oxygen species (ROS) that oxidize and injury neural cell components, such as lipids, proteins and DNA, leading to the appearance and prevalence of Alzheimer´s or Parkinson´s diseases, aging and many others neural disorders (Uttara et al., 2009). In this way, flavonoids such as catechin, quercetin, rutin, have demonstrated that act as excellent free radical scavengers by rapid donation of a hydrogen atom to radicals (Seyoum et al., 2006). The work of Pool et al. (2012) was based in the development of polymeric nanoparticles (NPs) using Eudragit L30-D55 (Eud) polymers for the encapsulation, protection and target release within gastrointestinal tract of quercetin. In this study, the encapsulation of quercetin into Eud enhanced the protection of multi-lamellar vesicles of soybean phosphatidylcholine against lipid peroxidation of peroxyl and hydroxyl radicals. These results suggested that Eud NPs increased the lipid phase solubility of quercetin and also NPs act as protective agents against molecular oxygen, leading to a better free radical action that the free quercetin molecules used in the same experiments. Thus, these type or nanoparticulated systems might be suitable to across the BBB and deliver antioxidants to combat the OS in brain.

Studies have demonstrated that curcumin exert a net protective effect against oxidative damages initiated by free radicals, Despite, curcumin is a highly lipophilic compound and have the ability to cross the BBB, the poor bioavailability represent a limiting to its use in a functional product to combat neurological diseases. Accordingly, NPs decorated with appropriate ligands for curcuin brain delivery were developed lately (Mulik et al., 2010). These approaches were based in the preparation of curcumin-loaded PnBCA NPs decorated with ApoE3 ligand to exploit LDL-r- mediated transcytosis across the BBB and through SH-SY5Y cells. Results showed a reduction of Aβ1-42 related toxicity on cells treated with the functional-

Figure 8. Microscopic examination of Alginate: Curcumin nanocomposites

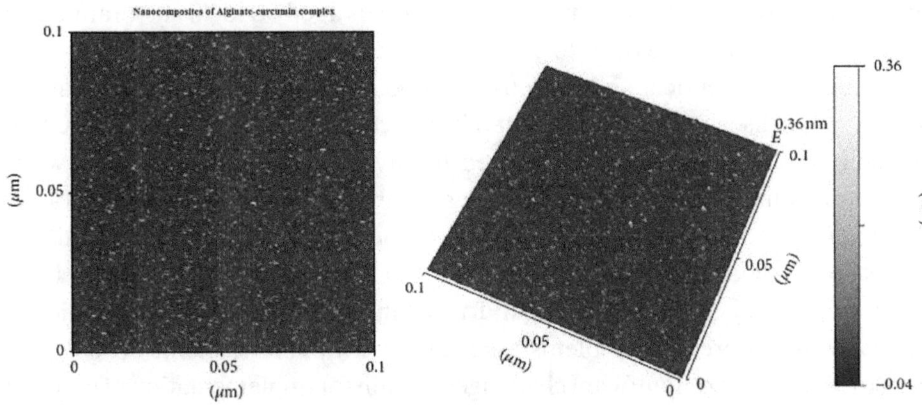

ized nanospheres along with a reduction of reactive oxygen species formation than the free curcumin used in this study. Another work demonstrated that curcumin-loaded PLGA nanoparticles obtained by an emulsion-diffusion evaporation method increase the neuroprotective effect on human SK-N-SH cells (neuroblastoma cell line) against the OS effect by reducing the elevation of ROS and the consumption of gluthatione induced by H_2O_2. Also the uptake of the encapsulated curcumin into SK-N-SH cells was significantly improved than the free curcumin, which impact on the effects found them by researchers. Overall, this study demonstrated that the PLGA NPs have the capacity to protect human neuronal cells against oxidative damages, having a great potential to be used as bioactive compounds carriers in the application to treat several neurodegenerative disorders (Doggui et al., 2012). Further alginate-curcumin nanocomposites (Yasir Hasan Siddique et al., 2013) reduced the oxidative stress as well as apoptosis in the brain of Parkinsons Disease model drosophila flies (Figure 8).

FUTURE RESEARCH DIRECTIONS

Neurodegenerative diseases a are characterized by a progressive and irreversible damage in the brain, which are the fourth leading cause of death in the developed world that precedes heart diseases, cancer, and stroke (Plassman et al., 2007). The most common neurodegenerative diseases are Alzheimer's disease, Parkinson's disease, Lewy body dementia, fronto temporal dementia, amyotrophic lateral sclerosis, Huntington's disease, and prion diseases (Bertram y Tanzi, 2005). However, the most widely recognized are Alzheimer's disease and Parkinson's disease, which are among the principal debilitating conditions of the current century. Approximately

24 million people worldwide suffer from dementia, 60% of cases being due to Alzheimer's disease, which occurs in 1% of individuals aged 50 to 70 and dramatically increases to 50% for those over 70 years (Cacciatore et al., 2012).

Several pharmaceutical and bioactive compounds have been used to treat neurodegenerative diseases, however, most of these components do not reach the brain completely and are, instead, metabolized totally or partially by the liver with a difficulty in solubility. Therefore, there is a necessity for the development of new technologies that help to protect and maintain the integrity of the drugs until they are released at specific site of action in the brain (Nazem Mansoori, 2011). Nano-sized drug delivery systems for herbal drugs can potentially enhance the biological activity and overcome problems associated with plant medicines (Pool et al., 2012;2013). However, significant challenges remain for implementation of clinically viable therapies in this field. Employing herbal drugs as nano formulations would aggravate the importance of ancient Ayurvedic medicine. The effectiveness of these potent phytocompounds as a synergetic effect with polymers can be further endorsed by performing clinical studies which would elevate the incredible competence of ayurvedic medicine on par with western medicine. The vital compounds present in the plant extract are further analysed towards the lysosomotrophic chronic effects along with the standard synthetic drugs by *in vitro* study using cell culture models. The regulation studies of the (neurotransmitters through receptors) β-Adrenoceptors and G protein would endorse the efficacy of the herbal drugs. Receptors stimulate the protein AC (Adenylate Cyclase) which there by induces CAMP (Secondary messenger) signalling the neurotransmitters which can authenticate the efficacy of drug released by the polymeric carriers. Experimental modes such as *in silico, in vitro* and *in vivo* studies with Cell lines and animal models can validate the therapeutic index of the herbal formulation. The cytotoxicity assay is the fore most parameter of in the process of drug development and in order to understand the *in vivo* mechanisms of ayurvedic nano formulations there is a need to propose good model system. This would propagate the therapeutic value of neurological diseases and disorders.

CONCLUSION

Herbal medicines are asserting importance in the current global scenario as the severity of pathogenesis is ascending. The advancement of modern science is elucidating the capacity several medicinal plant products. In order to compete with the rate of diseases and disorders there has to be an extensive rise in the therapeutic index. Multiple and synchronized therapy of combining the active ingredients of herbal extract would be an ideal stoppage solution. Polymeric nanocomposites augment the efficacy of these herbal formulations thereby causing less side effects. Nanosystems can deliver the active constituent at a sufficient concentration during the entire treatment period, directing it to the desired site of action which is failed by the conventional treatments. Brain drug delivery is a challenge as the rate of mental disorders are increasing hence there is stern constraint to exert more research to develop versatile drugs. An efficient herbal remedy without adverse effects is the need of the hour and smart herbal formulations can do this deed in a immaculately. This would rejuvenate the ancient riches present in the ayurvedic treatise.

REFERENCES

Ali, M. E., & Lamprecht, A. (2014). Spray freeze drying for dry powder inhalation of nanoparticles. *European Journal of Pharmaceutics and Biopharmaceutics*, *87*(3), 510–517. doi:10.1016/j.ejpb.2014.03.009 PMID:24657824

Alshamsan, A. (2014). Nanoprecipitation is more efficient than emulsion solvent evaporation method to encapsulate cucurbitacin I in PLGA nanoparticles. *Saudi Pharmaceutical Journal*, *22*(3), 2, 19–222. doi:10.1016/j.jsps.2013.12.002 PMID:25061407

Anita Margret, A., Nargis Begum, T., Parthasarathy, S., & Suvaithenamudhan, S. (2015). A Strategy to Employ *Clitoria ternatea* as a Prospective Brain Drug Confronting Monoamine Oxidase (MAO) Against Neurodegenerative Diseases and Depression. *Natural Products and Bioprospecting*, *5*(6), 293–306. doi:10.1007/s13659-015-0079-x PMID:26667936

Arora, S., Gupta, S., Narang, R. K., & Budhiraja, R. D. (2011). Amoxicillin loaded chitosan-alginate polyelectrolyte complex nanoparticles as mucopenetrating delivery system for *H. pylori*. *Scientia Pharmaceutica*, *79*(3), 673–694. doi:10.3797/scipharm.1011-05 PMID:21886911

Banerjee, S. K., & Chakravarti, R. N. (1963). Taraxerol from *Clitoria ternatea*. *Bulletin of the Calcutta School of Tropical Medicine*, *11*(1), 106–107. PMID:14068655

Bhattacharya, S. K., Goel, R. K., Kaur, R., & Ghosal, S. (1987). Anti - stress activity of Sitoindosides VII and VIII. New Acylsterylglucosides from *Withania somnifera*. *Phytotherapy Research*, *6*(1), 32–37. doi:10.1002/ptr.2650010108

Biber, A., Fischer, H., Romer, A., & Chatterjee, S. S. (1998). Oral bioavailability of hyperforin from *Hypericum* extracts in rats and human volunteers. *Pharmacopsychiatry*, *31*(1), 36–43. doi:10.1055/s-2007-979344 PMID:9684946

Bisht, S., Feldmann, G., & Soni, S. (2007). Polymeric nanoparticle-encapsulated curcumin ("nanocurcumin"): A novel strategy for human cancer therapy. *Journal of Nanobiotechnology*, *5*(3), 1–18. PMID:17439648

Brigger, I., Dubernet, C., & Couvreur, P. (2002). Nanoparticles in Cancer Therapy and Diagnosis. *Advanced Drug Delivery Reviews*, *5*(4), 631–651. doi:10.1016/S0169-409X(02)00044-3 PMID:12204596

Cacciatore, I., Baldassarre, L., Fornasari, E., Mollica, A Pinnen F. (2012). Recent Advances in the Treatment of Neurodegenerative Diseases Based on GSH Delivery Systems. *Oxidative Medicine Cell Longevity*. (2), 140-146.

Chu, K. S., Hasan, W., & Rawal, S. (2013). Plasma, tumor and tissue pharmacokinetics of Docetaxel delivered via nanoparticles of different sizes and shapes in mice bearing SKOV-3 human ovarian carcinoma xenograft. *Nanomedicine (London)*, *9*(5), 686–693. PMID:23219874

Cipriani, A., Furukawa, T. A., Salanti, G., Geddes, J. R., Higgins, J. P., Churchill, R., & Barbui, C. et al. (2009). Comparative efficacy and acceptability of 12 new-generation antidepressants: A multiple-treatments meta-analysis. *Lancet*, *373*(9665), 746–758. doi:10.1016/S0140-6736(09)60046-5 PMID:19185342

Das, J., Das, S., Samadder, A., Bhadra, K., & Khuda-Bukhsh, A. R. (2012). Poly (lactide-co-glycolide) encapsuled extract of Phytolacca decandra demonstrates better intervention against induced lung adenocarcinoma in mice and on A549 cells. *European Journal of Pharmaceutical Sciences*, *7*(2), 313–324. doi:10.1016/j.ejps.2012.06.018 PMID:22771545

Deepak, M., & Amit, A. (2004). The need for establishing identities of bacoside A and B, the putative major bioactive saponins of Indian medicinal plant *Bacopa monnieri*. *Phytomedicine*, *11*(2-3), 264–268. doi:10.1078/0944-7113-00351 PMID:15070183

Doggui, S., Sahni, J. K., Arseneault, M., Dao, L., & Ramassamy, C. (2012). Neuronal uptake and neuroprotective effect of curcumin-loaded PLGA nanoparticles on the human SK-N-SH cell line. *Journal of Alzheimer's Disease*, *30*(2), 377–392. PMID:22426019

Dong, Y., Ng, W. K., Shen, S., Kim, S., & Tan, R. B. (2013). Scalable ionic gelation synthesis of chitosan nanoparticles for drug delivery in static mixers. *Carbohydrate Polymers*, *94*(2), 940–945. doi:10.1016/j.carbpol.2013.02.013 PMID:23544653

Gaillard, P. J. (2010). Crossing barriers from blood-to-brain and academia-to-industry. *Therapeutic Delivery*, *1*(4), 495–500. doi:10.4155/tde.10.43 PMID:22833962

Galloway, A. L., Murphy, A., & De Simone, J. M. (2013). Development of a nanoparticle-based influenza vaccine using the PRINT technology. *Nanomedicine (London)*, *9*(4), 523–531. PMID:23178283

Gillman, P. K. (2007). Tricyclic antidepressant pharmacology and therapeutic drug interactions updated. *British Journal of Pharmacology*, *151*(6), 737–748. doi:10.1038/sj.bjp.0707253 PMID:17471183

Grady, M. M., & Stahl, S. M. (2012). Practical guide for prescribing MAOIs: Debunking myths and removing barriers. *CNS Spectrums*, *17*(1), 2–10. doi:10.1017/S109285291200003X PMID:22790112

Hardman, J. G., Limbird, L. E., & Goodman Gilman, A. (2007). *Goodman Gilman's; The Pharmacological Basis of Therapeutics* (11th ed.). New York: The McGraw Hill Companies, Inc.

Hu, X., Guo, Y., Wang, L., Hua, D., Hong, Y., & Li, J. (2011). Coenzyme Q10 nanoparticles prepared by a supercritical fluid-based method. *The Journal of Supercritical Fluids*, *57*(1), 66–72. doi:10.1016/j.supflu.2011.01.007

Ibrahim, E., Mustafa, N., & Arif, D. (2007). Venlafaxine modulates depression–induced oxidative stress in brain and medulla of rats. *Neurochemical Research*, *32*(3), 497–505. doi:10.1007/s11064-006-9258-9 PMID:17268845

Illum, L., Jabbal-Gill, I., Hinchcliffe, M., Fisher, A. N., & Davis, S. S. (2001). Chitosan as a novel nasal delivery system for vaccines. *Advanced Drug Delivery Reviews*, *51*(1), 81–96. doi:10.1016/S0169-409X(01)00171-5 PMID:11516781

Jain, K. K. (2012). Nanobiotechnology-based strategies for crossing the blood-brain barrier. *Nanomedicine (London)*, *7*(8), 1225–1233. doi:10.2217/nnm.12.86 PMID:22931448

Julien, R. M. (2004). *A Primer Of Drug Action: A Comprehensive Guide To The Actions, Uses, And Side Effects of Psychoactive Drugs*. London: Macmillan.

Kesarwani, K., & Gupta, R. (2013). Bioavailability enhancers of herbal origin: An overview. *Asian Pacific Journal of Tropical Biomedicine*, *3*(4), 253–266. doi:10.1016/S2221-1691(13)60060-X PMID:23620848

Kreuter, J. (2014). Drug delivery to the central nervous system by polymeric nanoparticles: What do we know? *Advanced Drug Delivery Reviews*, *71*, 2–14. doi:10.1016/j.addr.2013.08.008 PMID:23981489

Labhasetwar, V., Song, C., Humphrey, W., Shebuski, R., & Levy, R. J. (1998). Arterial uptake of biodegradable nanoparticles: Effect of surface modifications. *Journal of Pharmaceutical Sciences*, *87*(10), 1229–1234. doi:10.1021/js980021f PMID:9758682

Liu, Y., Tan, J., Thomas, A., Ou-Yang, D., & Muzykantov, V. R. (2012). The shape of things to come: Importance of design in nanotechnology for drug delivery. *Therapeutic Delivery*, *3*(2), 181–194. doi:10.4155/tde.11.156 PMID:22834196

Mainardes, R. M., & Gremião, M. P. D. (2012). Nanoencapsulation and characterization of zidovudine on poly(L-lactide) and poly(L-lactide)-poly(ethylene glycol)-blend nanoparticles. *Journal of Nanoscience and Nanotechnology*, *12*(11), 8513–8521. doi:10.1166/jnn.2012.6638 PMID:23421238

Mazzarino, L., Coche-Guérente, L., Labbé, P., Lemos-Senna, E., & Borsali, R. (2014). On the mucoadhesive properties of chitosan-coated polycaprolactone nanoparticles loaded with curcumin using quartz crystal microbalance with dissipation monitoring. *Journal of Biomedical Nanotechnology*, *10*(5), 787–794. doi:10.1166/jbn.2014.1768 PMID:24734531

Medina, C., Santos, M. J., Radomski, A., Corrigan, O., & Radomski, M. W. (2007). Nanoparticles: Pharmacological and Toxicological Significance. *British Journal of Pharmacology*, *150*(5), 552–558. doi:10.1038/sj.bjp.0707130 PMID:17245366

Min, K. H., Park, K., Kim, Y. S., Bae, S. M., Lee, S., Jo, H. G., & Kim, K. et al. (2008, May). Kwon, I.C., Hydrophobically modified glycol chitosan nanoparticles-encapsulated camptothecin enhance the drug stability and tumor targeting in cancer therapy. *Journal of Controlled Release*, *127*(3), 208–218. doi:10.1016/j.jconrel.2008.01.013 PMID:18336946

Mishra, L. C., Singh, B. B., & Dagenais, S. (2000). Scientific basis for the therapeutic use of Withania somnifera. (Ashwagandha): A review. *Alternative Medicine Review*, *5*, 334–346. PMID:10956379

Mukerjee, A., & Vishwanatha, J. K. (2009). Formulation, characterization and evaluation of Curcumin-loaded PLGA nanospheres for cancer therapy. *Anticancer Research*, *29*(10), 3867–3876. PMID:19846921

Mukherjee, P. K., Kumar, V., Mal, M., & Houghton, P. J. (2007). Acetyl cholinesterase inhibitors from plants. *Phytomedicine*, *14*(4), 289–300. doi:10.1016/j.phymed.2007.02.002 PMID:17346955

Mulik, R. S., Monkkonen, J., Juvonen, R. O., Mahadik, K. R., & Paradkar, A. R. (2010). ApoE3 mediated poly(butyl) cyanoacrylate nanoparticles containing curcumin: Study of enhanced activity of curcumin against beta amyloid induced cytotoxicity using *in vitro* cell culture model. *Molecular Pharmaceutics*, *7*(3), 815–825. doi:10.1021/mp900306x PMID:20230014

Murray, C. J., & Lopez, A. D. (1996). Evidence-based health policy--lessons from the Global Burden of Disease Study. *Science*, *274*(5288), 740–743. doi:10.1126/science.274.5288.740 PMID:8966556

Murray, C. J. L., & Lopez, A. D. (1996). *The Global Burden of Disease: A Comprehensive Assessment of Mortality and Disability from Diseases, Injuries, and Risk Factors in 1990 and Projected to 2020.* Cambridge: MA Harvard University Press.

Nahata, A., Patil, U. K., & Dixit, V. K. (2008). Effect of *Convolvulus pluricaulis* Choisy. on learning behaviour and memory enhancing activity in rodents. *Natural Product Research*, *22*(16), 1472–1482. doi:10.1080/14786410802214199 PMID:19023811

Nahata, A., Patil, U. K., & Dixit, V. K. (2010). Effect of *Evolvulus alsinoides* Linn. on learning behaviour and memory enhancing activity in rodents. *Phytotherapy Research*, *24*, 486–493. PMID:19610035

National Institute for Health and Clinical Excellence (NICE). *The treatment and management of depression in adults* (updated edition - 2010). Retrieved from http://www.nice.org.uk/nicemedia/live/12329/45896/45896.pdf

Nazem, A., & Mansoori, G. A. (2011). Nanotechnology for Alzheimer's disease detection and treatment. *Insciences Journal*, *14*(4), 169–193. doi:10.5640/insc.0104169

Panyam, J., & Labhasetwar, V. (2003). Biodegradable Nanoparticle from Drug and Gene Delivery to Cells and Tissue. *Advanced Drug Delivery Reviews*, *55*(3), 329–347. doi:10.1016/S0169-409X(02)00228-4 PMID:12628320

Perinchery. (n. d.). A History of Indian healing traditions: A science and society initiative. Retrieved from http://news.ncbs.res.in/story/history-indian-healing-traditions

Pool, H., Quintanar, D., Figueroa, J. D., Bechara, J. E. H., Mc Clements, D. J., & Mendoza, S. (2012). Polymeric Nanoparticles as Oral Delivery Systems for Encapsulation and Release of Polyphenolic Compounds: Impact on Quercetin Antioxidant Activity & Bioaccessibility. *Food Biophysics*, *7*, 276–288. doi:10.1007/s11483-012-9266-z

Pravina, K., Ravindra, K. R., Goudar, K. S., Vinod, D. R., Joshua, A. J., Wasim, P., & Amit, A. et al. (2007). Safety evaluation of BacoMind in healthy volunteers: A phase I study. *Phytomedicine*, *14*(5), 301–308. doi:10.1016/j.phymed.2007.03.010 PMID:17442556

Rai, K. S., Murthy, K. D., Karanth, K. S., Nalini, K., Rao, M. S., & Srinivasan, K. K. (2002). *Clitoria ternatea* root extract enhances acetylcholine content in rat hippocampus. *Fitoterapia*, *73*(7-8), 685–689. doi:10.1016/S0367-326X(02)00249-6 PMID:12490229

Rai, K. S., Murthy, K. D., Rao, M. S., & Karanth, K. S. (2005). Altered dendritic arborisation of amygdala neurons in young adult rats orally intubated with *Clitoria ternatea* aqueous root extract. *Phytotherapy Research*, *19*(7), 592–598. doi:10.1002/ptr.1657 PMID:16161034

Rajendran, R., Radhai, R., Kotresh, T. M., & Csiszar, E. (2013). Development of antimicrobial cotton fabrics using herb loaded nanoparticles. *Carbohydrate Polymers*, *91*(2), 613–617. doi:10.1016/j.carbpol.2012.08.064 PMID:23121954

Ramaswami, V., Varghese, N., & Simon, A. (2011). An investigation on cytotoxic and antioxidant properties of *Clitoria ternatea* L. *International Journal of Drug Discovery.*, *3*(1), 74–77. doi:10.9735/0975-4423.3.1.74-77

Richards, D. A. (2012). Stepped care: A method to deliver increased access to psychological therapies. *Canadian Journal of Psychiatry*, *57*(4), 210–215. PMID:22480585

Richelson, E. (2010). Pharmacology of antidepressants. *Mayo Clinic Proceedings*, *76*(5), 511–527. doi:10.4065/76.5.511 PMID:11357798

Ringsdorf, H. (1975). Structure and properties of pharmacologically active polymers. *Journal* of *Polymer Science. Polymer Symposia*, *51*(1), 135–153. doi:10.1002/polc.5070510111

Royal Pharmaceutical Society of Great Britain. (2010). British National Formulary (59th ed.). BMJ Group, London.

Sarandol, A., Sarandol, E., Eker, S. S., Erdinc, S., Vatansever, E., & Kirli, S. (2007). Major depressive disorder is accompanied with oxidative stress: Short-term antidepressant treatment does not alter oxidative–antioxidative systems. *Human Psychopharmacology Clinical and Experimental*, *22*(2), 67–73. doi:10.1002/hup.829 PMID:17299810

Sethiya, N. K., & Mishra, S. H. (2010). Review on ethanomedicinal uses and phyto-pharmacology of memory boosting herb Convolvulus pluricaulis Choisy. *The AustralianJournal of Herbal Medicine*, *22*(1), 19–25.

Sethiya, N. K., Patel, M. B., & Mishra, S. H. (2010). Phyto-pharmacologic aspects of *Canscora decussata* Roem and Schult. *Pharmacognosy Reviews*, *4*(7), 49–57. doi:10.4103/0973-7847.65326 PMID:22228941

Seyoum, A., Asres, K., & El-Fiky, F. K. (2006). Structure–radical scavenging activity relationships of flavonoids. *Phytochemistry*, *67*(18), 2058–2070. doi:10.1016/j.phytochem.2006.07.002 PMID:16919302

Sharma, R., Chaturvedi, C., & Tewari, P. V. (1987). Efficacy of *Bacopa monnieri* in revitalizing intellectual functions in children. *The Indian Journal of Medical Research*, *6*, 1–10.

Shinji, S., Yasukazu, T., Hatsue, W., Kazuo, K., Machiko, I., & Naoki, M. (2011). Analysis of brain cell activation by nano sized particles of Ginkgo biloba extract. *International Journal of Plant Physiology and Biochemistry*, *3*(3), 28–33.

Siddique, Y. H., Khan, W., Singh, B. R., & Naqvi, A. H. (2013). Synthesis of Alginate-Curcumin Nanocomposite and Its Protective Role in Transgenic *Drosophila* Model of Parkinson's Disease. *ISRN Pharmacology*, *2*, 1–8. doi:10.1155/2013/794582 PMID:24171120

Singh, R. H., & Udupa, K. N. (1993). *Clinical and experimental studies on rasayana drugs and rasayana therapy. Special Research Monograph, Central Council for Research in Ayurveda and Siddha (CCRAS)*. New Delhi: Ministry of Health and Family Welfare.

Sinico, C., De Logu, A., Lai, F., Valenti, D., Manconi, M., Loy, G., & Fadda, A. M. et al. (2005). Liposomal incorporation of Artemisia arborescens L. essential oil and in vitro antiviral activity. *European Journal of Pharmaceutics and Biopharmaceutics*, *59*(1), 161–168. doi:10.1016/j.ejpb.2004.06.005 PMID:15567314

Sivaranjan V.V. & Indira Balachandran. (1994). *Ayuruedic drugs and their plant sources*. Oxford and IBH Publishing Company, New Delhi, India.

Tripathi, K. D. (2008). Essentials of medical Pharmacology (6th ed.). New Delhi, India: Medical Publishers Ltd. doi:10.5005/jp/books/10282

Uttara, B., Singh, A. V., Zamboni, P., & Mahajan, R. (2009). Oxidative stress and neurodegenerative diseases: A review of upstream and downstream antioxidant therapeutic options. *Current Neuropharmacology*, *7*(1), 65–74. doi:10.2174/157015909787602823 PMID:19721819

Valiathan, S. (2006). Ayurveda – putting the house in order (Guest Editorial). *Current Science*, *90*, 5–6.

Vinogradov, S. V., Bronich, T. K., & Kabanov, A. V. (2002). Nanosized cationic hydrogels for drug delivery: Preparation, properties and interactions with cells. *Advanced Drug Delivery Reviews*, *54*(1), 135–147. doi:10.1016/S0169-409X(01)00245-9 PMID:11755709

WHO. (n. d.). Depression fact sheet. Retrieved from http://www.who.int/mediacentre/factsheets/fs369/en/index.html

Williams, B., Watanabe, C. M., Schultz, P. G., Rimbach, G., & Krucker, T. (2004). Age-related effects of *Ginkgo biloba* extract on synaptic plasticity and excitability. *Neurobiology of Aging*, *25*(7), 955–962. doi:10.1016/j.neurobiolaging.2003.10.008 PMID:15212849

Wu, T. Z., Yen, F. L., Lin, L. T., Tsai, T. R., Lin, C. C., & Cham, T. M. (2008). Preparation, physicochemical characterization, and antioxidant effects of quercetin nanoparticles. *International Journal of Pharmaceutics*, *346*(1–2), 160–168. doi:10.1016/j.ijpharm.2007.06.036 PMID:17689897

Yadava, R. N., & Verma, V. (2003). Antimicrobial activity of a novel flavonol glycosideisolated from the roots of *Clitoria ternatea* Linn. *Asian Journal of Chemistry*, *15*, 842–846.

Yen, F. L., Wu, T. Z., Lin, L. T., Cham, T. M., & Lin, C. C. (2008). Nanoparticles formulation of *Cuscuta chinensis* prevents acetaminophen-induced hepatotoxicity in rats. *Food and Chemical Toxicology*, *46*(5), 1771–1777. doi:10.1016/j.fct.2008.01.021 PMID:18308443

Zheng, X., Kan, B., Gou, M., Fu, S. Z., Zhang, J., Men, K., & Zhao, X. (2010). Preparation of MPEG-PLA nanoparticle for honokiol delivery in vitro. *International Journal of Pharmaceutics*, *386*(1–2), 262–267. doi:10.1016/j.ijpharm.2009.11.014 PMID:19932160

Compilation of References

Acosta-Torres, L. S., López-Marín, L. M., Núñez-Anita, R. E., Hernández-Padrón, G., & Castaño, V. M. (2011). Biocompatible Metal-Oxide Nanoparticles: Nanotechnology Improvement of Conventional Prosthetic Acrylic Resins. *Journal of Nanomaterials*, *2011*, 8. doi:10.1155/2011/941561

Acosta-Torres, L. S., Mendieta, I., Nunez-Anita, R. E., Cajero-Juarez, M., & Castano, V. M. (2012). Cytocompatible antifungal acrylic resin containing silver nanoparticles for dentures. *International Journal of Nanomedicine*, *7*, 4777–4786. PMID:22969297

Adams, L. P., Jooste, C. H., Thomas, C. J., & Harris, A. M. (1996). Biostereometric quantification of clinical denture tooth wear. *Journal of Oral Rehabilitation*, *23*(10), 667–674. doi:10.1046/j.1365-2842.1996.d01-178.x PMID:8933383

Afergan, E., Epstein, H., Dahan, R., Koroukhov, N., Rohekar, K., Danenberg, H. D., & Golomb, G. (2008). Delivery of serotonin to the brain by monocytes following phagocytosis of liposomes. *Journal of Controlled Release*, *132*(2), 84–90. doi:10.1016/j.jconrel.2008.08.017 PMID:18805446

Afrasiabi, E., Hietamäki, M., Viitanen, T., Sukumaran, P., Bergelin, N., & Törnquist, K. (2010). Expression and significance of HERG (KCNH2) potassium channels in the regulation of MDAMB-435S melanoma cell proliferation and migration. *Cellular Signalling*, *22*(1), 57–64. doi:10.1016/j.cellsig.2009.09.010 PMID:19765650

Agerer, F., Michel, A., Ohlsen, K., & Hauck, C. R. (2003). Integrin-mediated Invasion of *Staphylococcus aureus* into Human Cells Requires Src Family Protein-tyrosine Kinases. *The Journal of Biological Chemistry*, *278*(43), 42524–42531. doi:10.1074/jbc.M302096200 PMID:12893831

Ahn, J.-H., Choi, S.-J., Han, H. W., Park, T. J., Lee, S. Y., & Choi, Y.-K. (2011). Investigation of size dependence on sensitivity for nanowire FET biosensors. *IEEE Transactions on Nanotechnology*, *6*, 1405–1411.

Ai, H., Flask, C., Weinberg, B., Shuai, X., Pagel, M. D., & Farrell, D. (2005). Magnetite-loaded polymeric micelles as ultrasensitive magnetic-resonance probes. *Advanced Materials*, *17*(16).

Aishwarya, S., Mahalakshmi, S., & Sehgal, P. K. (2008). Collagen-coated polycaprolactone microparticles as a controlled drug delivery system. *Journal of Microencapsulation, 25*(5), 298–306. doi:10.1080/02652040801972004 PMID:18465301

Ajayan, P. (1999). Nanotubes from Carbon. *Chemical Reviews, 99*(7), 1787–1800. doi:10.1021/cr970102g PMID:11849010

Ajayan, P. M., Charlier, J. C., & Rinzler, A. G. (1999). Carbon nanotubes: From macromolecules to nanotechnology. *Proceedings of the National Academy of Sciences of the United States of America, 96*(25), 14199–14200. doi:10.1073/pnas.96.25.14199 PMID:10588679

Alasmari, W., Costello, S., Correia, J., Oxenham, S. K., Morris, J., Fernandes, L., & Barratt, C. L. et al. (2013). Ca2+ signals generated by CatSper and Ca2+ stores regulate different behaviors in human sperm. *The Journal of Biological Chemistry, 288*(9), 6248–6258. doi:10.1074/jbc.M112.439356 PMID:23344959

Al-Hardan, , Abdul Hamid, M., Ahmed, N., Jalar, A., Shamsudin, R., & Othman, N. et al. (2016). High Sensitivity pH Sensor Based on Porous Silicon (PSi) Extended Gate Field-Effect Transistor. *Sensors (Basel, Switzerland), 16*(6), 839. doi:10.3390/s16060839 PMID:27338381

Ali, M. E., & Lamprecht, A. (2014). Spray freeze drying for dry powder inhalation of nanoparticles. *European Journal of Pharmaceutics and Biopharmaceutics, 87*(3), 510–517. doi:10.1016/j.ejpb.2014.03.009 PMID:24657824

AlJehani, Y. A. (2014). Risk factors of periodontal disease: Review of the literature. *International Journal of Dentistry, 2014*, 182513. doi:10.1155/2014/182513 PMID:24963294

Allinger, N. L., Yuh, Y. H., & Lii, J. H. (1989). Molecular mechanics. The MM3 force field for hydrocarbons. 1. *Journal of the American Chemical Society, 111*(23), 8551–8566. doi:10.1021/ja00205a001

Alshamsan, A. (2014). Nanoprecipitation is more efficient than emulsion solvent evaporation method to encapsulate cucurbitacin I in PLGA nanoparticles. *Saudi Pharmaceutical Journal, 22*(3), 2, 19–222. doi:10.1016/j.jsps.2013.12.002 PMID:25061407

Alsmari, W., Barratt, C. L., Publicover, S. J., Whalley, K. M., Foster, E., Kay, V., & Oxenham, S. K. et al. (2013). The clinical significance of calcium-signaling pathways mediation human sperm hyperactivation. *Human Reproduction (Oxford, England), 28*(4), 866–876. doi:10.1093/humrep/des467 PMID:23406974

Álvarez, A. L., Espinar, F. O., & Méndez, J. B. (2011). The Application of Microencapsulation Techniques in the Treatment of Endodontic and Periodontal Disease. *Pharmaceutics, 3*(3), 538–571. doi:10.3390/pharmaceutics3030538 PMID:24310596

Alves, N. M., Leonor, I. B., Azevedo, H. S., Reis, R. L., & Mano, J. F. (2010). Designing biomaterials based on biomineralization of bone. *Journal of Materials Chemistry, 20*(15), 2911–2921. doi:10.1039/b910960a

Alyautdin, R. N., Petrov, V. E., Langer, K., Berthold, A., Kharkevich, D. A., & Kreuter, J. (1997). Delivery of loperamide across the blood–brain barrier with poly-sorbate 80-coated polybutylcyanoacrylate nanoparticles. *Pharmaceutical Research*, *14*(3), 325–328. doi:10.1023/A:1012098005098 PMID:9098875

Alyautdin, R. N., Tezikov, E. B., Ramge, P., Kharkevich, D. A., Begley, D. J., & Kreuter, J. (1998). Significant entry of tubocurarine into the brain of rats by adsorption to polysorbate 80-coated polybutylcyanoacrylate nanoparticles: An in situ brain perfusion study. *Journal of Microencapsulation*, *15*(1), 67–74. doi:10.3109/02652049809006836 PMID:9463808

Alyautdin, R., Khalin, I., Nafeeza, M. I., Haron, M. H., & Kuznetsov, D. (2014). Nanoscale drug delivery systems and the blood–brain barrier. *International Journal of Nanomedicine*, *9*, 795–811. PMID:24550672

Ambrosio, A., & Payne, J. (2013). Method of promoting new tissue growth and/or wound healing using bioresorbable dressing comprising microspheres or microparticles: Google Patents.

Amos, L. A., van den Ent, F., & Löwe, J. (2004). Structural/functional homology between the bacterial and eukaryotic cytoskeletons. *Current Opinion in Cell Biology*, *16*(1), 24–31. doi:10.1016/j.ceb.2003.11.005 PMID:15037301

Andersen, O. S. (2008). Perspectives on how to drug an ion channel. *The Journal of General Physiology*, *131*(5), 395–397. doi:10.1085/jgp.200810012 PMID:18411334

Anita Margret, A., Nargis Begum, T., Parthasarathy, S., & Suvaithenamudhan, S. (2015). A Strategy to Employ *Clitoria ternatea* as a Prospective Brain Drug Confronting Monoamine Oxidase (MAO) Against Neurodegenerative Diseases and Depression. *Natural Products and Bioprospecting*, *5*(6), 293–306. doi:10.1007/s13659-015-0079-x PMID:26667936

Aral, A., Yalçin, S., Karabuda, Z. C., Anil, A., Jansen, J. A., & Mutlu, Z. (2008). Injectable calcium phosphate cement as a graft material for maxillary sinus augmentation: An experimental pilot study. *Clinical Oral Implants Research*, *19*, 612–617. doi:10.1111/j.1600-0501.2007.01518.x PMID:18474064

Arcangeli, A., & Becchetti, A. (2010). New Trends in Cancer Therapy. *Targeting Ion Channels and Transporters*, *3*, 1202–1224.

Armitage, G. C. (1999). Development of a classification system for periodontal diseases and conditions. *Annals of Periodontology*, *4*(1), 1–6. doi:10.1902/annals.1999.4.1.1 PMID:10863370

Arora, S., Gupta, S., Narang, R. K., & Budhiraja, R. D. (2011). Amoxicillin loaded chitosan-alginate polyelectrolyte complex nanoparticles as mucopenetrating delivery system for *H. pylori*. *Scientia Pharmaceutica*, *79*(3), 673–694. doi:10.3797/scipharm.1011-05 PMID:21886911

Arruebo, M., Fernández-Pacheco, R., Ibarra, M. R., & Santamaría, J. (2007). Magnetic nanoparticles for drug delivery. *Nano Today*, *2*(3), 22–32. doi:10.1016/S1748-0132(07)70084-1

Atai, M., Solhi, L., Nodehi, A., Mirabedini, S. M., Kasraei, S., Akbari, K., & Babanzadeh, S. (2009). PMMA-grafted nanoclay as novel filler for dental adhesives. *Dental Materials*, *25*(3), 339–347. doi:10.1016/j.dental.2008.08.005 PMID:18829096

Atanasijevic, T., Shusteff, M., Fam, P., & Jasanoff, A. (2006). Calcium-sensitive MRI contrast agents based on superparamagnetic iron oxide nanoparticles and calmodulin. *Proceedings of the National Academy of Sciences of the United States of America*, *103*(40), 14707–14712. doi:10.1073/pnas.0606749103 PMID:17003117

Atri, A., Amundson, J., Clapham, D. E., & Sneyd, J. (1993). A single-pool model for intracellular calcium oscillations and waves in the Xenopus laevis oocyte. *Biophysical Journal*, *65*(4), 1727–1739. doi:10.1016/S0006-3495(93)81191-3 PMID:8274661

Austin, B., & Adams, C. (1996). Fish pathogens. In B. Austin, M. Altwegg, P. J. Gosling, & S. Joseph (Eds.), *The genus Aeromonas*. Chichester: John Wiley & Sons.

Austin, C. R. (1951). Observations on the penetration of the sperm in the mammalian egg. *Australian Journal of Scientific Research, Series B: Biological Sciences*, *4*, 581–596. PMID:14895481

Avila, M., Ojcius, D. M., & Yilmaz, Ö. (2009). The Oral Microbiota: Living with a Permanent Guest. *DNA and Cell Biology*, *28*(8), 405–411. doi:10.1089/dna.2009.0874 PMID:19485767

Avinoam, O., Schorb, M., Beese, C. J., Briggs, J. A. G., & Kaksonen, M. (2015). Endocytic sites mature by continuous bending and remodeling of the clathrin coat. *Science*, *348*(6241), 1369–1372. doi:10.1126/science.aaa9555 PMID:26089517

Aznar, M., Vera, P., Canellas, E., Nerín, C., Mercea, P., & Störmer, A. (2011). Composition of the adhesives used in food packaging multilayer materials and migration studies from packaging to food. *Journal of Materials Chemistry*, *21*(12), 4358–4370. doi:10.1039/c0jm04136j

Baaziz, W., Pichon, B. P., Fleutot, S., Liu, Y., Lefevre, C., Greneche, J.-M., & Begin-Colin, S. et al. (2014). Magnetic iron oxide nanoparticles: Reproducible tuning of the size and nanosized-dependent composition, defects, and spin canting. *The Journal of Physical Chemistry C*, *118*(7), 3795–3810. doi:10.1021/jp411481p

Bae, K. H., Park, M., Do, M. J., Lee, N., Ryu, J. H., Kim, G. W., & Hyeon, T. et al. (2012). Chitosan oligosaccharide-stabilized ferrimagnetic iron oxide nanocubes for magnetically modulated cancer hyperthermia. *ACS Nano*, *6*(6), 5266–5273. doi:10.1021/nn301046w PMID:22588093

Banerjee, S. K., & Chakravarti, R. N. (1963). Taraxerol from *Clitoria ternatea*. *Bulletin of the Calcutta School of Tropical Medicine*, *11*(1), 106–107. PMID:14068655

Banfi, S., Caruso, E., Buccafurni, L., Battini, V., Zazzaron, S., Barbieri, P., & Orlandi, V. (2006). Antibacterial activity of tetraaryl-porphyrin photosensitizers: An in vitro study on Gram negative and Gram positive bacteria. *Journal of Photochemistry and Photobiology. B, Biology*, *85*(1), 28–38. doi:10.1016/j.jphotobiol.2006.04.003 PMID:16737820

Batlle, X., & Labarta, A. (2002). Finite-size effects in fine particles: Magnetic and transport properties. *Journal of Physics. D, Applied Physics*, *35*(6), R15–R42. doi:10.1088/0022-3727/35/6/201

Batrakova, E. V., Gendelman, H. E., & Kabanov, A. V. (2011). Cell-mediated drugs delivery. *Expert Opinion on Drug Delivery*, *8*(4), 415–443. doi:10.1517/17425247.2011.559457 PMID:21348773

Batrakova, E. V., Li, S., Reynolds, A. D., Mosley, R. L., Bronich, T. K., Kabanov, A. V., & Gendelman, H. E. (2007). A macrophage-nanozyme delivery system for Parkinson's disease. *Bioconjugate Chemistry*, *18*(5), 1498–1506. doi:10.1021/bc700184b PMID:17760417

Bauer, A. W., Kirby, W. M., Sherris, J. C., & Turck, M. (1966). Antibiotic susceptibility testing by a standardized single disk method. *American Journal of Clinical Pathology*, *45*(4), 493–496. PMID:5325707

Baughman, R. H., Cui, C., Zakhidov, A. A., Iqbal, Z., Barisci, J. N., Spinks, G. M., & Jaschinski, O. et al. (1999). Carbon nanotube actuators. *Science*, *284*(5418), 1340–1344. doi:10.1126/science.284.5418.1340 PMID:10334985

Beachley, V., & Wen, X. J. (2010). Polymer nanofibrous structures: Fabrication, biofunctionalization, and cell interactions. *Progress in Polymer Science*, *35*(7), 868–892. doi:10.1016/j.progpolymsci.2010.03.003 PMID:20582161

Beall, J. R. (1943). Wear of Acrylic Resin Teeth (Progress Report). *The Journal of the American Dental Association*, *30*(3), 252–256. doi:10.14219/jada.archive.1943.0043

Bedran-Russo, A. K., Pashley, D. H., Agee, K., Drummond, J. L., & Miescke, K. J. (2007). Changes in stiffness of demineralized dentin following application of collagen crosslinkers. *Journal of Biomedical Materials Research. Part B, Applied Biomaterials*, *86*(2), 330–334. PMID:18161815

Bedran-Russo, A. K., Pereira, P. N., Duarte, W. R., Drummond, J. L., & Yamauchi, M. (2006). Application of crosslinkers to dentin collagen enhances the ultimate tensile strength. *Journal of Biomedical Materials Research. Part B, Applied Biomaterials*, *80*(1), 268–272. PMID:16767722

Bedran-Russo, A. K., Yoo, K. J., Ema, K. C., & Pashley, D. H. (2009). Mechanical properties of tannic-acid-treated dentin matrix. *Journal of Dental Research*, *88*(9), 807–811. doi:10.1177/0022034509342556 PMID:19767576

Berret, J.-F., Schonbeck, N., Gazeau, F., El Kharrat, D., Sandre, O., Vacher, A., & Airiau, M. (2006). Controlled clustering of superparamagnetic nanoparticles using block copolymers: Design of new contrast agents for magnetic resonance imaging. *Journal of the American Chemical Society*, *128*(5), 1755–1761. doi:10.1021/ja0562999 PMID:16448152

Bertrand, N., Simard, P., & Leroux, J. C. (2010). Serum-stable long-circulating pH-sensitive PEGylated liposomes. *Methods in Molecular Biology (Clifton, N.J.)*, *605*, 545–558. doi:10.1007/978-1-60327-360-2_36 PMID:20072905

Besinis, A., De Peralta, T., Tredwin, C. J., & Handy, R. D. (2015). Review of Nanomaterials in Dentistry: Interactions with the Oral Microenvironment, Clinical Applications, Hazards, and Benefits. *ACS Nano*, *9*(3), 2255–2289. doi:10.1021/nn505015e PMID:25625290

Beyth, N., Yudovin-Farber, I., Bahir, R., Domb, A. J., & Weiss, E. I. (2006). Antibacterial activity of dental composites containing quaternary ammonium polyethylenimine nanoparticles against Streptococcus mutans. *Biomaterials*, *27*(21), 3995–4002. doi:10.1016/j.biomaterials.2006.03.003 PMID:16564083

Beyth, N., Yudovin-Farber, I., Perez-Davidi, M., Domb, A. J., & Weiss, E. I. (2010). Polyethyleneimine nanoparticles incorporated into resin composite cause cell death and trigger biofilm stress in vivo. *Proceedings of the National Academy of Sciences of the United States of America*, *107*(51), 22038–22043. doi:10.1073/pnas.1010341107 PMID:21131569

Bhattacharya, S. K., Goel, R. K., Kaur, R., & Ghosal, S. (1987). Anti - stress activity of Sitoindosides VII and VIII. New Acylsterylglucosides from *Withania somnifera. Phytotherapy Research*, *6*(1), 32–37. doi:10.1002/ptr.2650010108

Biber, A., Fischer, H., Romer, A., & Chatterjee, S. S. (1998). Oral bioavailability of hyperforin from *Hypericum* extracts in rats and human volunteers. *Pharmacopsychiatry*, *31*(1), 36–43. doi:10.1055/s-2007-979344 PMID:9684946

Bield, A., & Kraus, R. (1898). U¨ ber eine bisher unbekannte toxische Wirkung der Gallensaurenauf das Zentralnervensystem. *Zhl Inn Med*, *19*, 1185–1200.

Bisht, S., Feldmann, G., & Soni, S. (2007). Polymeric nanoparticle-encapsulated curcumin ("nanocurcumin"): A novel strategy for human cancer therapy. *Journal of Nanobiotechnology*, *5*(3), 1–18. PMID:17439648

Blevins, R. D., & Plunkett, R. (1980). Formulas for natural frequency and mode shape. *Journal of Applied Mechanics*, *47*(2), 461. doi:10.1115/1.3153712

Blum, J. J., Reed, M. C., Janovick, J. A., & Conn, P. M. (2000). A mathematical model quantifying GnRH- induced LH secretion from gonadotropes. *American Journal of Physiology. Endocrinology and Metabolism*, *278*, 263–272. PMID:10662710

Bödding, M. (2007). TRP proteins and cancer. *Cellular Signalling*, *19*(3), 617–624. doi:10.1016/j.cellsig.2006.08.012 PMID:17029734

Boden, S. D., Martin, G. J. Jr, Morone, M. A., Ugbo, J. L., & Moskovitz, P. A. (1999). Posterolateral lumbar intertransverse process spine arthrodesis with recombinant human bone morphogenetic protein 2/hydroxyapatite–tricalcium phosphate after laminectomy in the nonhuman primate. *Spine*, *24*(12), 1179–1185. doi:10.1097/00007632-199906150-00002 PMID:10382242

Bohner, M., & Baroud, G. (2005). Injectability of calcium phosphate pastes. *Biomaterials*, *26*(13), 1553–1563. doi:10.1016/j.biomaterials.2004.05.010 PMID:15522757

Bohner, M., Lemaitre, J., & Ring, T. A. (1996). Effects of sulfate, pyrophosphate and citrate ions on the physico chemical properties of cements made of beta tricalcium phosphate – phosphoric acid – water mixtures. *Journal of the American Ceramic Society*, *79*(6), 1427–1434. doi:10.1111/j.1151-2916.1996.tb08746.x

Bohren, C. F., & Huffman, D. R. (1983). Absorption and scattering by a sphere. In *Absorption and Scattering of Light by Small Particles* (pp. 82–129).

Bordelon, D. E., Cornejo, C., Grüttner, C., Westphal, F., DeWeese, T. L., & Ivkov, R. (2011). Magnetic nanoparticle heating efficiency reveals magneto-structural differences when characterized with wide ranging and high amplitude alternating magnetic fields. *Journal of Applied Physics*, *109*(12), 124904. doi:10.1063/1.3597820

Branda, S. S., Vik, A., Friedman, L., & Kolter, R. (2005). Biofilms: The matrix revisited. *Trends in Microbiology*, *13*(1), 20–26. doi:10.1016/j.tim.2004.11.006 PMID:15639628

Brenker, C., Goodwin, N., Weyand, I., Kashikar, N. D., Naruse, M., Krähling, M., & Strünker, T. et al. (2012). The CatSper channel: A polymodal chemosensor in human sperm. *The EMBO Journal*, *31*(7), 1654–1665. doi:10.1038/emboj.2012.30 PMID:22354039

Brenner, C., & Moulin, M. (2012). Physiological roles of the permeability transition pore. *Circulation Research*, *111*(9), 1237–1247. doi:10.1161/CIRCRESAHA.112.265942 PMID:23065346

Brenner, W., Shenderova, O. A., Harrison, J. A., Stuart, S. J., Ni, B., & Sinnott, S. B. (2002). Second generation reactive empirical bonds order (rebo) potential energy expression for hydrocarbons. *Journal of Physics*, *14*, 783–802.

Brigger, I., Dubernet, C., & Couvreur, P. (2002). Nanoparticles in Cancer Therapy and Diagnosis. *Advanced Drug Delivery Reviews*, *5*(4), 631–651. doi:10.1016/S0169-409X(02)00044-3 PMID:12204596

Brown, W. E., & Chow, L. C. (1986). US Patent, 4612053, 16 September.

Brown, L., Wolf, J. M., Prados-Rosales, R., & Casadevall, A. (2015). Through the wall: Extracellular vesicles in Gram-positive bacteria, mycobacteria and fungi. *Nature Reviews. Microbiology*, *13*, 620–630. PMID:26324094

Brown, S. S. (1999). Cooperation between microtubule- and actin-based motor proteins. *Annual Review of Cell and Developmental Biology*, *15*(1), 63–80. doi:10.1146/annurev.cellbio.15.1.63 PMID:10611957

Bruschi, M. L., de Freitas, O., Lara, E. H., Panzeri, H., Gremiao, M. P., & Jones, D. S. (2008). Precursor system of liquid crystalline phase containing propolis microparticles for the treatment of periodontal disease: Development and characterization. *Drug Development and Industrial Pharmacy*, *34*(3), 267–278. doi:10.1080/03639040701655911 PMID:18363142

Budtz-Jörgensen, E. (1996). Restoration of the partially edentulous mouth — a comparison of overdentures, removable partial dentures, fixed partial dentures and implant treatment. *Journal of Dentistry*, *24*(4), 237–244. doi:10.1016/0300-5712(95)00075-5 PMID:8783527

Burg, K. J. L., Porter, S., & Kellam, J. F. (2000). Biomaterial developments for bone tissue engineering. *Biomaterials*, *21*(23), 2347–2359. doi:10.1016/S0142-9612(00)00102-2 PMID:11055282

Bustos-Martinez, J. A., Hamdan-Partida, A., & Gutierrez-Cardenas, M. (2006). *Staphylococcus aureus*: La reemergencia de un patógeno en la comunidad. *Rev Biomed*, *17*, 287–305.

Cacciatore, I., Baldassarre, L., Fornasari, E., Mollica, A Pinnen F. (2012). Recent Advances in the Treatment of Neurodegenerative Diseases Based on GSH Delivery Systems. *Oxidative Medicine Cell Longevity*. (2), 140-146.

Cai, B., Huang, L., Zhang, H., Sun, Z., Zhang, Z., & Zhang, G.-J. (2015). Gold nanoparticles-decorated graphene field-effect transistor bio-sensor for femtomolar microRNA detection. *Biosensors & Bioelectronics*, *74*, 329–334. doi:10.1016/j.bios.2015.06.068 PMID:26159152

Cai, B., Wang, S., Huang, L., Ning, Y., Zhang, Z., & Zhang, G.-J. (2014). Ultrasensitive label-free detection of PNA-DNA hybridization by reduced graphene oxide field-effect transistor biosensor. *ACS Nano*, *8*(3), 2632–2638. doi:10.1021/nn4063424 PMID:24528470

Cai, Q., Wang, L., Deng, G., Liu, J., Chen, Q., & Chen, Z. (2016). Systemic delivery to central nervous system by engineered PLGA nanoparticles. *American Journal of Translational Research*, *8*(2), 749–764. PMID:27158367

Cai, Q., Xu, Q. Q., Feng, Q. F., Cao, X. Y., Yang, X. P., & Deng, X. L. (2011). Biomineralization of electrospun poly(L-lactic acid)/gelatin composite fibrous scaffold by using a supersaturated simulated body fluid with continuous CO2 bubbling. *Applied Surface Science*, *257*(23), 10109–10118. doi:10.1016/j.apsusc.2011.06.157

Cai, S., Liang, G., Zhang, P., Chen, H., Zhang, S., Liu, B., & Kong, J. (2011). Rational strategy of magnetic relaxation switches for glycoprotein sensing. *Analyst (London)*, *136*(1), 201–204. doi:10.1039/C0AN00362J PMID:20949140

Campus, G., Lallai, M. R., & Carboni, R. (2003). Fluoride concentration in saliva after use of oral hygiene products. *Caries Research*, *37*(1), 66–70. doi:10.1159/000068220 PMID:12566642

Carballido-López, R., & Errington, J. (2003). A dynamic bacterial cytoskeleton. *Trends in Cell Biology*, *13*(11), 577–583. doi:10.1016/j.tcb.2003.09.005 PMID:14573351

Carroll, J. A., Coleman, S. A., Smitherman, L. S., & Minnick, M. F. (2000). Hemin-binding surface protein fromBartonella quintana. *Infection and Immunity*, *68*(12), 6750–6757. doi:10.1128/IAI.68.12.6750-6757.2000 PMID:11083791

Cedervall, T., Lynch, I., Lindman, S., Berggård, T., Thulin, E., Nilsson, H., & Linse, S. et al. (2007). Understanding the nanoparticle–protein corona using methods to quantify exchange rates and affinities of proteins for nanoparticles. *Proceedings of the National Academy of Sciences of the United States of America*, *104*(7), 2050–2055. doi:10.1073/pnas.0608582104 PMID:17267609

Center for Disease Control and Prevention. (n. d.) Biological effects of exposure. Retrieved from www.cdc.gov/niosh/pdfs/7311022b.pdf

Chang, S-P., Yang, T.-H. (2012). Sensing performance of EGFET pH sensors with Cuo nanowires fabricated on glass substrate. *International Journal of Electrochemical Sensing*, 7(6), 5020-5027.

Chang, H., & Suarez, S. S. (2011). Two distinct Ca2+ signaling pathways modulate sperm flagellar beating patterns in mice. *Biology of Reproduction*, *85*(2), 296–305. doi:10.1095/biolreprod.110.089789 PMID:21389347

Chang, M. C. (1951). Fertilizing capacity of spermatozoa deposited into the fallopian tubes. *Nature*, *168*(4277), 697–698. doi:10.1038/168697b0 PMID:14882325

Chang, P. C., Dovban, A. S., Lim, L. P., Chong, L. Y., Kuo, M. Y., & Wang, C. H. (2013). Dual delivery of PDGF and simvastatin to accelerate periodontal regeneration in vivo. *Biomaterials*, *34*(38), 9990–9997. doi:10.1016/j.biomaterials.2013.09.030 PMID:24079892

Chanteau, B., Fresnais, J., & Berret, J.-F. (2009). Electrosteric enhanced stability of functional sub-10 nm cerium and iron oxide particles in cell culture medium. *Langmuir*, *25*(16), 9064–9070. doi:10.1021/la900833v PMID:19572532

Chávez de Paz, L. E., Resin, A., Howard, K. A., Sutherland, D. S., & Wejse, P. L. (2011). Antimicrobial Effect of Chitosan Nanoparticles on Streptococcus mutans Biofilms. *Applied and Environmental Microbiology*, *77*(11), 3892–3895. doi:10.1128/AEM.02941-10 PMID:21498764

Chen, J. & Lee., J.E. (2010). Atomistic analysis of nano/micro biosensors. *Interaction and Multiscale mechanics*, 3(2), 111-121.

Chen, J., Fang, Z., Lie, P., & Zeng, L. (2012). Computational lateral flow biosensor for proteins and small Molecules: A new class of strip logic gates. *Analytical Chemistry*, *84*(15), 6321–6325. doi:10.1021/ac301508b PMID:22780691

Cheon, J., Kang, N.-J., Lee, S.-M., Lee, J.-H., Yoon, J.-H., & Oh, S. J. (2004). Shape evolution of single-crystalline iron oxide nanocrystals. *Journal of the American Chemical Society*, *126*(7), 1950–1951. doi:10.1021/ja038722o PMID:14971924

Chikazumi, S., Taketomi, S., Ukita, M., Mizukami, M., Miyajima, H., Setogawa, M., & Kurihara, Y. (1987). Physics of magnetic fluids. *Journal of Magnetism and Magnetic Materials*, *65*(2), 245–251. doi:10.1016/0304-8853(87)90043-6

Chirio, D., Gallarate, M., Peira, E., Battaglia, L., Muntoni, E., & Riganti, C. et al. (2014). Positive-charged solid lipid nanoparticles as paclitaxel drug delivery system in glioblastoma treatment. *European Journal of Pharmaceutics and Biopharmaceutics*, *88*(3), 746–758. doi:10.1016/j.ejpb.2014.10.017 PMID:25445304

Chodon, D., Guilbert, A., Dhennin-Duthille, I., Gautier, M., Telliez, M. S., Sevestre, H., & Ouadid-Ahidouch, H. (2010). Estrogen regulation of TRPM8 expression in breast cancer cells. *BMC Cancer*, *10*(1), 212. doi:10.1186/1471-2407-10-212 PMID:20482834

Cho, H., Yu, M.-F., Vakakis, A. F., Bergman, L. A., & McFarland, D. M. (2010). Tunable Broadband Nonlinear Nanomechanical Resonator. *Nano Letters*, *10*(5), 1793–1798. doi:10.1021/nl100480y PMID:20384349

Choi, S., Na, H. S., Kim, J., Lee, J., Lee, S., Kim, D., & Shin, H. S. et al. (2007). Attenua.ted pain responses in mice lacking CaV32 T-type. channels. *Genes Brain & Behavior, 6*(5), 425–431. doi:10.1111/j.1601-183X.2006.00268.x PMID:16939637

Chomoucka, J., Drbohlavova, J., Huska, D., Adam, V., Kizek, R., & Hubalek, J. (2010). Magnetic nanoparticles and targeted drug delivering. *Pharmacological Research, 62*(2), 144–149. doi:10.1016/j.phrs.2010.01.014 PMID:20149874

Chowdhury, R., & Adhikari, S. (2011). Boron nitride nanotubes as zeptogram-scale bionano sensors: Theoretical investigations. *IEEE Transactions on Nanotechnology, 10*(4), 659–667. doi:10.1109/TNANO.2010.2060492

Chowdhury, R., Adhikari, S., & Mitchell, J. (2009). Vibrating carbon nanotube based bio-sensors. *Physica E, Low-Dimensional Systems and Nanostructures, 42*(2), 104–109. doi:10.1016/j.physe.2009.09.007

Chow, L. C., Takagi, S., & Ishikawa, K. (1994). Formation of hydroxyapatite in cement systems. In W. E. Brown & E. Constanz (Eds.), *Hydroxyapatite and related materials* (pp. 127–137). Boca Raton, FL: CRC Press.

Chua, K. N., Lim, W. S., Zhang, P. C., Lu, H. F., Wen, J., Ramakrishna, S., & Mao, H. Q. et al. (2005). Stable immobilization of rat hepatocyte spheroids on galactosylated nanofiber scaffold. *Biomaterials, 26*(15), 2537–2547. doi:10.1016/j.biomaterials.2004.07.040 PMID:15585256

Chu, K. S., Hasan, W., & Rawal, S. (2013). Plasma, tumor and tissue pharmacokinetics of Docetaxel delivered via nanoparticles of different sizes and shapes in mice bearing SKOV-3 human ovarian carcinoma xenograft. *Nanomedicine (London), 9*(5), 686–693. PMID:23219874

Chung, J. J., Navarro, B., Krapivinsky, G., Krapivinsky, L., & Clapham, D. E. (2011). A novel gene required for male fertility and functional CATSPER channel formation in spermatozoa. *Nature Communications, 2*(1), 153. doi:10.1038/ncomms1153 PMID:21224844

Ciampolini, J., & Harding, K. G. (2000). Pathophysiology of chronic bacterial osteomyelitis. Why do antibiotics fail so often? *Postgraduate Medical Journal, 76*(898), 479–483. doi:10.1136/pmj.76.898.479 PMID:10908375

Ciancio, S. G., Mather, M. L., & McMullen, J. A. (1980). An Evaluation of Minocycline in Patients With Periodontal Disease*. *Journal of Periodontology, 51*(9), 530–534. doi:10.1902/jop.1980.51.9.530 PMID:6999148

Cifra, M., Fields, J. Z., & Farhadi, A. (2011). Electromagnetic cellular interactions. *Progress in Biophysics and Molecular Biology, 105*(3), 223–246. doi:10.1016/j.pbiomolbio.2010.07.003 PMID:20674588

Cioffi, N., Torsi, L., Ditaranto, N., Tantillo, G., Ghibelli, L., Sabbatini, L., & Traversa, E. et al. (2005). Copper nanoparticle/polymer composites with antifungal and bacteriostatic properties. *Chemistry of Materials, 17*(21), 5255–5262. doi:10.1021/cm0505244

Cipriani, A., Furukawa, T. A., Salanti, G., Geddes, J. R., Higgins, J. P., Churchill, R., & Barbui, C. et al. (2009). Comparative efficacy and acceptability of 12 new-generation antidepressants: A multiple-treatments meta-analysis. *Lancet*, *373*(9665), 746–758. doi:10.1016/S0140-6736(09)60046-5 PMID:19185342

Clemens, H., Liu, Y., Mauser, N. J., Ringhofer, C., & Dutton, R. W. (2010). Calculation of fluctuations in boundary layers of nanowire Field effect biosensors. *Journal of Computational and Theoretical Nanoscience*, *7*, 1–7.

Colombelli, A., Manera, M.G., Taurino, A., Catalano, M., Convertino, A. Au nanoparticles decoration of silica nanowires for improved optical bio-sensing. *Sensor actuators B: Chemical*, 226, 589-597.

Colombo, A. V., Silva, C. M., Haffajee, A., & Colombo, A. P. (2006). Identification of oral bacteria associated with crevicular epithelial cells from chronic periodontitis lesions. *Journal of Medical Microbiology*, *55*(Pt 5), 609–615. doi:10.1099/jmm.0.46417-0 PMID:16585650

Colombo, M., Carregal-Romero, S., Casula, M. F., Gutiérrez, L., Morales, M. P., Bohm, I. B., & Parak, W. J. et al. (2012). Biological applications of magnetic nanoparticles. *Chemical Society Reviews*, *41*(11), 4306–4334. doi:10.1039/c2cs15337h PMID:22481569

Colombo, M., Ronchi, S., Monti, D., Corsi, F., Trabucchi, E., & Prosperi, D. (2009). Femtomolar detection of autoantibodies by magnetic relaxation nanosensors. *Analytical Biochemistry*, *392*(1), 96–102. doi:10.1016/j.ab.2009.05.034 PMID:19481052

Constantz, B. R., Ison, I. C., Fulmer, M. T., Poser, R. D., Smith, S. T., VanWagoner, M., & Rosenthal, D. et al. (1995). Skeletal repair by in situ formation of the mineral phase of bone. *Science*, *267*(5205), 1796–1799. doi:10.1126/science.7892603 PMID:7892603

Cook, A. D., Hrkach, J. S., Gao, N. N., Johnson, I. M., Pajvani, U. B., Cannizzaro, S. M., & Langer, R. (1997). Characterization and development of RGD-peptide-modified poly(lactic acid-co-lysine) as an interactive, resorbable biomaterial. *Journal of Biomedical Materials Research*, *35*(4), 513–523. doi:10.1002/(SICI)1097-4636(19970615)35:4<513::AID-JBM11>3.0.CO;2-C PMID:9189829

Cooper, C. A., Ravich, D., Lips, D., Mayer, J., & Wagner, H. D. (2002). Distribution and alignment of carbon nanotubes and nanofibrils in a polymer matrix. *Composites Science and Technology*, *62*(7–8), 1105–1112. doi:10.1016/S0266-3538(02)00056-8

Corchero, J. L., & Villaverde, A. (2009). Biomedical applications of distally controlled magnetic nanoparticles. *Trends in Biotechnology*, *27*(8), 468–476. doi:10.1016/j.tibtech.2009.04.003 PMID:19564057

Cornell, R. M., & Schwertmann, U. (2003). The Iron Oxides. Wiley.

Cornell, W. D., Cieplak, P., Bayly, C. I., Gould, I. R., Merz, K. M., Ferguson, D. M., & Kollman, P. A. (1996). A second generation force field for the simulation of proteins, nucleic acids, and organic molecules j. am. chem. soc. 1995, 117, 5179-5197. *Journal of the American Chemical Society*, *118*(9), 2309–2309. doi:10.1021/ja955032e

Cowan, C. M., Shi, Y. Y., Aalami, O. O., Chou, Y. F., Mari, C., Thomas, R., & Longaker, M. T. et al. (2004). Adipose-derived adult stromal cells heal critical-size mouse calvarial defects. *Nature Biotechnology*, *22*(5), 560–567. doi:10.1038/nbt958 PMID:15077117

Craparo, E. F., Bond, M. L., Pitarresi, G., & Cavallaro, G. (2011). Nanoparticulate systems for drug delivery and targeting to the central nervous system. *CNS Neuroscience & Therapeutics*, *17*(6), 670–677. doi:10.1111/j.1755-5949.2010.00199.x PMID:20950327

Cristina Buzea, I. I. P. B., & Kevin, R. (2007). Nanomaterials and nanoparticles: Sources and toxicity. *Biointerphases*, *2*, MR17–MR172. PMID:20419892

Cullity, B. D. (1978). *Elements of X-ray Diffraction* (2nd ed.). USA: Edison-Wesley Publishing Company Inc.

Curia, R., Erega, A., & Milani, M. (2015). Study of the Formation, Secretion and Trafficking of Membrane Vesicles Loaded with Polyurethane Nanoparticles in Staphylococcus aureus via Electron Microscopy. *HSOA Journal of Nanotechnology: Nanomedicine & Nanobiotechnology*, 2.

Curia, R., Milani, M., Didenko, L. V., Avtandilov, G. A., Shevlyagina, N. V., & Smirnova, T. A. (2014). Beyond the biodestruction of polyurethane: *S. aureus* uptake of nanoparticles is a challenge for toxicology. In A. Méndez-Vilas (Ed.), *Microscopy: advances in scientific research and education* (pp. 16–23). Spain: Formatex Research Center.

Curia, R., Milani, M., Didenko, L. V., & Shevlyagina, N. V. (2013). Electron microscopy broadens toxicology's horizons: The role of nanoparticles vehiculated by bacteria. *Current Topics in Toxicology*, *9*, 93–98.

Dabbagh, A., Kasim, N. H. A., Bakri, M. M., Wakily, H., Ramasindarum, C., & Abdullah, B. J. J. (2014). Polyethylene-glycol coated maghemite nanoparticles for treatment of dental hypersensitivity. *Materials Letters*, *121*, 89–92. doi:10.1016/j.matlet.2014.01.120

Damien, C. J., & Parsons, J. R. (1991). Bone-Graft and Bone-Graft Substitutes - a Review of Current Technology and Applications. *Journal of Applied Biomaterials*, *2*(3), 187–208. doi:10.1002/jab.770020307 PMID:10149083

Damm, C., Münstedt, H., & Rösch, A. (2007). Long-term antimicrobial polyamide 6/silver-nanocomposites. *Journal of Materials Science*, *42*(15), 6067–6073. doi:10.1007/s10853-006-1158-5

Daood, U., Iqbal, K., Nitisusanta, L. I., & Fawzy, A. S. (2012). Effect of chitosan/riboflavin modification on resin/dentin interface: Spectroscopic and microscopic investigations. *Journal of Biomedical Materials Research. Part A*, *101*(7), 1846–1856. PMID:23184366

Daood, U., Swee Heng, C., Neo Chiew Lian, J., & Fawzy, A. S. (2015). In vitro analysis of riboflavin-modified, experimental, two-step etch-and-rinse dentin adhesive: Fourier transform infrared spectroscopy and micro-Raman studies. *Journal of Oral Science*, *7*(2), 110–124. doi:10.1038/ijos.2014.49 PMID:25257880

Darkrim, F. L., Malbrunot, P., & Tartaglia, G. P. (2002). Review of hydrogen storage by adsorption in carbon nanotubes. *International Journal of Hydrogen Energy*, *27*(2), 193–202. doi:10.1016/S0360-3199(01)00103-3

Darszon, A., Acevedo, J. J., Galindo, B. E., Hernandez-Gonzalez, E. O., Nishigaki, T., Trevino, C. L., & Beltran, C. et al. (2006). Sperm channel diversity and functional multiplicity. *Reproduction (Cambridge, England)*, *131*(6), 977–988. doi:10.1530/rep.1.00612 PMID:16735537

Darszon, A., Martínez, L. P., Acevedo, J. J., Cruz, H. A., & Treviño, C. L. (2006). T-type Ca2+ channels in sperm function. *Cell Calcium*, *40*(2), 241–252. doi:10.1016/j.ceca.2006.04.028 PMID:16797697

Darszon, A., Nishiqaki, T., Wood, C., Trevino, C. L., Felix, R., & Beltran, C. (2005). Calcium channels and Ca2+ fluctuations in sperm physiology. *International Review of Cytology*, *243*, 79–172. doi:10.1016/S0074-7696(05)43002-8 PMID:15797459

Das, J., Das, S., Samadder, A., Bhadra, K., & Khuda-Bukhsh, A. R. (2012). Poly (lactide-co-glycolide) encapsuled extract of Phytolacca decandra demonstrates better intervention against induced lung adenocarcinoma in mice and on A549 cells. *European Journal of Pharmaceutical Sciences*, *7*(2), 313–324. doi:10.1016/j.ejps.2012.06.018 PMID:22771545

Das, , Ko, D. H., Chen, C.-H., Chang, L.-B., Lai, C.-S., & Chu, F.-C. et al. (2014). Highly sensitive palladium oxide thin film extended gate FETs as pH sensor. *Sensors and Actuators. B, Chemical*, *205*, 199–205. doi:10.1016/j.snb.2014.08.057

Davis, J. J., Green, M. L., Hill, H. A. O., Leung, Y. C., Sadler, P. J., Sloan, J., & Tsang, S. C. (1998). The immobilisation of proteins in carbon nanotubes. *Inorganica Chimica Acta*, *272*(1), 261–266. doi:10.1016/S0020-1693(97)05926-4

Dayeh, V. R., Chow, S. L., Schirmer, K., Lynn, D. H., & Bols, N. C. (2004). Evaluating the toxicity of Triton X-100 to protozoan, fish, and mammalian cells using fluorescent dyes as indicators of cell viability. *Ecotoxicology and Environmental Safety*, *57*(3), 375–382. doi:10.1016/S0147-6513(03)00083-6 PMID:15041260

de Francisco, L. M., Cerquetani, J. A., & Bruschi, M. L. (2012). Development and characterization of gelatin and ethylcellulose microparticles designed as platforms to delivery fluoride. *Drug Development and Industrial Pharmacy*, *39*(11), 1644–1650. doi:10.3109/03639045.2012.728610 PMID:23034061

de Oliveira, M., & Mendes, G. S. (2008). Modulation of calcium signaling in human sperm by nitric oxide [Ph.D. Thesis]. University of Birmingham.

De Vico, L., Sørensen, M. H., Iversen, L., Rogers, D. M., Sørensen, B. S., & Brandbyge, M. et al. (2010). Quantifying signal changes in nano-wire based biosensors. *Nanoscale*, *3*(2), 706–717. doi:10.1039/C0NR00442A PMID:21173975

Deatherage, B. L., & Cookson, B. T. (2012). Membrane Vesicle Release in Bacteria, Eukaryotes, and Archaea: A Conserved yet Underappreciated Aspect of Microbial Life. *Infection and Immunity*, *80*(6), 1948–1957. doi:10.1128/IAI.06014-11 PMID:22409932

Deepak, M., & Amit, A. (2004). The need for establishing identities of bacoside A and B, the putative major bioactive saponins of Indian medicinal plant *Bacopa monnieri. Phytomedicine*, *11*(2-3), 264–268. doi:10.1078/0944-7113-00351 PMID:15070183

Del Real, R. P., Wolke, J. G. C., Vallet-Regí, M., & Jansen, J. A. (2002). A new method to produce macropores in calcium phosphate cements. *Biomaterials*, *23*(17), 3673–3680. doi:10.1016/S0142-9612(02)00101-1 PMID:12109693

Denora, N., Trapani, A., Laquintana, V., Lopedota, A., & Tropani, G. (2009). Recent advances in medicinal chemistry and pharmaceutical technology-strategies for drug delivery to the brain. *Current Topics in Medicinal Chemistry*, *9*(2), 182–196. doi:10.2174/156802609787521571 PMID:19200004

Dereli, G., & Ozdogan, C. (2003). Structural stability and energetics of single walled carbon nanotubes under uniaxial strain. *Physical Review B: Condensed Matter and Materials Physics*, *67*(3), 035416. doi:10.1103/PhysRevB.67.035416

Destache, C. J., Belgum, T., Goede, M., Shibata, A., & Belshan, M. A. (2010). Antiretroviral release from poly(DL-lactide-co-glycolide) nanoparticles in mice. *The Journal of Antimicrobial Chemotherapy*, *65*(10), 2183–2187. doi:10.1093/jac/dkq318 PMID:20729545

Dhadse, P., Gattani, D., & Mishra, R. (2010). The link between periodontal disease and cardiovascular disease: How far we have come in last two decades? *Journal of Indian Society of Periodontology*, *14*(3), 148–154. doi:10.4103/0972-124X.75908 PMID:21760667

Diarra, M., Pourroy, G., Boymond, C., & Muster, D. (2003). Fluoride controlled release tablets for intrabuccal use. *Biomaterials*, *24*(7), 1293–1300. doi:10.1016/S0142-9612(02)00444-1 PMID:12527271

Didenko, L. V., Avtandilov, G. A., Shevlyagina, N. V., Smirnova, T. A., Lebedenko, I. Y., Tatti, F., & Milani, M. (2012). Biodestruction of polyurethane by *Staphylococcus aureus* (an investigation by SEM, TEM and FIB). In A. Méndez-Vilas (Ed.), *Current Microscopy Contributions to Advances in Science and Technology* (pp. 323–334). Spain: Formatex Research Center.

Didenko, L. V., Shevlyagina, N. V., Curia, R., Erega, A., & Milani, M. (2014). Electron Microscopy Furthers the Investigation of Bacteria-Nanoparticles Interactions Subcellular Dynamics. Global Journal of Medical Research: D Radiology. *Diagnostic Imaging and Instrumentation*, *14*(4), 13–23.

Didenko, L., Avtandilov, G., Shevlyagina, N., Shustrova, N., Smirnova, T., Lebedenko, I., & Milani, M. et al. (2013). Nanoparticles production and inclusion in *S. aureus* incubated with polyurethane: An electron microscopy analysis. *Open Journal of Medical Imaging*, *3*(2), 69–73. doi:10.4236/ojmi.2013.32010

Ding, F., Jiao, K., Wu, M., & Yakobson, B. I. (2007). Pseudoclimb and dislocation dynamics in superplastic nanotubes. *Physical Review Letters*, *98*(7), 075503. doi:10.1103/PhysRevLett.98.075503 PMID:17359035

Doadrio, A. L., Sousa, E. M., Doadrio, J. C., Pérez Pariente, J., Izquierdo-Barba, I., & Vallet-Regí, M. (2004). Mesoporous SBA-15 HPLC evaluation for controlled gentamicin drug delivery. *Journal of Controlled Release*, *97*(1), 125–132. doi:10.1016/j.jconrel.2004.03.005 PMID:15147810

Dobson, J. (2006). Magnetic nanoparticles for drug delivery. *Drug Development Research*, *67*(1), 55–60. doi:10.1002/ddr.20067

Doggui, S., Sahni, J. K., Arseneault, M., Dao, L., & Ramassamy, C. (2012). Neuronal uptake and neuroprotective effect of curcumin-loaded PLGA nanoparticles on the human SK-N-SH cell line. *Journal of Alzheimer's Disease*, *30*(2), 377–392. PMID:22426019

Döhner, K., Nagel, C., & Sodeik, B. (2005). Viral stop-and-go along microtubules: Taking a ride with dynein and kinesins. *Trends in Microbiology*, *13*(7), 320–327. doi:10.1016/j.tim.2005.05.010 PMID:15950476

Dong, Y., Ng, W. K., Shen, S., Kim, S., & Tan, R. B. (2013). Scalable ionic gelation synthesis of chitosan nanoparticles for drug delivery in static mixers. *Carbohydrate Polymers*, *94*(2), 940–945. doi:10.1016/j.carbpol.2013.02.013 PMID:23544653

Dougherty, D. P., Wright, G. A., & Yew, A. C. (2005). Computational model of the cAMP-mediated sensory response and calcium-dependent adaptation in vertebrate olfactory receptor neurons. *Proceedings of the National Academy of Sciences of the United States of America*, *102*(30), 10415–10420. doi:10.1073/pnas.0504099102 PMID:16027364

Duan, X., Li, Y., Rajan, N. K., Routenberg, D. A., Modis, Y., & Reed, M. A. (2012). Quantification of the affinities and kinetics of protein interactions using silicon nanowire biosensors. *Nature Nanotechnology*, *7*(6), 401–407. doi:10.1038/nnano.2012.82 PMID:22635097

Du, G. J., Li, J.-H., Liu, W.-J., Liu, Y.-H., Zhao, B., Li, H.-R., & Duan, Y.-J. et al. (2014). The combination of TRPM8 and TRPA1 expression causes an invasive phenotype in lung cancer. *Tumour Biology*, *35*(2), 1251–1261. doi:10.1007/s13277-013-1167-3 PMID:24037916

Dung, T. H., Lee, S. R., Han, S. D., Kim, S. J., Ju, Y. M., Kim, M. S., & Yoo, H. (2007). Chitosan-TPP nanoparticle as a release system of antisense oligonucleotide in the oral environment. *Journal of Nanoscience and Nanotechnology*, *7*(11), 3695–3699. doi:10.1166/jnn.2007.041 PMID:18047039

Duran, N., Marcato, P. D., Conti, R. D., Alves, O. L., Costa, F. T. M., & Brocchi, M. (2010). Potential use of silver nanoparticles on pathogenic bacteria, their toxicity and possible mechanisms of action. *Journal of the Brazilian Chemical Society*, *21*(6), 949–959. doi:10.1590/S0103-50532010000600002

Ehrlich, P. (1885). Oxygen necessity of the organism. A colourful Analytical Study. August Hirschwald, Berlin, Germany.

Eidelman, N., Chow, L. C., & Brown, W. E. (1987). Calcium phosphate saturation levels in ultrafiltered serum. *Calcified Tissue International*, *40*(2), 71–78. doi:10.1007/BF02555708 PMID:3105836

El-Boubbou, K., Zhu, D. C., Vasileiou, C., Borhan, B., Prosperi, D., Li, W., & Huang, X. (2010). Magnetic glyco-nanoparticles: A tool to detect, differentiate, and unlock the glyco-codes of cancer via magnetic resonance imaging. *Journal of the American Chemical Society*, *132*(12), 4490–4499. doi:10.1021/ja100455c PMID:20201530

Elishakoff, I., Versaci, C., Maugeri, N., & Muscolino, G. (2011). Clamped-free single-walled carbon nanotube-based mass sensor treated as Bernoulli–Euler beam. *Journal of nanotechnology in engineering and medicine, 2*.

Elliott, D. W., & Zhang, W.-X. (2001). Field assessment of nanoscale bimetallic particles for groundwater treatment. *Environmental Science & Technology*, *35*(24), 4922–4926. doi:10.1021/es0108584 PMID:11775172

El-Sayed, M. A. (2001). Some interesting properties of metals confined in time and nanometer space of different shapes. *Accounts of Chemical Research*, *34*(4), 257–264. doi:10.1021/ar960016n PMID:11308299

Elsdale, T., & Bard, J. (1972). Collagen Substrata for Studies on Cell Behavior. *Journal of Cell Biology, 54*(3), 626. Doi:10.1083/Jcb.54.3.626

Engqvist-Goldstein, A. E. Y., & Drubin, D. G. (2003). Actin Assembly and Endocytosis: From Yeast to Mammals. *Annual Review of Cell and Developmental Biology*, *19*(1), 287–332. doi:10.1146/annurev.cellbio.19.111401.093127 PMID:14570572

Fang, C., Wang, K., Stephen, Z. R., Mu, Q., Kievit, F. M., & Chiu, D. T. et al. (2015). Temozolomide Nanoparticles for Targeted Glioblastoma Therapy. *ACS Applied Materials & Interfaces*, *7*(12), 6674–6682. doi:10.1021/am5092165 PMID:25751368

Fang, J. A., Wang, H. X., Niu, H. T., Lin, T., & Wang, X. G. (2010). Evolution of Fiber Morphology During Electrospinning. *Journal of Applied Polymer Science*, *118*(5), 2553–2561. doi:10.1002/app.32569

Fantner, G. E., Birkedal, H., Kindt, J. H., Hassenkam, T., Weaver, J. C., Cutroni, J. A., & Hansma, P. K. et al. (2004). Influence of the degradation of the organic matrix on the microscopic fracture behavior of trabecular bone. *Bone*, *35*(5), 1013–1022. doi:10.1016/j.bone.2004.05.027 PMID:15542025

Farhadi, J., Jaquiery, C., Haug, M., Pierer, G., Zeilhofer, H. F., & Martin, I. (2006). Bone and cartilage tissue engineering for facial reconstructive surgery. *IEEE Engineering in Medicine and Biology Magazine*, *25*(1), 106–109. doi:10.1109/MEMB.2006.1578673 PMID:16485400

Fauconnier, N., Bee, A., Roger, J., & Pons, J. N. (1999). Synthesis of aqueous magnetic liquids by surface complexation of maghemite nanoparticles. *Journal of Molecular Liquids*, *83*(1), 233–242. doi:10.1016/S0167-7322(99)00088-4

Fawzy, A. S., Nitisusanta, L. I., Iqbal, K., Daood, U., Beng, L. T., & Neo, J. (2012). Chitosan/Riboflavin-modified demineralized dentin as a potential substrate for bonding. *Journal of the Mechanical Behavior of Biomedical Materials*, *17*, 278–289. doi:10.1016/j.jmbbm.2012.09.008 PMID:23127636

Fawzy, A. S., Nitisusanta, L. I., Iqbal, K., Daood, U., & Neo, J. (2012). Riboflavin as a dentin crosslinking agent: Ultraviolet A versus blue light. *Dental Materials*, *28*(12), 1284–1291. doi:10.1016/j.dental.2012.09.009 PMID:23031483

Fennimore, A. M., Yuzvinsky, T. D., Han, W. Q., Fuhrer, M. S., Cumings, J., & Zettl, A. (2003). Rotational actuators based on carbon nanotubes. *Nature*, *424*(6947), 408–410. doi:10.1038/nature01823 PMID:12879064

Fernandes, J. C., Raphael, R. A. S., & Mulato, M. (2016). Effects of measurements conditions on an extended gate FET used as pH sensor. *Materials Research*, *19*(1), 267–272. doi:10.1590/1980-5373-MR-2015-0248

Fernández-Pacheco, R., Marquina, C., Valdivia, J. G., Gutiérrez, M., Romero, M. S., Cornudella, R., & García, A. et al. (2007). Magnetic nanoparticles for local drug delivery using magnetic implants. *Journal of Magnetism and Magnetic Materials*, *311*(1), 318–322. doi:10.1016/j.jmmm.2006.11.192

Florman, H. M., Arnoult, C., Kazam, I. G., Li, C., & O'Toole, C. M. (1998). A perspective on the control of mammalian fertilization by egg-activated ion channels in sperm: A tale of two channels. *Biology of Reproduction*, *59*(1), 12–16. doi:10.1095/biolreprod59.1.12 PMID:9674987

Foster, T. J. (2005). Immune evasion by Staphylococci. *Nature Reviews. Microbiology*, *3*(12), 948–958. doi:10.1038/nrmicro1289 PMID:16322743

Freskgard, P. O., Niewoehner, J., & Urich, E. (2014). Time to Open the Blood–Brain Barrier Gate for Biologics? *Future Neurology*, *9*(3), 243–245. doi:10.2217/fnl.14.15

Friedman, C. D., Costantino, P. D., Takagi, S., & Chow, L. C. (1998). Bone Source hydroxyapatite cement: A novel biomaterial for craniofacial skeletal tissue engineering and reconstruction. *Journal of Biomedical Materials Research*, *43*(4), 428–432. doi:10.1002/(SICI)1097-4636(199824)43:4<428::AID-JBM10>3.0.CO;2-0 PMID:9855201

Frohlich, M., Grayson, W. L., Wan, L. Q., Marolt, D., Drobnic, M., & Vunjak-Novakovic, G. (2008). Tissue Engineered Bone Grafts: Biological Requirements, Tissue Culture and Clinical Relevance. *Current Stem Cell Research & Therapy*, *3*(4), 254–264. doi:10.2174/157488808786733962 PMID:19075755

Fujisawa, R., Wada, Y., Nodasaka, Y., & Kuboki, Y. (1996). Acidic amino acid-rich sequences as binding sites of osteonectin to hydroxyapatite crystals. *Biochimica Et Biophysica Acta-Protein Structure and Molecular Enzymology*, *1292*(1), 53–60. doi:10.1016/0167-4838(95)00190-5 PMID:8547349

Fujishima, A., Rao, T. N., & Tryk, D. A. (2000). Titanium dioxide photocatalysis. *Journal of Photochemistry and Photobiology C, Photochemistry Reviews*, *1*(1), 1–21. doi:10.1016/S1389-5567(00)00002-2

Fux, C. A., Costerton, J. W., Stewart, P. S., & Stoodley, P. (2005). Survival strategies of infectious biofilms. *Trends in Microbiology*, *13*(1), 34–40. doi:10.1016/j.tim.2004.11.010 PMID:15639630

Gaillard, P. J. (2010). Crossing barriers from blood-to-brain and academia-to-industry. *Therapeutic Delivery*, *1*(4), 495–500. doi:10.4155/tde.10.43 PMID:22833962

Galler, K. M., & D'Souza, R. N. (2010). Tissue engineering approaches for regenerative dentistry. *Regenerative Medicine*, *6*(1), 111–124. doi:10.2217/rme.10.86 PMID:21175291

Galloway, A. L., Murphy, A., & De Simone, J. M. (2013). Development of a nanoparticle-based influenza vaccine using the PRINT technology. *Nanomedicine (London)*, *9*(4), 523–531. PMID:23178283

Gao, A., Lu, N., Wang, Y., & Li, T. (2016). Robust ultrasensitive tunneling-FET biosensor for point-of care diagnostics. *Scientific Reports*, *2*, 22554. doi:10.1038/srep22554 PMID:26932158

Gao, J. Q., Lv, Q., Li, L. M., Tang, X. J., Li, F. Z., Hu, L. Y., & Han, M. (2013). Glioma targeting and blood-brain barrier penetration by dual-targeting doxorubincin liposomes. *Biomaterials*, *34*(22), 5628–5639. doi:10.1016/j.biomaterials.2013.03.097 PMID:23628475

Gao, J., Gu, H., & Xu, B. (2009). Multifunctional Magnetic Nanoparticles: Design, Synthesis, and Biomedical Applications. *Accounts of Chemical Research*, *42*(8), 1097–1107. doi:10.1021/ar9000026 PMID:19476332

Gao, J., Li, L., Ho, P., Mak, G. C., Gu, H., & Xu, B. (2006). Combining fluorescent probes and biofunctional magnetic nanoparticles for rapid detection of bacteria in human blood. *Advanced Materials*, *18*(23), 3145–3148. doi:10.1002/adma.200601058

Gao, J., & Xu, B. (2009). Applications of nanomaterials inside cells. *Nano Today*, *4*(1), 37–51. doi:10.1016/j.nantod.2008.10.009

Garcia, P. S. et al. (2015). A Nanobiosensor based on 4-hydrosyphenylpyruvate dioxygenase Enzyme for mesotrione detection. *IEEE Sensors Journal*, *15*(4), 2106–2113. doi:10.1109/JSEN.2014.2371773

Garvin, K. A., Dalecki, D., & Hocking, D. C. Vascularization of Three-Dimensional Collagen Hydrogels Using Ultrasound Standing Wave Fields. *Ultrasound in Medicine & Biology, 37*(11), 1853-1864. doi:10.1016/j.ultrasmedbio.2011.07.008

Gelin, B. R. (1994). *Molecular modeling of polymer structures and properties*. Cincinnati, OH: Hanser/Gardner Publishers.

Gendelman, H. E., Kabanov, A., & Linder, J. (2008). The promise and perils of CNS drug delivery: A video debate. *Journal of Neuroimmune Pharmacology*, *3*(2), 58–64. doi:10.1007/s11481-008-9103-9 PMID:18322803

Georgantzinos, S. K., & Anifantis, N. K. (2010). Carbon nanotube-based resonant nanomechanical sensors: A computational investigation of their behavior. *Physica E, Low-Dimensional Systems and Nanostructures*, *42*(5), 1795–1801. doi:10.1016/j.physe.2010.02.002

Ghannoum, M. A., & O'Toole, G. A. (2004). *Microbial Biofilms*. Washington, DC, USA: Amer. Society for Microbiology Press.

Giannopoulos, G. I., Kakavas, P. A., & Anifantis, N. K. (2008). Evaluation of the effective mechanical properties of single walled carbon nanotubes using a spring based finite element approach. *Computational Materials Science*, *41*(4), 561–569. doi:10.1016/j.commatsci.2007.05.016

Gibbons, R. J., & Houte, J. V. (1975). Dental Caries. *Annual Review of Medicine*, *26*(1), 121–136. doi:10.1146/annurev.me.26.020175.001005 PMID:1096752

Gillman, P. K. (2007). Tricyclic antidepressant pharmacology and therapeutic drug interactions updated. *British Journal of Pharmacology*, *151*(6), 737–748. doi:10.1038/sj.bjp.0707253 PMID:17471183

Ginebra, M. P., Traykova, T., Planell, J. A. (2006). Calcium phosphate cements as bone drug delivery systems: a review. *Journal of controlled release*, 113, 102–110.

Ginebra, M. P., Traykova, T., & Planell, J. A. (2006). Calcium phosphate cements: Competitive drug carriers for the musculoskeletal system. *Biomaterials*, *27*(10), 2171–2177. doi:10.1016/j.biomaterials.2005.11.023 PMID:16332349

Ginebra, M.-P., Traykova, T., & Planell, J. (2006). Calcium phosphate cements as bone drug delivery systems: A review. *Journal of Controlled Release*, *113*(2), 102–110. doi:10.1016/j.jconrel.2006.04.007 PMID:16740332

Go, J., & Alam, M. A. (2009). Statistical interpretation of "femtomolar" detection. *Applied Physics Letters*, *95*(3), 033110. doi:10.1063/1.3176017 PMID:19690630

Gokel, G. W., & Negin, S. (2013). Synthetic Ion Channels: From Pores to Biological Applications. *Accounts of Chemical Research*, *46*(12), 2824–2833. doi:10.1021/ar400026x PMID:23738778

Gong, P., Li, H., He, X., Wang, K., Hu, J., Tan, W., & Yang, X. et al. (2007). Preparation and antibacterial activity of FeO @Ag nanoparticles. *Nanotechnology*, *18*(28), 604–611. doi:10.1088/0957-4484/18/28/285604

Goodson, J. M., Gunsolley, J. C., Grossi, S. G., Bland, P. S., Otomo-Corgel, J., Doherty, F., & Comiskey, J. (2007). Minocycline HCl microspheres reduce red-complex bacteria in periodontal disease therapy. *Journal of Periodontology*, *78*(8), 1568–1579. doi:10.1902/jop.2007.060488 PMID:17668977

Gopal, V., Xavier, J., Dar, G. H., Jafurulla, M., Chattopadhyay, A., & Rao, N. M. (2011). Targeted liposomes to deliver DNA to cells expressing 5-HT receptors. *International Journal of Pharmaceutics*, *419*(1-2), 347–354. doi:10.1016/j.ijpharm.2011.08.004 PMID:21855617

Gopinath, V., Ramakrishnan, T., Emmadi, P., Ambalavanan, N., Mammen, B., & Vijayalakshmi, . (2009). Effect of a controlled release device containing minocycline microspheres on the treatment of chronic periodontitis: A comparative study. *Journal of Indian Society of Periodontology*, *13*(2), 79–84. doi:10.4103/0972-124X.55844 PMID:20407655

Gordon, R. J., & Lowy, F. D. (2008). Pathogenesis of Methicillin-Resistant *Staphylococcus aureus* Infection. *Clinical Infectious Diseases*, *46*(Suppl. 5), S350–S359. doi:10.1086/533591 PMID:18462090

Gould, T., Westbury, L., & Tillman, J. (1982). Dentin matrix gelatin (DMG) as a possible "universal" grafting material in periodontics. *Journal of Periodontology*, *53*(1), 22–25. doi:10.1902/jop.1982.53.1.22 PMID:6948946

Grady, M. M., & Stahl, S. M. (2012). Practical guide for prescribing MAOIs: Debunking myths and removing barriers. *CNS Spectrums*, *17*(1), 2–10. doi:10.1017/S109285291200003X PMID:22790112

Grainger, D. W., & Okano, T. (2003). Biomedical micro and nano-technology. *Advanced Drug Delivery Reviews*, *55*(3), 311–313. doi:10.1016/S0169-409X(02)00222-3

Granger, E., McNee, G., Allan, V., & Woodman, P. (2014). The role of the cytoskeleton and molecular motors in endosomal dynamics. *Seminars in Cell & Developmental Biology*, *31*, 20–29. doi:10.1016/j.semcdb.2014.04.011 PMID:24727350

Graves, D. (2008). Cytokines that promote periodontal tissue destruction. *Journal of Periodontology*, *79*(8), 1585–1591. doi:10.1902/jop.2008.080183 PMID:18673014

Greenwald, S., & Boden, S.D., & Barrack, R.L. (2010). The Evolving Role of Bone-Graft Substitutes. *Paper presented at the American Academy of Orthopaedic Surgeons*, New Orleans, Louisiana.

Grüttner, C., Müller, K., Teller, J., Westphal, F., Foreman, A., & Ivkov, R. (2007). Synthesis and antibody conjugation of magnetic nanoparticles with improved specific power absorption rates for alternating magnetic field cancer therapy. *Journal of Magnetism and Magnetic Materials*, *311*(1), 181–186. doi:10.1016/j.jmmm.2006.10.1151

Guarnieri, D., Falanga, A., Muscetti, O., Tarallo, R., Fusco, S., & Galdiero, M. et al. (2013). Shuttle-mediated nanoparticle delivery to the blood brain barrier. *Small*, *9*(6), 853–862. doi:10.1002/smll.201201870 PMID:23135878

Gu, H., Ho, P.-L., Tsang, K. W. T., Wang, L., & Xu, B. (2003). Using biofunctional magnetic nanoparticles to capture vancomycin-resistant enterococci and other gram-positive bacteria at ultralow concentration. *Journal of the American Chemical Society*, *125*(51), 15702–15703. doi:10.1021/ja0359310 PMID:14677934

Gu, H., Xu, K., Xu, C., & Xu, B. (2006). Biofunctional magnetic nanoparticles for protein separation and pathogen detection. *Chemical Communications*, (9): 941–949. doi:10.1039/b514130c PMID:16491171

Gu, L., Elkin, T., Jiang, X., Li, H., Lin, Y., Qu, L., & Sun, Y.-P. et al. (2005). Single-walled carbon nanotubes displaying multivalent ligands for capturing pathogens. *Chemical Communications*, (7): 874. doi:10.1039/b415015e PMID:15700066

Gulyaev, A. E., Gelperina, S. E., Skidan, I. N., Antropov, A. S., Kivman, G. Y., & Kreuter, J. (1999). Significant transport of doxorubicin into the brain with polysorbate 80-coated nanoparticles. *Pharmaceutical Research*, *16*(10), 1564–1569. doi:10.1023/A:1018983904537 PMID:10554098

Gupta, A. K., & Gupta, M. (2005a). Synthesis and surface engineering of iron oxide nanoparticles for biomedical applications. *Biomaterials*, *26*(18), 3995–4021. doi:10.1016/j.biomaterials.2004.10.012 PMID:15626447

Gurung, M., Moon, D. C., Choi, C. W., Lee, J. H., Bae, Y. C., Kim, J., & Lee, J. C. (2011). *Staphylococcus aureus* Produces Membrane-Derived Vesicles That Induce Host Cell Death. *PLoS ONE*, *6*(11), e27958. doi:10.1371/journal.pone.0027958 PMID:22114730

Gutiérrez, L., Costo, R., Grüttner, C., Westphal, F., Gehrke, N., Heinke, D., & Morales, M. P. et al. (2015). Synthesis methods to prepare single- and multi-core iron oxide nanoparticles for biomedical applications. *Dalton Transactions (Cambridge, England)*, *44*(7), 2943–2952. doi:10.1039/C4DT03013C PMID:25564784

Gutiérrez, L., Lázaro, F. J., Abadía, A. R., Romero, M. S., Quintana, C., Morales, M. P., & Arranz, R. et al. (2006). Bioinorganic transformations of liver iron deposits observed by tissue magnetic characterisation in a rat model. *Journal of Inorganic Biochemistry*, *100*(11), 1790–1799. doi:10.1016/j.jinorgbio.2006.06.010 PMID:16899298

Gutiérrez, L., Mejías, R., Barber, D. F., Veintemillas-Verdaguer, S., Serna, C. J., Lázaro, F. J., & Morales, M. P. (2011). Ac magnetic susceptibility study of in vivo nanoparticle biodistribution. *Journal of Physics. D, Applied Physics*, *44*(25), 255002. doi:10.1088/0022-3727/44/25/255002

Guzman, M. G., Dille, J., & Godet, S. (2008). Synthesis of silver nanoparticles by chemical reduction method and their antibacterial activity. *World Acad Sci Eng Technol*, *43*, 357–364.

Haddon, R. C. (2002). Carbon nanotubes. *Accounts of Chemical Research*, *35*(12), 997–997. doi:10.1021/ar020259h PMID:12484786

Hafaiedh, I., Baccar, H., Ktari, T., & Abdelghani, A. (2012). Electrochemical chacterization of Streptavidin-HRP Immobilized on Multiwall carbon Nanotubes for Biosensor Applications. *Journal of Biomaterials and Nanobiotechnology*, *3*(01), 31–36. doi:10.4236/jbnb.2012.31005

Hahn, H., Logas, J., & Averback, R. S. (1990). Sintering characteristics of nanocrystalline TiO 2. *Journal of Materials Research*, *5*(03), 609–614. doi:10.1557/JMR.1990.0609

Haitham, M. A., & Almajhdi, F. N. (2011). Development of a SYBR Green I based real-time RT-PCR assay for detection and quantification of bovine coronavirus. *Molecular and Cellular Probes*, *25*(2-3), 101–107. doi:10.1016/j.mcp.2011.03.001 PMID:21419844

Hannig, M., & Hannig, C. (2010). Nanomaterials in preventive dentistry. *Nature Nanotechnology*, *5*(8), 565–569. doi:10.1038/nnano.2010.83 PMID:20581832

Han, Y., Zeng, Q., Lingling, E., Wang, D., He, H., & Liu, H. (2012). Sustained Topical delivery of insulin from fibrin gel loaded with poly (lactic-co-glycolic acid) microspheres improves the biomechanical retention of titanium implants in type 1 diabetic rats. *Journal of Oral and Maxillofacial Surgery*, *70*(10), 2299–2308. doi:10.1016/j.joms.2012.05.028 PMID:22871310

Hao, R., Xing, R., Xu, Z., Hou, Y., Gao, S., & Sun, S. (2010). Synthesis, functionalization, and biomedical applications of multifunctional magnetic nanoparticles. *Advanced Materials*, *22*(25), 2729–2742. doi:10.1002/adma.201000260 PMID:20473985

Hardman, J. G., Limbird, L. E., & Goodman Gilman, A. (2007). *Goodman Gilman's; The Pharmacological Basis of Therapeutics* (11th ed.). New York: The McGraw Hill Companies, Inc.

Harris, L. G., Foster, S. J., & Richards, R. G. (2002). An introduction to Staphylococcus aureus, and techniques for identifying and quantifying S. aureus adhesins in relation to adhesion to biomaterials: review. *Eur. Cell. Mater.*, 4, 39–60.

Hayashi, K., Pu, H., Tian, J., Andras, I. E., Lee, Y. W., Hennig, B., & Toborek, M. (2005). HIV-Tat protein induces P-glycoprotein expression in brain microvascular endothelial cells. *Journal of Neurochemistry*, *93*(5), 1231–1241. doi:10.1111/j.1471-4159.2005.03114.x PMID:15934943

Heath, J. R. (1995). The chemistry of size and order on a nanometer scale. *Science*, *270*(5240), 1315–1316. doi:10.1126/science.270.5240.1315

Heitzinger, C., Kennell, R., Klimeck, G., Mauser, N., Mclennan, M., & Ringhofer, C. (2008). Modeling and simulation of field effect biosensors and their deployment on the nanoHUB. *Journal of Physics: Conference Series*, *107*, 012004. doi:10.1088/1742-6596/107/1/012004

Heitzinger, C., & Klimeck, G. (2007). Computational Aspects of the three dimensional feature-scale simulation of silicon nanowire field-effect sensors for DNA detection. *Journal of Computational Electronics*, *6*(1-3), 387–390. doi:10.1007/s10825-006-0139-x

Heitzinger, C., Mauser, N. J., & Ringhofer, C. (2010). Multiscale modeling of planar and nanowire field-effect biosensors. *SIAM Journal on Applied Mathematics*, *70*(5), 1634–1654. doi:10.1137/080725027

He, J., Söderling, E., Vallittu, P. K., & Lassila, L. V. J. (2013). Preparation and Evaluation of Dental Resin with Antibacterial and Radio-Opaque Functions. *International Journal of Molecular Sciences*, *14*(3), 5445–5460. doi:10.3390/ijms14035445 PMID:23470923

Hellstrom, M. K., McClain, P. K., Schallhorn, R. G., Bellis, L., Hanlon, A. L., & Ramberg, P. (2008). Local minocycline as an adjunct to surgical therapy in moderate to severe, chronic periodontitis. *Journal of Clinical Periodontology*, *35*(6), 525–531. doi:10.1111/j.1600-051X.2008.01219.x PMID:18384391

Hertel, T., Walkup, R. E., & Avouris, P. (1998). Deformation of carbon nanotubes by surface van der waals forces. *Physical Review B: Condensed Matter and Materials Physics*, *58*(20), 13870–13873. doi:10.1103/PhysRevB.58.13870

He, X. Q., Kitipornchai, S., & Liew, K. M. (2005). Buckling analysis of multi-walled carbon nanotubes: A continuum model accounting for van der Waals interaction. *Journal of the Mechanics and Physics of Solids*, *53*(2), 303–326. doi:10.1016/j.jmps.2004.08.003

He, X. Z., Ma, J. Y., & Jabbari, E. (2008). Effect of Grafting RGD and BMP-2 Protein-Derived Peptides to a Hydrogel Substrate on Osteogenic Differentiation of Marrow Stromal Cells. *Langmuir*, *24*(21), 12508–12516. doi:10.1021/la802447v PMID:18837524

Hierold, C., Helbling, T., Roman, C., Durrer, L., Jungen, A., & Stampfer, C. (2008). CNT based sensors. In Advances in Science and Technology (Vol. 54, pp. 343-349). Trans Tech Publications. doi:10.4028/www.scientific.net/AST.54.343

Hille, B. (2001). *Ion channels of excitable membranes* (3rd ed.). Sunderland, Massachusetts, USA: Sinauer Associates, Inc.

Hladky, S. B., & Hadlay, D. A. (1972). Ion transfer across lipid membranes in the presence of gramicidin A. I. Studies of the unit conductance channel. *Biochimica et Biophysica Acta*, *274*(2), 294–312. doi:10.1016/0005-2736(72)90178-2 PMID:5048999

Hodgkin, A. L. & Huxley, A. F. (1952b). The components of membrane conductance in the giant axon of Loligo. *Journal of Physiology*, 116(4), 473-496.

Hodgkn, A. L., & Huxley, A. F. (1952a). Currents carried by sodium and potassium ions through the membrane of the giant axon of Loligo. *The Journal of Physiology*, *116*(4), 449–472. doi:10.1113/jphysiol.1952.sp004717 PMID:14946713

Hod, O., Rabani, E., & Baer, R. (2003). Carbon nanotube closed-ring structures. *Physical Review B: Condensed Matter and Materials Physics*, *67*(19), 195408. doi:10.1103/PhysRevB.67.195408

Hoffmann, E. K., Lambert, I. H., & Pedersen, S. F. (2009). Physiology of cell volume regulation in vertebrates. *Physiological Reviews*, *89*(1), 193–277. doi:10.1152/physrev.00037.2007 PMID:19126758

Hoffmann, M. R., Martin, S. T., Choi, W., & Bahnemann, D. W. (1995). Environmental Applications of Semiconductor Photocatalysis. *Chemical Reviews*, *95*(1), 69–96. doi:10.1021/cr00033a004

Hong, L., Krishnamachari, Y., Seabold, D., Joshi, V., Schneider, G., & Salem, A. K. (2011). Intracellular Release of 17-β Estradiol from Cationic Polyamidoamine Dendrimer Surface-Modified Poly (Lactic-co-Glycolic Acid) Microparticles Improves Osteogenic Differentiation of Human Mesenchymal Stromal Cells. *Tissue Engineering. Part C, Methods*, *17*(3), 319–325. doi:10.1089/ten.tec.2010.0388 PMID:20883116

Hong, R., Cima, M. J., Weissleder, R., & Josephson, L. (2008). Magnetic microparticle aggregation for viscosity determination by MR. *Magnetic Resonance in Medicine*, *59*(3), 515–520. doi:10.1002/mrm.21526 PMID:18306403

Hong, Y. S., Legge, R. L., Zhang, S., & Chen, P. (2003). Effect of amino acid sequence and pH on nanofiber formation of self-assembling peptides EAK16-II and EAK16-IV. *Biomacromolecules*, *4*(5), 1433–1442. doi:10.1021/bm0341374 PMID:12959616

Hood, J. D., Bednarski, M., Frausto, R., Guccione, S., Reisfeld, R. A., Xiang, R., & Cheresh, D. A. (2002). Tumor regression by targeted gene delivery to the neovasculature. *Science*, *296*(5577), 2404–2407. doi:10.1126/science.1070200 PMID:12089446

Hood, L., Heath, J. R., Phelps, M. E., & Lin, B. (2004). Systems biology and new technologies enable predictive and preventative medicine. *Science*, *306*(5696), 640–643. doi:10.1126/science.1104635 PMID:15499008

Hoomann, T., Jahnke, N., Horner, A., Keller, S., & Pohl, P. (2013). Filter gate closure inhibits ion but not water transport through potassium channels. *Proceedings of the National Academy of Sciences of the United States of America*, *10*(26), 10842–10847. doi:10.1073/pnas.1304714110 PMID:23754382

Hoopes, P. J., Tate, J. A., Ogden, J. A., Strawbridge, R. R., Fiering, S. N., & Petryk, A. A. … Ivkov, R. (2009). Assessment of intratumor non-antibody directed iron oxide nanoparticle hyperthermia cancer therapy and antibody directed IONP uptake in murine and human cells. In SPIE BiOS: Biomedical Optics (p. 71810P).

Horev, B., Klein, M. I., Hwang, G., Li, Y., Kim, D., Koo, H., & Benoit, D. S. W. (2015). pH-activated nanoparticles for controlled topical delivery of farnesol to disrupt oral biofilm virulence. *ACS Nano*, *9*(3), 2390–2404. doi:10.1021/nn507170s PMID:25661192

Horstmann, W. G., Verheyen, C. C. P. M., & Leemans, R. (2003). An injectable calcium phosphate cement as a bone-graft substitute in the treatment of displaced lateral tibial plateau fractures. *Injury*, *34*(2), 141–144. doi:10.1016/S0020-1383(02)00105-5 PMID:12565022

Howard, G. T. (2011). Microbial biodegradation of polyurethane. In A. Fainleib & O. Grigoryeva (Eds.), *Recent Developments in Polymer Recycling* (pp. 215–238). Kerala, India: Transworld Research Network.

Howell, D. (2016). Tooth decay still on the rise among English children. from http://www.bbc.com/news/health-35672775

Hsiao, F.-L., & Lee, C. (2011). Nanophotonic biosensors using hexagonal nanoring resonators: Computational study. Journal of micro/nanolithography. *MEMS and MOEMS*, *10*(1), 013001. doi:10.1117/1.3532834

Huang, Z.-M., Zhang, Y. Z., Kotaki, M., & Ramakrishna, S. (2003). A review on polymer nanofibers by electrospinning and their applications in nanocomposites. *Composites Science and Technology*, *63*(15), 2223–2253. doi:10.1016/S0266-3538(03)00178-7

Hugounenq, P., Levy, M., Alloyeau, D., Lartigue, L., Dubois, E., Cabuil, V., & Gazeau, F. et al. (2012). Iron oxide monocrystalline nanoflowers for highly efficient magnetic hyperthermia. *The Journal of Physical Chemistry C*, *116*(29), 15702–15712. doi:10.1021/jp3025478

Hung, S.-C., Cheng, N. J., Yang, C.-F., & Lo, Y.-P. (2014). Investigation of extended-gate field-effect transistor pH sensors based on different-temperature-ennealed bi-layer MWCNTs-In_2O_3 films. *Nanoscale Research Letters*, *9*(1), 502. doi:10.1186/1556-276X-9-502 PMID:25288911

Huotari, J., & Helenius, A. (2011). Endosome maturation. *The EMBO Journal*, *30*(17), 3481–3500. doi:10.1038/emboj.2011.286 PMID:21878991

Huse, J. T., Holland, E., & DeAngelis, L. M. (2013). Glioblastoma: Molecular Analysis and Clinical Implications. *Annual Review of Medicine*, *64*(1), 59–70. doi:10.1146/annurev-med-100711-143028 PMID:23043492

Hutter, E., & Maysinger, D. (2013). Gold-nanoparticle-based biosensors for detection of enzyme activity. *Trends in Pharmacological Sciences*, *34*(9), 497–507. doi:10.1016/j.tips.2013.07.002 PMID:23911158

Hu, X., Guo, Y., Wang, L., Hua, D., Hong, Y., & Li, J. (2011). Coenzyme Q10 nanoparticles prepared by a supercritical fluid-based method. *The Journal of Supercritical Fluids*, *57*(1), 66–72. doi:10.1016/j.supflu.2011.01.007

Hu, Y. H., Winn, S. R., Krajbich, I., & Hollinger, J. O. (2003). Porous polymer scaffolds surface-modified with arginine-glycine-aspartic acid enhance bone cell attachment and differentiation in vitro. *Journal of Biomedical Materials Research. Part A*, *64A*(3), 583–590. doi:10.1002/jbm.a.10438 PMID:12579573

Hyeon, T. (2003). Chemical synthesis of magnetic nanoparticles. *Chemical Communications*, (8): 927–934. doi:10.1039/b207789b PMID:12744306

Hyeon, T., Lee, S. S., Park, J., Chung, Y., & Na, H. (2001). Synthesis of highly crystalline and monodisperse maghemite nanocrystallites without a size-selection process. *Journal of the American Chemical Society*, *123*(51), 12798–12801. doi:10.1021/ja016812s PMID:11749537

Ibrahim, E., Mustafa, N., & Arif, D. (2007). Venlafaxine modulates depression–induced oxidative stress in brain and medulla of rats. *Neurochemical Research*, *32*(3), 497–505. doi:10.1007/s11064-006-9258-9 PMID:17268845

Iijima, S. (1991). Helical microtubules of graphitic carbon. *Nature, 354,* 6348, 56.

Iijima, S., Brabec, C., Maiti, A., & Bernholc, J. (1996). Structural flexibility of carbon nanotubes. *The Journal of Chemical Physics*, *104*(5), 2089–2092. doi:10.1063/1.470966

Iijima, S., & Ichihashi, T. (1993). Single-shell carbon nanotubes of 1-nm diameter. *Nature*, *363*(6430), 603–605. doi:10.1038/363603a0

Illum, L., Jabbal-Gill, I., Hinchcliffe, M., Fisher, A. N., & Davis, S. S. (2001). Chitosan as a novel nasal delivery system for vaccines. *Advanced Drug Delivery Reviews*, *51*(1), 81–96. doi:10.1016/S0169-409X(01)00171-5 PMID:11516781

Iyer, S., Romanowicz, B., & Laudon, M. (2004). *Biomolecular Motors, - A DARPA commissioned overview with highlights from the special session at Nanotech 2004*. Boston, MA, USA. Retrieved fromhttp://www.nsti.org/outreach/Biomolecular_Motors/

Jaberi-Ansari, Z., Ekrami, M., & Nojehdehian, H. (2013). Controlled Release of Drugs for Management of Pulpitis. *Avicenna Journal of Dental Research*, *6*(1).

Jain, A., Nair, P. R., & Alam, M. A. (2012). Flexure-fet biosensor to break the fundamental sensitivity limits of biosensors using nonlinear electromechanical coupling. *Proceedings of the National Academy of Sciences of the United States of America*, *109*(24), 9304–9308. doi:10.1073/pnas.1203749109 PMID:22623527

Jain, K. K. (2009). Current status and future prospects of nanoneurology. *Journal of Nanoneuroscience*, *1*(1), 56–64. doi:10.1166/jns.2009.006

Jain, K. K. (2012). Nanobiotechnology-based strategies for crossing the blood-brain barrier. *Nanomedicine (London)*, *7*(8), 1225–1233. doi:10.2217/nnm.12.86 PMID:22931448

Jamal, T., Rahman, M. A., Mirza, M. A., Panda, A. K., Talegaonkar, S., & Iqbal, Z. (2012). Formulation, antimicrobial and toxicity evaluation of bioceramic based ofloxacin loaded biodegradable microspheres for periodontal infection. *Current Drug Delivery*, *9*(5), 515–526. PMID:22758655

Jang, H.-J., & Cho, W.-J. (2016). Performance enhancement of capacitive-coupling dual-gate ion-sensitive field-effect transistor in ultra-thin-body. *Scientific Reports*, *4*, 5284. PMID:24923751

Jayesh, P., Chatterjee, A. K. P., Duttagupta, S., & Mukherji, S. (2007). Strain specificity in antimicrobial activity of silver and copper nanoparticles. *Nanomedicine (London)*, *3*, 95–101.

Jensen, K., Kim, K., & Zettl, A. (2008). An atomic-resolution nanomechanical mass sensor. *Nature Nanotechnology*, *3*(9), 533–537. doi:10.1038/nnano.2008.200 PMID:18772913

Jeun, M., Bae, S., Tomitaka, A., Takemura, Y., Park, K. H., Paek, S. H., & Chung, K.-W. (2009). Effects of particle dipole interaction on the ac magnetically induced heating characteristics of ferrite nanoparticles for hyperthermia. *Applied Physics Letters*, *95*(8), 82501. doi:10.1063/1.3211120

Jhajharia, K., Parolia, A., Shetty, K. V., & Mehta, L. K. (2015). Biofilm in endodontics: A review. *Journal of International Society of Preventive & Community Dentistry*, *5*(1), 1–12. doi:10.4103/2231-0762.151956 PMID:25767760

Jiles, D. C. (1998). *Introduction to magnetism and magnetic materials.* CRC press.

Jin, J. L., O'Doherty, A. M., Wang, S., Zheng, H., Sanders, K. M., & Yan, W. (2005). CatSper3 and CatSper4 encode two cation channel-like proteins exclusively expressed in the testis. *Biology of Reproduction, 73*(6), 1235–1242. doi:10.1095/biolreprod.105.045468 PMID:16107607

Johari, R., Kumar, G., Kumar, D., Singh, S. (2009). Synthesis and antibacterial activity of m (ii) schiff base complex. *J Ind Coun chem.*, 26(1), 23-27.

Jorgensen, J., Shariati, M., Shields, C. P., Durr, D. P., & Proskin, H. M. (1989). Fluoride uptake into demineralized primary enamel from fluoride-impregnated dental floss in vitro. *Pediatric Dentistry, 11*(1), 17–20. PMID:2626334

Josephson, L., Tung, C.-H., Moore, A., & Weissleder, R. (1999). High-efficiency intracellular magnetic labeling with novel superparamagnetic-Tat peptide conjugates. *Bioconjugate Chemistry, 10*(2), 186–191. doi:10.1021/bc980125h PMID:10077466

Joshi Anand, Y., Harsha, S. P., & Sharma Satish, C. (2010a). Vibration signature analysis of single walled carbon nanotube based nano mechanical sensors. *Physica E, Low-Dimensional Systems and Nanostructures, 42*(8), 2115–2123. doi:10.1016/j.physe.2010.03.033

Joshi Anand, Y., Sharma Satish, C., & Harsha, S. P. (2008). Vibration Analysis of Pre-Stressed Single Walled CNT Based Mass Sensor. *International Journal of Electrospun Nanofibers and Applications, 2*, 161–170.

Joshi Anand, Y., Sharma Satish, C., & Harsha, S. P. (2010b). The Effect of Pinhole Defect on Vibration Characteristics of Single Walled Carbon Nanotube. *Physica E, Low-Dimensional Systems and Nanostructures.*

Joshi Anand, Y., Sharma Satish, C., & Harsha, S. P. (2010d). Dynamic Analysis of a Clamped Wavy Single Walled Carbon Nanotube based Nano Mechanical Sensors. *ASME Journal of Nanotechnology in Engineering & Medicine, 1*, 031007-1–7.

Joshi, A. Y., Harsha, S. P., & Sharma Satish, C. (2010c). Vibration Response Analysis of Doubly Clamped Single Walled Wavy Carbon Nanotube Based Nano Mechanical Sensors. *ASME Journal of Nanotechnology in Engineering & Medicine, 1*, 031004-1–5.

Joshi, A. Y., Sharma, S. C., & Harsha, S. P. (2011). Effect of chirality and atomic vacancies on dynamics of nanoresonators based on SWCNT. *Sensor Review, 31*(1), 47–57. doi:10.1108/02602281111099080

Julien, M., Khairoun, I., LeGeros, R. Z., Delplace, S., Pilet, P., Weiss, P., & Guicheux, J. et al. (2007). Physico-chemical mechanical and in vitro biological properties of calcium phosphate cements with doped amorphous calcium phosphates. *Biomaterials, 28*(6), 956–965. doi:10.1016/j.biomaterials.2006.10.018 PMID:17123598

Julien, R. M. (2004). *A Primer Of Drug Action: A Comprehensive Guide To The Actions, Uses, And Side Effects of Psychoactive Drugs.* London: Macmillan.

Jung, C. W., & Jacobs, P. (1995). Physical and chemical properties of superparamagnetic iron oxide MR contrast agents: Ferumoxides, ferumoxtran, ferumoxsil. *Magnetic Resonance Imaging*, *13*(5), 661–674. doi:10.1016/0730-725X(95)00024-B PMID:8569441

Jun, Y., Huh, Y.-M., Choi, J., Lee, J.-H., Song, H.-T., Kim, S., & Shin, J.-S. et al. (2005). Nanoscale size effect of magnetic nanocrystals and their utilization for cancer diagnosis via magnetic resonance imaging. *Journal of the American Chemical Society*, *127*(16), 5732–5733. doi:10.1021/ja0422155 PMID:15839639

Jun, Y., Seo, J., & Cheon, J. (2008). Nanoscaling laws of magnetic nanoparticles and their applicabilities in biomedical sciences. *Accounts of Chemical Research*, *41*(2), 179–189. doi:10.1021/ar700121f PMID:18281944

Kamal, A., & Goldstein, L. S. B. (2000). Connecting vesicle transport to the cytoskeleton. *Current Opinion in Cell Biology*, *12*(4), 503–508. doi:10.1016/S0955-0674(00)00123-X PMID:10873823

Kamerer, D. B., Hirsch, B. E., Snyderman, C. H., Costantino, P., & Friedman, C. D. (1994). Hydroxyapatite cement: A new method for achieving watertight closure in transtemporal surgery. *The American Journal of Otology*, *15*, 47–49. PMID:8109630

Kam, H., Lee, D. T., Howe, R. T., & King, T.-J. (2005). A new nano-electro-mechanical field effect transistor (NEMFET) design for low-power electronics.*Proceedings of IEEE International Electron Device Meeting* (pp. 463-466).

Karageorgiou, V., & Kaplan, D. (2005). Porosity of 3D biornaterial scaffolds and osteogenesis. *Biomaterials*, *26*(27), 5474–5491. doi:10.1016/j.biomaterials.2005.02.002 PMID:15860204

Karaman, O., Kumar, A., Moeinzadeh, S., He, X., Cui, T., & Jabbari, E. (2013). Effect of surface modification of nanofibres with glutamic acid peptide on calcium phosphate nucleation and osteogenic differentiation of marrow stromal cells. *Journal of Tissue Engineering and Regenerative Medicine*. doi:10.1002/term.1775 PMID:23897753

Karger, A. (2008). Theoretical investigation of silicon nanowire pH sensor. *Proceedings of the 2nd IEEE RAS & EMBS International Conference on Biomedical Robotics and Biomechatronics* (pp. 765-769). Retrieved from http://lammps.sandia.gov/

Kashi, T. S., Eskandarion, S., Esfandyari-Manesh, M., Marashi, S. M., Samadi, N., & Fatemi, S. M. et al.. (2012). Improved drug loading and antibacterial activity of minocycline-loaded PLGA nanoparticles prepared by solid/oil/water ion pairing method. *International Journal of Nanomedicine*, *7*, 221–234. PMID:22275837

Kawaguchi, H. (2000). Functional polymer microspheres. *Progress in Polymer Science*, *25*(8), 1171–1210. doi:10.1016/S0079-6700(00)00024-1

Kawano, F., Ohguri, T., Ichikawa, T., Mizuno, I., & Hasegawa, A. (2002). Shock absorbability and hardness of commercially available denture teeth. *The International Journal of Prosthodontics*, *15*(3), 243–247. PMID:12066486

Keegan, G. M., Smart, J. D., Ingram, M. J., Barnes, L. M., Burnett, G. R., & Rees, G. D. (2012). Chitosan microparticles for the controlled delivery of fluoride. *Journal of Dentistry*, *40*(3), 229–240. doi:10.1016/j.jdent.2011.12.012 PMID:22212236

Kell, A. J., Stewart, G., Ryan, S., Peytavi, R., Boissinot, M., Huletsky, A., & Simard, B. et al. (2008). Vancomycin-modified nanoparticles for efficient targeting and preconcentration of Gram-positive and Gram-negative bacteria. *ACS Nano*, *2*(9), 1777–1788. doi:10.1021/nn700183g PMID:19206416

Kesarwani, K., & Gupta, R. (2013). Bioavailability enhancers of herbal origin: An overview. *Asian Pacific Journal of Tropical Biomedicine*, *3*(4), 253–266. doi:10.1016/S2221-1691(13)60060-X PMID:23620848

Khodashenas, B., & Ghorbani, H. R. Synthesis of silver nanoparticles with different shapes. *Arabian Journal of Chemistry*.

Kim, D. K., Mikhaylova, M., Wang, F. H., Kehr, J., Bjelke, B., Zhang, Y., & Muhammed, M. et al. (2003). Starch-coated superparamagnetic nanoparticles as MR contrast agents. *Chemistry of Materials*, *15*(23), 4343–4351. doi:10.1021/cm031104m

Kim, D., Lee, N., Park, M., Kim, B. H., An, K., & Hyeon, T. (2008). Synthesis of uniform ferrimagnetic magnetite nanocubes. *Journal of the American Chemical Society*, *131*(2), 454–455. doi:10.1021/ja8086906 PMID:19099480

Kim, H. M., Miyaji, F., Kokubo, T., & Nakamura, T. (1996). Preparation of bioactive Ti and its alloys via simple chemical surface treatment. *Journal of Biomedical Materials Research*, *32*(3), 409–417. doi:10.1002/(SICI)1097-4636(199611)32:3<409::AID-JBM14>3.0.CO;2-B PMID:8897146

Kim, H. W., Lee, H. H., & Knowles, J. C. (2006). Electrospinning biomedical nanocomposite fibers of hydroxyapaite/poly(lactic acid) for bone regeneration. *Journal of Biomedical Materials Research. Part A*, *79A*(3), 643–649. doi:10.1002/jbm.a.30866 PMID:16826596

Kim, J. S., Valencia, C. A., Liu, R., & Lin, W. (2007). Highly-efficient purification of native polyhistidine-tagged proteins by multivalent NTA-modified magnetic nanoparticles. *Bioconjugate Chemistry*, *18*(2), 333–341. doi:10.1021/bc0601951 PMID:17311440

Kim, J., Piao, Y., & Hyeon, T. (2009). Multifunctional nanostructured materials for multimodal imaging, and simultaneous imaging and therapy. *Chemical Society Reviews*, *38*(2), 372–390. doi:10.1039/B709883A PMID:19169455

Kim, M. M., & Kim, S. K. (2006). Chitooligosaccharides inhibit activation and expression of matrix metalloproteinase-2 in human dermal fibroblasts. *FEBS Letters*, *580*(11), 2661–2666. doi:10.1016/j.febslet.2006.04.015 PMID:16647062

Kim, S. E., Jeon, O., Lee, J. B., Bae, M. S., Chun, H. J., Moon, S. H., & Kwon, I. K. (2008). Enhancement of ectopic bone formation by bone morphogenetic protein-2 delivery using heparin-conjugated PLGA nanoparticles with transplantation of bone marrow-derived mesenchymal stem cells. *Journal of Biomedical Science*, *15*(6), 771–777. doi:10.1007/s11373-008-9277-4 PMID:18773307

Kinane, D. L. J. (2003). Clinical Periodontology and Implant Dentistry (4 ed.). Copenhagen: Blackwell Munksgaard.

King, R. W., & Kulkarni, R. (2013). Osteomyelitis in Emergency Medicine. *Emedicine.* Retrieved from http://emedicine.medscape.com/article/785020-overview#showall

Kirichok, Y., & Lishko, P. V. (2011). Rediscovering sperm ion channels with the patch-clamp technique. *Molecular Human Reproduction*, *17*(8), 478–499. doi:10.1093/molehr/gar044 PMID:21642646

Kishen, A., & Shrestha, A. (2013). Multi-functional micro and nanoparticles for use in root canal therapies: Google Patents.

Kishen, A., Shi, Z., Shrestha, A., & Neoh, K. G. (2008). An investigation on the antibacterial and antibiofilm efficacy of cationic nanoparticulates for root canal disinfection. *Journal of Endodontics*, *34*(12), 1515–1520. doi:10.1016/j.joen.2008.08.035 PMID:19026885

Klepac-Ceraj, V., Patel, N., Song, X., Holewa, C., Patel, C., Kent, R., & Soukos, N. S. et al. (2011). Photodynamic effects of methylene blue-loaded polymeric nanoparticles on dental plaque bacteria. *Lasers in Surgery and Medicine*, *43*(7), 600–606. doi:10.1002/lsm.21069 PMID:22057487

Koert, U. (2005). Synthetic ion channels: Functional analysis and structural studies. *Physical Chemistry Chemical Physics*, *7*(7), 1501–1506. doi:10.1039/b418561g PMID:19787974

Koh, I., Hong, R., Weissleder, R., & Josephson, L. (2008). Sensitive NMR sensors detect antibodies to influenza. *Angewandte Chemie*, *120*(22), 4187–4189. doi:10.1002/ange.200800069 PMID:18428168

Kokubo, T. (2008). Bioceramics and their clinical applications. Cambridge: Woodhead Pub. and Maney Pub.

Kokubo, T., & Takadama, H. (2006). How useful is SBF in predicting in vivo bone bioactivity? *Biomaterials*, *27*(15), 2907–2915. doi:10.1016/j.biomaterials.2006.01.017 PMID:16448693

Kretlow, J. D., Young, S., Klouda, L., Wong, M., & Mikos, A. G. (2009). Injectable Biomaterials for Regenerating Complex Craniofacial Tissues. *Advanced Materials*, *21*(32-33), 3368–3393. doi:10.1002/adma.200802009 PMID:19750143

Kreuter, J. & Gelperina, S (2008). Use of nanoparticles for cerebral cancer. *Tumori journal* 94, 271–277.

Kreuter, J. (2001). Nanoparticulate systems for brain delivery of drugs. *Advanced Drug Delivery*, *47*(1), 65–81. doi:10.1016/S0169-409X(00)00122-8 PMID:11251246

Kreuter, J. (2004). Influence of the surface properties on nanoparticle-mediated transport of drugs to the brain. *Journal of Nanoscience and Nanotechnology*, *4*(5), 484–488. doi:10.1166/jnn.2003.077 PMID:15503433

Kreuter, J. (2013). Mechanism of polymeric nanoparticle-based drug transport across the blood-brain barrier (BBB). *Journal of Microencapsulation: Micro and Nano Carriers*, *30*(1), 49–54. doi:10.3109/02652048.2012.692491 PMID:22676632

Kreuter, J. (2014). Drug delivery to the central nervous system by polymeric nanoparticles: What do we know? *Advanced Drug Delivery Reviews*, *71*, 2–14. doi:10.1016/j.addr.2013.08.008 PMID:23981489

Kreuter, J., & Alyautdin, R. N. (2000). Using nanoparticles to target drugs to the central nervous system. In D. J. Begley, M. W. Bradbury, & J. Kreuter (Eds.), *The Blood–Brain Barrier and Drug Delivery to the CNS* (pp. 205–223). New York, NY: Marcel Dekker, Inc. doi:10.1201/9780824741990

Kreuter, J., Alyautdin, R., Kharkevich, D., & Ivanov, A. A. (1995). Passage of peptides through the blood–brain barrier with colloidal polymer particles (nanoparticles). *Brain Research*, *674*(1), 171–174. doi:10.1016/0006-8993(95)00023-J PMID:7773690

Kreuter, J., Ramge, P., Petrov, V., Hamm, S., Gelperina, S. E., & Engelhardt, B. et al. (2003). Direct evidence that polysorbate-80-coated poly (butylcyanoacrylate) nanoparticles deliver drugs to the CNS via specific mechanisms requiring prior binding of drug to the nanoparticles. *Pharmaceutical Research*, *20*(3), 409–416. doi:10.1023/A:1022604120952 PMID:12669961

Kreuter, J., Shamenkov, D., Petrov, V., Ramge, P., Cychutek, K., Koch-Brandt, C., & Alyautdin, R. (2002). Apolipoprotein mediated transport of nanoparticle-bound drugs across the blood–brain barrier. *Journal of Drug Targeting*, *10*(4), 317–325. doi:10.1080/10611860290031877 PMID:12164380

Krumeick, F. (n. d.) Properties of electrons, their interactions with matter and applications in electron microscopy. *Microcopy.ethz.ch.* Retrieved from http://www.microscopy.ethz.ch/downloads/Interactions.pdf

Krycka, K. L., Jackson, A. J., Borchers, J. A., Shih, J., Briber, R., Ivkov, R., … Dennis, C. L. (2011). Internal magnetic structure of dextran coated magnetite nanoparticles in solution using small angle neutron scattering with polarization analysis. *Journal of Applied Physics, 109*(7).

Kulkarni, P. V., Roney, C. A., Antich, P. P., Bonte, F. J., Raghu, A. V., & Aminabhavi, T. M. (2010). Quinoline-*n*-butylcyanoacrylate-based nanoparticles for brain targeting for the diagnosis of Alzheimer's disease. *WIREs Nanomed Nanobiotechnol*, *2*(1), 35–47. doi:10.1002/wnan.59 PMID:20049829

Kumari, A., Yadav, S. K., & Yadav, S. C. (2010). Biodegradable polymeric nanoparticles based drug delivery systems. *Colloids and Surfaces*, *75*(1), 1–18. doi:10.1016/j.colsurfb.2009.09.001 PMID:19782542

Kumar, O., Singh, Y., Rao, V. K., & Vijayaraghavan, R. (2008). Carbon nanotubes: Detection of chemical and biological warfare agents. *Defence Science Journal*, *58*(5), 617–625. doi:10.14429/dsj.58.1684

Kung, H., & Foecke, T. (1999). Mechanical behavior of nanostructured materials. *MRS Bulletin*, *24*(2), 14–15.

Kunzelmann, K. (2005). Ion channels and cancer. *The Journal of Membrane Biology*, *205*(3), 159–173. doi:10.1007/s00232-005-0781-4 PMID:16362504

Kurakhmaeva, K. B., Kijindijkhasvili, I. A., Petrov, V. E., Balabanyan, V. U., Voronina, T. A., & Trofimov, S. S. et al. (2009). Brain targeting of nerve growth factor using poly(butyl cyanoacrylate) nanoparticles. *Journal of Drug Targeting*, *17*(8), 564–574. doi:10.1080/10611860903112842 PMID:19694610

Labhasetwar, V., Song, C., Humphrey, W., Shebuski, R., & Levy, R. J. (1998). Arterial uptake of biodegradable nanoparticles: Effect of surface modifications. *Journal of Pharmaceutical Sciences*, *87*(10), 1229–1234. doi:10.1021/js980021f PMID:9758682

Lacroix, L.-M., Ho, D., & Sun, S. (2010). Magnetic nanoparticles as both imaging probes and therapeutic agents. *Current Topics in Medicinal Chemistry*, *10*(12), 1184–1197. doi:10.2174/156802610791384207 PMID:20388109

Lamont, R. J., & Yilmaz, O. (2002). In or out: The invasiveness of oral bacteria. *Periodontology 2000*, *30*(1), 61–69. doi:10.1034/j.1600-0757.2002.03006.x PMID:12236896

Lang, F., Ritter, M., Gamper, N., Huber, S., Fillon, S., Tanneur, V., & Gulbins, E. et al. (2000). Cell volume in the regulation of cell proliferation and apoptotic cell death. *Cellular Physiology and Biochemistry*, *10*(5-6), 417–428. doi:10.1159/000016367 PMID:11125224

Laquintana, V., Trapani, A., Denora, N., Wang, F., Gallo, G. M., & Tripani, G. (2009). New strategies to deliver anticancer drugs to brain tumors. *Expert Opinion on Drug Delivery*, *6*(10), 1017–1032. doi:10.1517/17425240903167942 PMID:19732031

Lartigue, L., Hugounenq, P., Alloyeau, D., Clarke, S. P., Lévy, M., Bacri, J.-C., & Gazeau, F. et al. (2012). Cooperative organization in iron oxide multi-core nanoparticles potentiates their efficiency as heating mediators and MRI contrast agents. *ACS Nano*, *6*(12), 10935–10949. PMID:23167525

Lassagne, B., Garcia-Sanchez, D., Aguasca, A., & Bachtold, A. (2008). Ultrasensitive mass sensing with a nanotube electromechanical resonator. *Nano Letters*, *8*(11), 3735–3738. doi:10.1021/nl801982v PMID:18939809

Laurent, S., Forge, D., Port, M., Roch, A., Robic, C., Vander Elst, L., & Muller, R. N. (2008). Magnetic iron oxide nanoparticles: Synthesis, stabilization, vectorization, physicochemical characterizations, and biological applications. *Chemical Reviews*, *108*(6), 2064–2110. doi:10.1021/cr068445e PMID:18543879

Leanza, L., Biasutto, L., Managò, A., Gulbins, E. Zoratti, M. and Szabò, I. (2013). Intracellular ion channels and cancer. *Frontiers in Physiology*, 4, 227(1-7)

Lee, J.U., Nguyen, A.H. & Sim, S.J. (2015). A nanoplasmonic biosensor for label-free multiplex detection of cancer biomarkers. *Biosensors and Bioelectronics*, 74, 341-346.

Lee, E., Choi, D., Kim, D., Kim, J., Park, J. O., Kim, S., & Gho, Y. S. et al. (2009). Gram-positive bacteria produce membrane vesicles: Proteomics-based characterization of *Staphylococcus aureus*-derived membrane vesicles. *Proteomics*, *9*(24), 5425–5436. doi:10.1002/pmic.200900338 PMID:19834908

Lee, H., Sun, E., Ham, D., & Weissleder, R. (2008). Chip–NMR biosensor for detection and molecular analysis of cells. *Nature Medicine*, *14*(8), 869–874. doi:10.1038/nm.1711 PMID:18607350

Lee, H., Yoon, T., & Weissleder, R. (2009). Ultrasensitive Detection of Bacteria Using Core–Shell Nanoparticles and an NMR-Filter System. *Angewandte Chemie International Edition*, *48*(31), 5657–5660. doi:10.1002/anie.200901791 PMID:19554581

Lee, I.-K., Bae, T.-E., Lau, H.-C., Lim, J.-O., & Cho, W.-J. (2014). A study on a differential-measurement biosensor using high performance extended-gate field-effect transistors with TiO_2 sensing membranes. *Sensor Letters*, *12*(6), 1003–1007. doi:10.1166/sl.2014.3193

Lee, J. C. (2012). *Staphylococcus aureus* Membrane Vesicles and Its Potential Role in Bacterial Pathogenesis. *Journal of Bacteriology and Virology*, *42*(3), 181–188. doi:10.4167/jbv.2012.42.3.181

Lee, J.-H., Huh, Y.-M., Jun, Y., Seo, J., Jang, J., Song, H.-T., & Suh, J.-S. et al. (2007). Artificially engineered magnetic nanoparticles for ultra-sensitive molecular imaging. *Nature Medicine*, *13*(1), 95–99. doi:10.1038/nm1467 PMID:17187073

Lee, M., & Fauchet, P. M. (2007). Two dimensional silicon photonic crystal based biosensing platform for protein detection. *Optics Express*, *15*(8), 4530. doi:10.1364/OE.15.004530 PMID:19532700

Lee, P.-Y., Chang, S.-P., Kuo, P. J., & Shei, S. C. (2013). Sensing performance of EGFET pH sensors with CZTse nanoparticles fabricated on glass substrates. *International Journal of Electrochemical Sensing*, *8*, 3866–3875.

Lee, S., & Yoon, D. S. (2007). Self-Assembly of Colloidal Particles on a Patterned Surface with Wettability. *Biochip Journal*, *1*, 193.

Lehen'kyi, V., Shapovalov, C., Skryma, R., & Prevarskaya, N. (2011). Ion channnels and transporters in cancer. 5. Ion channels in control of cancer and cell apoptosis. *American Journal of Physiology. Cell Physiology*, *30*(6), C1281–C1289. doi:10.1152/ajpcell.00249.2011 PMID:21940667

Leppard, K., Dimmock, N., & Easton, A. (2007). *Introduction to Modern Virology* (p. 450). Blackwell Publishing Limited.

Lewandowsky, M. (1900). Zur Lehre der Zerebrospinalflussigkeit. *Zeitschrift fur Klinische Medizin*, *40*, 480–484.

Lew, D. P., & Waldvogel, F. A. (2004). Osteomyelitis. *Lancet*, *364*(9431), 369–379. doi:10.1016/S0140-6736(04)16727-5 PMID:15276398

Li, C., & Chou, T. W. (2003). A structural mechanics approach for the analysis of carbon nanotubes. *International Journal of Solids and Structures*, *40*(10), 2487–2499. doi:10.1016/S0020-7683(03)00056-8

Li, C., & Chou, T. W. (2004). Mass detection using carbon nanotube-based nanomechanical resonators. *Applied Physics Letters*, *84*(25), 5246–5248. doi:10.1063/1.1764933

Li, C., & Chou, T.-W. (2003b). Single-walled carbon nanotubes as ultrahigh frequency nano mechanical resonators. *Physical Review B: Condensed Matter and Materials Physics*, *68*, 734051–734053.

Li, C., & Chou, T.-W. (2004). Vibrational behaviors of multi-walled carbon nanotube-based nano mechanical resonators. *Applied Physics Letters*, *84*(1), 121–123. doi:10.1063/1.1638623

Li, G., Wang, Y., & Tao, Q. (2013). Modeling and simulation of silicon nanowire-based biosensors. In *Modeling and control for micro/nano devices and systems*. London: Taylor and Francis.

Li, L., Mak, K. Y., Leung, C. W., Chan, K. Y., Chan, W. K., Zhong, W., & Pong, P. W. T. (2013). Effect of synthesis conditions on the properties of citric-acid coated iron oxide nanoparticles. *Microelectronic Engineering*, *110*, 329–334. doi:10.1016/j.mee.2013.02.045

Lin, L. M., Skribner, J. E., & Gaengler, P. (1992). Factors associated with endodontic treatment failures. *Journal of Endodontics*, *18*(12), 625–627. doi:10.1016/S0099-2399(06)81335-X PMID:1298804

Lin, Y. E., Vidic, R. D., Stout, J. E., Mccartney, C. A., & Yu, V. L. (1998). Inactivation of Mycobacterium avium by copper and silver ions. *Water Research*, *32*(7), 1997–2000. doi:10.1016/S0043-1354(97)00460-0

Lin, Y. E., Vidic, R. D., Stout, J. E., & Yu, V. L. (1996). Individual and combined effects of copper and silver ions on inactivation of *Legionella pneumophila. Water Research*, *30*(8), 1905–1913. doi:10.1016/0043-1354(96)00077-2

Lin, Y., Taylor, S., Li, H., Fernando, K. S., Qu, L., Wang, W., & Sun, Y. P. (2004). Advances toward bioapplications of carbon nanotubes. *Journal of Materials Chemistry*, *14*(4), 527–541. doi:10.1039/b314481j

Lishko, P., Kirichok, Y., Ren, D., Navarro, B., Chung, J.J. & Clapham, D.E. (2012), The control of male fertility by spermatozoan ion channels. *Annual Review of Physiology*, 74, 1-10.23.

Lishko, P. V., Botchkina, I. L., & Kirichok, Y. (2011). Progesterone activates the principal Ca2+ channel of human sperm. *Letter to Nature*, *471*(7338), 387–391. doi:10.1038/nature09767 PMID:21412339

Listgarten, M. A. (1987). Nature of periodontal diseases: Pathogenic mechanisms. *Journal of Periodontal Research*, *22*(3), 172–178. doi:10.1111/j.1600-0765.1987.tb01560.x PMID:2955092

Litman, T., Brangi, M., Hudson, E., Fetsch, P., Abati, A., & Ross, D. D. et al. (2000). The multidrug-resistant phenotype associated with overexpression of the new ABC half-transporter, MXR (ABCG2). *Journal of Cell Science*, *113*(Pt 11), 2011–2021. PMID:10806112

Liu, D. Z., Chen, W. P., Lee, C. P., Wu, S. L., Wang, Y. C., & Chung, T. W. (2005). Effects of alginate coated on PLGA microspheres for delivery tetracycline hydrochloride to periodontal pockets. *Journal of Microencapsulation*, *21*(6), 643–652. doi:10.1080/02652040400000512 PMID:15762321

Liu, H. L., & Lin, J. C. (2004). A set of homology models of pore loop domain of six eukaryotic voltage- gated potassium channels Kv1.1-Kv. Proteins: Structure. *Function and Ioinformatics*, *55*, 558–567. doi:10.1002/prot.20065

Liu, J., Xia, J., Cho, K. H., Clapham, D. E., & Ren, D. (2007). CatSperbeta, a novel transmembrane protein in the CatSper channel complex. *The Journal of Biological Chemistry*, *282*(26), 18945–18952. doi:10.1074/jbc.M701083200 PMID:17478420

Liu, W. Y., Yeh, Y. C., Lipner, J., Xie, J. W., Sung, H. W., Thomopoulos, S., & Xia, Y. N. (2011). Enhancing the Stiffness of Electrospun Nanofiber Scaffolds with a Controlled Surface Coating and Mineralization. *Langmuir*, *27*(15), 9088–9093. doi:10.1021/la2018105 PMID:21710996

Liu, X., Smith, L. A., Hu, J., & Ma, P. X. (2009). Biomimetic nanofibrous gelatin/apatite composite scaffolds for bone tissue engineering. *Biomaterials*, *30*(12), 2252–2258. doi:10.1016/j.biomaterials.2008.12.068 PMID:19152974

Liu, Y., Tan, J., Thomas, A., Ou-Yang, D., & Muzykantov, V. R. (2012). The shape of things to come: Importance of design in nanotechnology for drug delivery. *Therapeutic Delivery*, *3*(2), 181–194. doi:10.4155/tde.11.156 PMID:22834196

Liverneaux, P. A. (2004). Osteoporotic distal radius curettage-filling with an injectable calcium phosphate cement. A cadaveric study. *European Journal of Orthopaedic Surgery & Traumatology*, *15*(1), 1–6. doi:10.1007/s00590-004-0182-x

Li, X. R., Xie, J. W., Yuan, X. Y., & Xia, Y. N. (2008). Coating Electrospun Poly(ε-caprolactone) Fibers with Gelatin and Calcium Phosphate and Their Use as Biomimetic Scaffolds for Bone Tissue Engineering. *Langmuir*, *24*(24), 14145–14150. doi:10.1021/la802984a PMID:19053657

Li, Z., Kawashita, M., Araki, N., Mitsumori, M., Hiraoka, M., & Doi, M. (2010). Preparation of magnetic iron oxide nanoparticles for hyperthermia of cancer in a FeCl2-NaNO3-NaOH aqueous system. *Journal of Biomaterials Applications*. PMID:20207773

Lobley, A., Pierron, V., Reynolds, L., Allen, L., & Michalovich, D. (2003). Identification of human and mouse CatSper3 and CatSper4 genes: Characterisation of a common interaction domain and evidence for expression in testis. *Reproductive Biology and Endocrinology*, *1*(1), 53. doi:10.1186/1477-7827-1-53 PMID:12932298

Logeart-Avramoglou, D., Anagnostou, F., Bizios, R., & Petite, H. (2005). Engineering bone: Challenges and obstacles. *Journal of Cellular and Molecular Medicine*, *9*(1), 72–84. doi:10.1111/j.1582-4934.2005.tb00338.x PMID:15784166

Longo, J. P. F., Leal, S. C., Simioni, A. R., Almeida-Santos, M. F. M., Tedesco, A. C., & Azevedo, R. B. (2012). Photodynamic therapy disinfection of carious tissue mediated by aluminum-chloride-phthalocyanine entrapped in cationic liposomes: An in vitro and clinical study. *Lasers in Medical Science*, *27*(3), 575–584. doi:10.1007/s10103-011-0962-6 PMID:21809069

Löscher, W., & Potschka, H. (2005). Blood-Brain Barrier Active Efflux Transporters: ATP-Binding Cassette Gene Family. *NeuroRx*, *2*(1), 86–98. doi:10.1602/neurorx.2.1.86 PMID:15717060

Löwe, J., van den Ent, F., & Amos, L. A. (2004). Molecules of the bacterial cytoskeleton. *Annual Review of Biophysics and Biomolecular Structure*, *33*(1), 177–198. doi:10.1146/annurev.biophys.33.110502.132647 PMID:15139810

Lowy, F. D. (1998). *Staphylococcus aureus* infections. *The New England Journal of Medicine*, *339*(8), 520–532. doi:10.1056/NEJM199808203390806 PMID:9709046

Lowy, F. D. (2000). Is *Staphylococcus aureus* an intracellular pathogen? *Trends in Microbiology*, *8*(8), 341–343. doi:10.1016/S0966-842X(00)01803-5 PMID:10920387

Lu, A.-H., Salabas, E. L., & Schüth, F. (2007). Magnetic Nanoparticles: Synthesis, Protection, Functionalization, and Application. *Angewandte Chemie International Edition*, *46*(8), 1222–1244. doi:10.1002/anie.200602866 PMID:17278160

Lu, A., Schmidt, W., Matoussevitch, N., Bönnemann, H., Spliethoff, B., Tesche, B., & Schüth, F. et al. (2004). Nanoengineering of a magnetically separable hydrogenation catalyst. *Angewandte Chemie*, *116*(33), 4403–4406. doi:10.1002/ange.200454222 PMID:15368378

Luengo, Y., Nardecchia, S., Morales, M. P., & Serrano, M. C. (2013). Different cell responses induced by exposure to maghemite nanoparticles. *Nanoscale*, *5*(23), 11428–11437. doi:10.1039/c3nr02148c PMID:23963338

Lu, F., Gu, L., Meziani, M. J., Wang, X., Luo, P. G., Veca, L. M., & Sun, Y. P. (2009). Advances in bioapplications of carbon nanotubes. *Advanced Materials*, *21*(2), 139–152. doi:10.1002/adma.200801491

Lynch, I., Salvati, A., & Dawson, K. A. (2009). Protein-nanoparticle interactions: What does the cell see? *Nature Nanotechnology*, *4*(9), 546–547. doi:10.1038/nnano.2009.248 PMID:19734922

Madigan, M., & Martinko, J. (Eds.), (2005). *Brock Biology of Microorganisms* (11th ed.). Prentice Hall.

Magalh, xe, es, A. P. R., Santos, L. B., Lopes, L. G., Estrela, C. R. d. A., et al. (2012). Nanosilver Application in Dental Cements. *ISRN Nanotechnology, 2012*, 6.

Maguina C, Gotuzzo E. (2000). Bartonellosis: New and old. *Infect. Dis. Clin. North Am.*, 14, 1–22. Doi: 10.1016/S0891-5520(05)70215-4

Mahan, G. D. (2002). Oscillations of a thin hollow cylinder: Carbon nanotubes. *Physical Review B: Condensed Matter and Materials Physics, 65*(23), 235402. doi:10.1103/PhysRevB.65.235402

Mailander, V., & Landfester, K. (2009). Interaction of nanoparticles with cells. *Biomacromolecules, 10*(9), 2379–2400. doi:10.1021/bm900266r PMID:19637907

Mainardes, R. M., & Gremião, M. P. D. (2012). Nanoencapsulation and characterization of zidovudine on poly(L-lactide) and poly(L-lactide)-poly(ethylene glycol)-blend nanoparticles. *Journal of Nanoscience and Nanotechnology, 12*(11), 8513–8521. doi:10.1166/jnn.2012.6638 PMID:23421238

Mao, Y., Shin, K.S., Wang, X., Ji, Z., Meng., H. & Chui, C.O. (2016). Semiconductor electronic label-free assay for predictive toxicology. *Scientific Reports, 6*, 24982. doi:10.1038/srep24982

Ma, P. X. (2008). Biomimetic materials for tissue engineering. *Advanced Drug Delivery Reviews, 60*(2), 184–198. doi:10.1016/j.addr.2007.08.041 PMID:18045729

Ma, P. X., & Zhang, R. Y. (1999). Synthetic nano-scale fibrous extracellular matrix. *Journal of Biomedical Materials Research, 46*(1), 60–72. doi:10.1002/(SICI)1097-4636(199907)46:1<60::AID-JBM7>3.0.CO;2-H PMID:10357136

Ma, P. X., Zhang, R. Y., Xiao, G. Z., & Franceschi, R. (2001). Engineering new bone tissue in vitro on highly porous poly(alpha-hydroxyl acids)/hydroxyapatite composite scaffolds. *Journal of Biomedical Materials Research, 54*(2), 284–293. doi:10.1002/1097-4636(200102)54:2

Marra, K. G., Szem, J. W., Kumta, P. N., DiMilla, P. A., & Weiss, L. E. (1999). In vitro analysis of biodegradable polymer blend/hydroxyapatite composites for bone tissue engineering. *Journal of Biomedical Materials Research, 47*(3), 324–335. doi:10.1002/(SICI)1097-4636(19991205)47:3<324::AID-JBM6>3.0.CO;2-Y PMID:10487883

Martina, M.-S., Fortin, J.-P., Ménager, C., Clément, O., Barratt, G., Grabielle-Madelmont, C., & Lesieur, S. et al. (2005). Generation of superparamagnetic liposomes revealed as highly efficient MRI contrast agents for in vivo imaging. *Journal of the American Chemical Society, 127*(30), 10676–10685. doi:10.1021/ja0516460 PMID:16045355

Martin, D. H., & Martin, D. H. (1967). *Magnetism in solids.* MA: MIT Press Cambridge.

Mascarelli, A. (2011). Tooth decay is prevalent among poor children. Retrieved from http://articles.latimes.com/2011/jul/01/health/hk-he-dental-underserved-20110701

Massart, R. (1981). Preparation of aqueous magnetic liquids in alkaline and acidic media. *IEEE Transactions on Magnetics, 17*(2), 1247–1248. doi:10.1109/TMAG.1981.1061188

Masuzaki, T., Ayukawa, Y., Moriyama, Y., Jinno, Y., Atsuta, I., Ogino, Y., & Koyano, K. (2010). The effect of a single remote injection of statin-impregnated poly (lactic-co-glycolic acid) microspheres on osteogenesis around titanium implants in rat tibia. *Biomaterials, 31*(12), 3327–3334. doi:10.1016/j.biomaterials.2010.01.016 PMID:20129662

Mattson, M. P., Haddon, R. C., & Rao, A. M. (2000). Molecular functionalization of carbon nanotubes and use as substrates for neuronal growth. *Journal of Molecular Neuroscience*, *14*(3), 175–182. doi:10.1385/JMN:14:3:175 PMID:10984193

May, R. C., & Machesky, L. M. (2001). Phagocytosis and the actin cytoskeleton. *Journal of Cell Science*, *114*(6), 1061–1077. PMID:11228151

Ma, Z. W., Kotaki, M., Inai, R., & Ramakrishna, S. (2005). Potential of nanofiber matrix as tissue-engineering scaffolds. *Tissue Engineering*, *11*(1-2), 101–109. doi:10.1089/ten.2005.11.101

Mazzarino, L., Coche-Guérente, L., Labbé, P., Lemos-Senna, E., & Borsali, R. (2014). On the mucoadhesive properties of chitosan-coated polycaprolactone nanoparticles loaded with curcumin using quartz crystal microbalance with dissipation monitoring. *Journal of Biomedical Nanotechnology*, *10*(5), 787–794. doi:10.1166/jbn.2014.1768 PMID:24734531

McCommis, K. S., & Baines, C. P. (2012). The role of VDAC in cell death: friend or foe. *Biochimica et Biophysica Acta*, *1818*(6), 1444–1450. doi:10.1016/j.bbamem.2011.10.025 PMID:22062421

McCormick, F. (1991). GTP-binding proteins as oncogenes in human tumors. *Environmental Health Perspectives*, *93*, 17–18. PMID:1773789

McCurrie, R. A. (1994). *Ferromagnetic Materials: Structure and Properties. New York, Academic.*

McRae, M. (2016). HIV and viral protein effects on the blood brain barrier. *Tissue Barriers.* 28;4(1). doi:.10.1080/21688370.2016.1143543

Medina, C., Santos, M. J., Radomski, A., Corrigan, O., & Radomski, M. W. (2007). Nanoparticles: Pharmacological and Toxicological Significance. *British Journal of Pharmacology*, *150*(5), 552–558. doi:10.1038/sj.bjp.0707130 PMID:17245366

Mehra, M., Vahidi, F., & Berg, R. W. (2013). A complete denture impression technique survey of postdoctoral prosthodontic programs in the United States. *Journal of Prosthodontics*, *23*(4), 320–327. doi:10.1111/jopr.12099 PMID:24118134

Mei, L.-P., Feng, J.-J., Wu, L., Zhou, J.-Y., Chen, J.-R., & Wang, A. J. (2015). Novel phenol biosensor based on laccase immobilized on reduced graphene oxide supported palladium–copper alloyed nanocages. *Biosensors & Bioelectronics*, *74*, 347–352. doi:10.1016/j.bios.2015.06.060 PMID:26159155

Meirovitch, L. (1997). *Principles and techniques of vibrations* (Vol. 1). New Jersey: Prentice Hall.

Melo, M. A. S., Guedes, S. F. F., Xu, H. H. K., & Rodrigues, L. K. A. (2013). Nanotechnology-based restorative materials for dental caries management. *Trends in Biotechnology*, *31*(8), 459–467. doi:10.1016/j.tibtech.2013.05.010 PMID:23810638

Meolin, J., & Quake, S. R. (2007). Microfluidic large scale integration: The evolution of design rules for biological automation. *Annual Review of Biophysics and Biomolecular Structure*, *36*(1), 213–231. doi:10.1146/annurev.biophys.36.040306.132646 PMID:17269901

Michaelis, K., Hoffmann, M. M., Dreis, S., Herbert, E., Alyautdin, R. N., & Michaelis, M. et al. (2006). Covalent linkage of apolipoprotein E to albumin nanoparticles strongly enhances drug transport into the brain. *The Journal of Pharmacology and Experimental Therapeutics*, *317*(3), 1246–1253. doi:10.1124/jpet.105.097139 PMID:16554356

Michie, K. A., & Löwe, J. (2006). Dynamic Filaments of the Bacterial Cytoskeleton. *Annual Review of Biochemistry*, *75*(1), 467–492. doi:10.1146/annurev.biochem.75.103004.142452 PMID:16756499

Milani, M., Didenko, L. V., Avtandilov, G. A., Curia, R., Erega, A., & Shevlyagina, N. V. (2016). Electron Microscopy Documents the Microorganisms' Biodestructive Action on Polyurethane and the Production, Internalization and Vesicular Trafficking of Nanoparticles. *British Journal of Applied Science & Technology*, *12*(3), 1–19. doi:10.9734/BJAST/2016/20290

Milani, M., Drobne, D., & Tatti, F. (2007). How to study biological samples by FIB/SEM? In A. Méndez-Vilas & J. Díaz (Eds.), *Modern Research and Educational Topics in Microscopy* (pp. 787–794). Spain: Formatex Research Center.

Miller, D. S., Bauer, B., & Hartz, A. M. S. (2008). Modulation of P-glycoprotein at the blood–brain barrier: opportunities to improve central nervous system pharmacotherapy, *Pharmacol. Rev.* 60:196–209.

Minami, T., Sasaki, Y., Minamiki, T., Wakida, S., Kurita, R., Niwa, O., & Tokito, S. (2016). Selectrive nitrate detection by an enzymatic sensor based on an extended-gate type organic field-effect transistor. *Biosensors & Bioelectronics*, *81*, 87–91. doi:10.1016/j.bios.2016.02.036 PMID:26921557

Min, K. H., Park, K., Kim, Y. S., Bae, S. M., Lee, S., Jo, H. G., & Kim, K. et al. (2008, May). Kwon, I.C., Hydrophobically modified glycol chitosan nanoparticles-encapsulated camptothecin enhance the drug stability and tumor targeting in cancer therapy. *Journal of Controlled Release*, *127*(3), 208–218. doi:10.1016/j.jconrel.2008.01.013 PMID:18336946

Mirtchi, A. A., Lemaitre, J., & Terao, N. (1989). Calcium phosphate cements: Study of the beta-tricalcium phosphate-monocalcium phosphate system. *Biomaterials*, *10*(7), 475–480. doi:10.1016/0142-9612(89)90089-6 PMID:2804235

Mirtichi, A. A., Lemaitre, J., & Munting, E. (1989). Calcium phosphate cements: Action of setting regulators on the properties of the beta-tricalcium phosphate-monocalcium phosphate cements. *Biomaterials*, *10*(9), 634–638. doi:10.1016/0142-9612(89)90120-8 PMID:2611315

Mishra, B., Patel, B. B., & Tiwari, S. (2010). Colloidal nanocarriers: A review on formulation technology, types and applications toward targeted drug delivery. *Nanomedicine; Nanotechnology, Biology, and Medicine*, *6*(1), 9–24. doi:10.1016/j.nano.2009.04.008 PMID:19447208

Mishra, L. C., Singh, B. B., & Dagenais, S. (2000). Scientific basis for the therapeutic use of Withania somnifera. (Ashwagandha): A review. *Alternative Medicine Review*, *5*, 334–346. PMID:10956379

Molander, A., Reit, C., & Dahlén, G. (1999). The antimicrobial effect of calcium hydroxide in root canals pretreated with 5% iodine potassium iodide. *Dental Traumatology*, *15*(5), 205–209. doi:10.1111/j.1600-9657.1999.tb00775.x PMID:10825827

Møller-Jensen, J., & Löwe, J. (2005). Increasing complexity of the bacterial cytoskeleton. *Current Opinion in Cell Biology*, *17*(1), 75–81. doi:10.1016/j.ceb.2004.11.002 PMID:15661522

Monteiro, D. R., Gorup, L. F., Takamiya, A. S., de Camargo, E. R., Filho, A. C., & Barbosa, D. B. (2011). Silver distribution and release from an antimicrobial denture base resin containing silver colloidal nanoparticles. *Journal of Prosthodontics*, *21*(1), 7–15. doi:10.1111/j.1532-849X.2011.00772.x PMID:22050139

Moon, D., Peycelon, M., Kim, J.-Y., Ahn, J.-H., Park, T. J., & Choi, Y. K. (2013). A biristor based on a floating-body silicon nanowire for biosensor applications. *Applied Physics Letters*, *102*(4), 043701. doi:10.1063/1.4789904

Morães, R. R., Garcia, J. W., Wilson, N. D., Lewis, S. H., Barros, M. D., Yang, B., & Stansbury, J. W. et al. (2012). Improved Dental Adhesive Formulations Based on Reactive Nanogel Additives. *Journal of Dental Research*, *91*(2), 179–184. doi:10.1177/0022034511426573 PMID:22019910

Morales, M. P., Serna, C. J., Bødker, F., & Mørup, S. (1997). Spin canting due to structural disorder in maghemite. *Journal of Physics Condensed Matter*, *9*(25), 5461–5467. doi:10.1088/0953-8984/9/25/013

Morales, M., Veintemillas-Verdaguer, S., Montero, M. I., Serna, C. J., Roig, A., Casas, L., & Sandiumenge, F. et al. (1999). Surface and internal spin canting in γ-Fe2O3 nanoparticles. *Chemistry of Materials*, *11*(11), 3058–3064. doi:10.1021/cm991018f

Mora, M., Sagrista, M. L., Trombetta, D., Bonina, F. P., Pasquale, A. D., & Saija, A. (2002). Design and characterization of liposomes containing long-chain N-acylPEs for brain delivery: Penetration of liposomes incorporating GM1 into the rat brain. *Pharmaceutical Research*, *19*(10), 1430–1438.

Morones, J. R., Elechiguerra, J. L., Camacho, A., Holt, K., Kouri, J. B., Ramirez, J. T., & Yacaman, M. J. (2005). The bactericidal effect of silver nanoparticles. *Nanotechnology*, *16*(10), 2346–2353. doi:10.1088/0957-4484/16/10/059 PMID:20818017

Moulari, B., Lboutounne, H., Chaumont, J.-P., Guillaume, Y., Millet, J., & Pellequer, Y. (2006). Potentiation of the bactericidal activity of Harungana madagascariensis Lam. ex Poir.(Hypericaceae) leaf extract against oral bacteria using poly (D, L-lactide-co-glycolide) nanoparticles: In vitro study. *Acta Odontologica Scandinavica*, *64*(3), 153–158. doi:10.1080/00016350500483152 PMID:16809192

Mukerjee, A., & Vishwanatha, J. K. (2009). Formulation, characterization and evaluation of Curcumin-loaded PLGA nanospheres for cancer therapy. *Anticancer Research*, *29*(10), 3867–3876. PMID:19846921

Mukerji, G., Yiangou, Y., Corocran, S. L., Selmer, I., Smith, G. D., Benham, C. D., & Anand, R. et al. (2006). Cool and menthol receptor TRPM8 in human urinary bladder disorders and clinical correlations. *BMC Urology*, *6*(1), 6. doi:10.1186/1471-2490-6-6 PMID:16519806

Mukherjee, P. K., Kumar, V., Mal, M., & Houghton, P. J. (2007). Acetyl cholinesterase inhibitors from plants. *Phytomedicine*, *14*(4), 289–300. doi:10.1016/j.phymed.2007.02.002 PMID:17346955

Mu, L., Chang, Y., Sawtelle, S. D., Wipf, M., Duan, X., & Reed, M. (2015). Silicon nanowire field effect transistors-A versatile class of potentiometric nanobiosensors. *Access, IEEE*, *3*, 287–302. doi:10.1109/ACCESS.2015.2422842

Mulik, R. S., Monkkonen, J., Juvonen, R. O., Mahadik, K. R., & Paradkar, A. R. (2010). ApoE3 mediated poly(butyl) cyanoacrylate nanoparticles containing curcumin: Study of enhanced activity of curcumin against beta amyloid induced cytotoxicity using *in vitro* cell culture model. *Molecular Pharmaceutics*, *7*(3), 815–825. doi:10.1021/mp900306x PMID:20230014

Muller, L., & Muller, F. A. (2006). Preparation of SBF with different HCO3- content and its influence on the composition of biomimetic apatites. *Acta Biomaterialia*, *2*(2), 181–189. doi:10.1016/j.actbio.2005.11.001 PMID:16701876

Mundargi, R. C., Srirangarajan, S., Agnihotri, S. A., Patil, S. A., Ravindra, S., Setty, S. B., & Aminabhavi, T. M. (2007). Development and evaluation of novel biodegradable microspheres based on poly(d,l-lactide-co-glycolide) and poly(ε-caprolactone) for controlled delivery of doxycycline in the treatment of human periodontal pocket: In vitro and in vivo studies. *Journal of Controlled Release*, *119*(1), 59–68. doi:10.1016/j.jconrel.2007.01.008 PMID:17331611

Murphy, W. L., & Mooney, D. J. (2002). Bioinspired growth of crystalline carbonate apatite on biodegradable polymer substrata. *Journal of the American Chemical Society*, *124*(9), 1910–1917. doi:10.1021/ja012433n PMID:11866603

Murray, C. J. L., & Lopez, A. D. (1996). *The Global Burden of Disease: A Comprehensive Assessment of Mortality and Disability from Diseases, Injuries, and Risk Factors in 1990 and Projected to 2020*. Cambridge: MA Harvard University Press.

Murray, C. J., & Lopez, A. D. (1996). Evidence-based health policy--lessons from the Global Burden of Disease Study. *Science*, *274*(5288), 740–743. doi:10.1126/science.274.5288.740 PMID:8966556

Nahata, A., Patil, U. K., & Dixit, V. K. (2008). Effect of *Convolvulus pluricaulis* Choisy. on learning behaviour and memory enhancing activity in rodents. *Natural Product Research*, *22*(16), 1472–1482. doi:10.1080/14786410802214199 PMID:19023811

Nahata, A., Patil, U. K., & Dixit, V. K. (2010). Effect of *Evolvulus alsinoides* Linn. on learning behaviour and memory enhancing activity in rodents. *Phytotherapy Research*, *24*, 486–493. PMID:19610035

Nair, P. R., Go, J., Landells, G. J., Pandit, T. R., Alam, M., Jin, X., et al. (2014). BioSensorLab. Retrieved from https://nanohub.org/resources/senstran DOI:10.4231/D3000014H

Nair, P. R., & Alam, M. A. (2006). Performance limits of nanobiosensors. *Applied Physics Letters, 88*(23), 233120. doi:10.1063/1.2211310

Nair, P. R., & Alam, M. A. (2007). Dimensionally frustrated diffusion towards fractal absorbers. *Physical Review Letters, 99*(25), 256101. doi:10.1103/PhysRevLett.99.256101 PMID:18233533

Nair, P. R., & Alam, M. A. (2008). Screening limited response of nanobiosensors. *Nano Letters, 8*(5), 1281–1285. doi:10.1021/nl072593i PMID:18386914

Nair, P. R., & Alam, M. A. (2010). Theory of selectivity of label-free nanobiosensors: A geometro-physical perspective. *Journal of Applied Physics, 107*(6), 064701. doi:10.1063/1.3310531 PMID:20428486

Naito, Y., Terukina, T., Galli, S., Kozai, Y., Vandeweghe, S., Tagami, T., & Jimbo, R. et al. (2014). The effect of simvastatin-loaded polymeric microspheres in a critical size bone defect in the rabbit calvaria. *International Journal of Pharmaceutics, 461*(1–2), 157–162. doi:10.1016/j.ijpharm.2013.11.046 PMID:24296045

Nakabayashi. (1982). Bonding mechanism of resins and the tooth. *Kokubyo Gakkai Zasshi., 49*(410).

Nakabayashi, N., Kojima, K., & Masuhara, E. (1982). The promotion of adhesion by the infiltration of monomers into tooth substrates. *Journal of Biomedical Materials Research, 16*(3), 265–273. doi:10.1002/jbm.820160307 PMID:7085687

Nance, E. A., Woodworth, G. F., Sailor, K. A., Shih, T. Y., Xu, Q., & Swaminathan, G. et al. (2012). Dense Poly(Ethylene Glycol) Coating Improves Penetration of Large Polymeric Nanoparticles Within Brain Tissue. *Science Translational Medicine, 4*(149), 149ra119. doi:10.1126/scitranslmed.3003594 PMID:22932224

Nathanson, H. C., Newell, W. E., Wickstrom, R. A., & Davis, J. R. Jr. (1967). Resonant gate transistor. *IEEE Transactions on Electron Devices, 14*(3), 117–133. doi:10.1109/T-ED.1967.15912

National Institute for Health and Clinical Excellence (NICE). *The treatment and management of depression in adults* (updated edition - 2010). Retrieved from http://www.nice.org.uk/nicemedia/live/12329/45896/45896.pdf

Navarro, B., Kirichok, Y., Chung, J.-J., & Clapham, D. E. (2008). Ion channels that control fertility in mammalian spermatozoa. *The International Journal of Developmental Biology, 52*(5-6), 607–613. doi:10.1387/ijdb.072554bn PMID:18649274

Nawattanapaiboon, K., Kiatpathomchai, W., Santanirand, P., Vongsakulyanon, A., Amarit, R., & Somboonkaew, A. et al. (2015). SPR-DNA array for detection of methicillin-resistant Staphylococcus aureus (MRSA) in combination with loop-mediated isothermal amplification. *Biosensors & Bioelectronics, 74*, 335–340. doi:10.1016/j.bios.2015.06.038 PMID:26159153

Nazem, A., & Mansoori, G. A. (2011). Nanotechnology for Alzheimer's disease detection and treatment. *Insciences Journal, 14*(4), 169–193. doi:10.5640/insc.0104169

Nel, A. E., Mädler, L., Velegol, D., Xia, T., Hoek, E. M. V., Somasundaran, P., & Thompson, M. et al. (2009). Understanding biophysicochemical interactions at the nano–bio interface. *Nature Materials*, *8*(7), 543–557. doi:10.1038/nmat2442 PMID:19525947

Ngiam, M., Liao, S. S., Patil, A. J., Cheng, Z. Y., Chan, C. K., & Ramakrishna, S. (2009). The fabrication of nano-hydroxyapatite on PLGA and PLGA/collagen nanofibrous composite scaffolds and their effects in osteoblastic behavior for bone tissue engineering. *Bone*, *45*(1), 4–16. doi:10.1016/j.bone.2009.03.674 PMID:19358900

Nielsen, M. E., Høi, L., Schmidt, A. S., Qian, D., Shimada, T., Shen, J. Y., & Larsen, J. L. (2001). Is *Aeromonas hydrophila* the dominant motile Aeromonas species that causes disease outbreaks in aquaculture production in the Zhejiang Province of China? *Diseases of Aquatic Organisms*, *46*, 23–29. doi:10.3354/dao046023 PMID:11592699

Niemeyer, C. M. (2001). Nanoparticles, proteins, and nucleic acids: Biotechnology meets materials science. *Angewandte Chemie International Edition*, *40*(22), 4128–4158. doi:10.1002/1521-3773(20011119)40:22<4128::AID-ANIE4128>3.0.CO;2-S

Niihara, K. (1991). New design concept of structural ceramics—ceramic nanocomposites. *Nippon Seramikkusu Kyokai Gakujutsu Ronbunshi*, *99*(10), 974–982. doi:10.2109/jcersj.99.974

Nowacek, A. S., Miller, R. L., McMillan, J., Kanmogne, G., Kanmogne, M., & Mosley, R. L. et al. (2009). NanoART synthesis, characterization, uptake, release and toxicology for human monocyte-macrophage drug delivery. *Nanomedicine (London)*, *4*(8), 903–917. doi:10.2217/nnm.09.71 PMID:19958227

Obata, K., Tajima, H., Yohda, M., & Matsunaga, T. (2002). Recent developments in laboratory automation using magnetic particles for genome analysis. *Pharmacogenomics*, *3*(5), 697–708. doi:10.1517/14622416.3.5.697 PMID:12223054

Ogle, R. E., David, L. J., & Ortman, H. R. (1985). Clinical wear study of a new tooth material: Part II. *The Journal of Prosthetic Dentistry*, *54*(1), 67–75. doi:10.1016/S0022-3913(85)80073-1 PMID:3894642

Ohtsuki, S., & Terasaki, T. (2007). Contribution of carrier-mediated transport systems to the blood–brain barrier as a supporting and protecting interface for the brain; importance for CNS drug discovery and development. *Pharmaceutical Research*, *24*(9), 1745–1758. doi:10.1007/s11095-007-9374-5 PMID:17619998

Olivares, W., & McQuarrie, D. A. (1976). A vibrational approach to the theory of ionic solutions. *The Journal of Chemical Physics*, *65*(9), 3604. doi:10.1063/1.433545

Oliver, D., Baukrowitz, T., & Fakler, B. (2000). Polyamines as gating molecules of inward-rectifier K^+ channels. *European Journal of Biochemistry*, *267*(19), 5824–5829. doi:10.1046/j.1432-1327.2000.01669.x PMID:10998040

Olivier, J. C. (2005). Drug transport to brain with targeted nanoparticles. *NeuroRx*, *2*(1), 108–119. doi:10.1602/neurorx.2.1.108 PMID:15717062

Olson, S. D., Suarez, S. S., & Fauci, L. J. (2010). A model of CatSper channel mediated calcium dynamics in mammalian spermatozoa. *Bulletin of Mathematical Biology*, *72*(8), 1925–1946. doi:10.1007/s11538-010-9516-5 PMID:20169416

Oringer, R. J., Al-Shammari, K. F., Aldredge, W. A., Iacono, V. J., Eber, R. M., Wang, H. L., & Giannobile, W. V. et al. (2002). Effect of locally delivered minocycline microspheres on markers of bone resorption. *Journal of Periodontology*, *73*(8), 835–842. doi:10.1902/jop.2002.73.8.835 PMID:12211491

Osorio, R., Osorio, E., Medina-Castillo, A. L., & Toledano, M. (2014). Polymer Nanocarriers for Dentin Adhesion. *Journal of Dental Research*, *93*(12), 1258–1263. doi:10.1177/0022034514551608 PMID:25227634

Ousingsawat, J., Spitzer, M., Puntheeranurak, S., & Terracciano, L. (2007). Expression of voltage-gated potassium channels in human and mouse colonic carcinoma. *Clinical Cancer Research*, *13*(3), 824–831. doi:10.1158/1078-0432.CCR-06-1940 PMID:17289873

Oxenham (2013). The clinical significance of calcium-signaling pathways mediation human sperm hyperactivation. *Human Reproduction*, 28(4), 866-876.

Ozak, S. T., & Ozkan, P. (2013). Nanotechnology and dentistry. *European Journal of Dentistry*, *7*(1), 145–151. PMID:23408486

Pagonis, T. C., Chen, J., Fontana, C. R., Devalapally, H., Ruggiero, K., Song, X., & Yamazaki, H. et al. (2010). Nanoparticle-based endodontic antimicrobial photodynamic therapy. *Journal of Endodontics*, *36*(2), 322–328. doi:10.1016/j.joen.2009.10.011 PMID:20113801

Pahuja, R., Seth, K., Shukla, A., Shukla, R. K., Bhatnagar, P., & Chauhan, L. K. et al. (2015). Trans-blood brain barrier delivery of dopamine-loaded nanoparticles reverses functional deficits in Parkinsonian rats. *ACS Nano*, *9*(5), 4850–4871. doi:10.1021/nn506408v PMID:25825926

Pandey, R., & Khuller, G. K. (2006). Oral nanoparticle-based antituberculosis drug delivery to the brain in an experimental model. *The Journal of Antimicrobial Chemotherapy*, *57*(6), 1146–1152. doi:10.1093/jac/dkl128 PMID:16597631

Pankhurst, Q. A., Connolly, J., Jones, S. K., & Dobson, J. J. (2003). Applications of magnetic nanoparticles in biomedicine. *Journal of Physics. D, Applied Physics*, *36*(13), R167–R181. doi:10.1088/0022-3727/36/13/201

Panyam, J., & Labhasetwar, V. (2003). Biodegradable Nanoparticle from Drug and Gene Delivery to Cells and Tissue. *Advanced Drug Delivery Reviews*, *55*(3), 329–347. doi:10.1016/S0169-409X(02)00228-4 PMID:12628320

Paquette, D., Oringer, R., Lessem, J., Offenbacher, S., Genco, R., Persson, G. R., & Williams, R. C. et al. (2003). Locally delivered minocycline microspheres for the treatment of periodontitis in smokers. *Journal of Clinical Periodontology*, *30*(9), 787–794. doi:10.1034/j.1600-051X.2003.00375.x PMID:12956654

Pardo, L. A. (2004). Voltage-gated potassium channels in cell proliferation. *Physiology (Bethesda, MD)*, *19*(5), 285–292. doi:10.1152/physiol.00011.2004 PMID:15381757

Pardridge, W. (2006). Molecular Trojan horses for blood–brain barrier drug delivery. *Current Opinion in Pharmacology*, *6*(5), 494–500. doi:10.1016/j.coph.2006.06.001 PMID:16839816

Pardridge, W. M. (2001). Crossing the blood–brain barrier: Are we getting it right? *Drug Discovery Today*, *6*(1), 1–2. doi:10.1016/S1359-6446(00)01583-X PMID:11165157

Pardridge, W. M. (2015). Blood–brain barrier drug delivery of IgG fusion proteins with a transferrin receptor monoclonal antibody. *Expert Opinion on Drug Delivery*, *12*(2), 207–222. doi:10.1517/17425247.2014.952627 PMID:25138991

Parikh, T., Bommana, M. M., & Squillante, E. III. (2010). Efficacy of surface charge in targeting pegylated nanoparticles of sulpiride to the brain. *European Journal of Pharmaceutics and Biopharmaceutics*, *74*(3), 442–450. doi:10.1016/j.ejpb.2009.11.001 PMID:19941957

Park, J., Kang, N., Jun, Y., Oh, S. J., Ri, H., & Cheon, J. (2002). Superlattice and magnetism directed by the size and shape of nanocrystals. *ChemPhysChem*, *3*(6), 543–547. doi:10.1002/1439-7641(20020617)3:6<543::AID-CPHC543>3.0.CO;2-E PMID:12465496

Park, Y. J., Lee, J. Y., Yeom, H. R., Kim, K. H., Lee, S. C., Shim, I. K., & Lee, S. J. et al. (2005). Injectable polysaccharide microcapsules for prolonged release of minocycline for the treatment of periodontitis. *Biotechnology Letters*, *27*(22), 1761–1766. doi:10.1007/s10529-005-3550-7 PMID:16314967

Patel, A. M., & Joshi, A. Y. (2014b). Vibration Analysis of Defective Double Walled Carbon Nanotube Based Nano Resonators. *Proceedings of ASME 2014 International Mechanical Engineering Congress and Exposition* (pp. V010T13A024-V010T13A024). American Society of Mechanical Engineers. doi:10.1115/IMECE2014-36454

Patel, Ajay M., Joshi A.Y. (2014e). Effect of Waviness on the Dynamic Characteristics of Double Walled Carbon Nanotubes, *Nanoscience Nano technology Letter*, 6, 1-9.

Patel, A. M., Gosai, N., & Joshi, A. Y. (2015c). A Review on Defects in Carbon Nanotubes. *Applied Mechanics and Materials, Trans Tech Publications*, *813-814*, 145–150. doi:10.4028/www.scientific.net/AMM.813-814.145

Patel, A. M., & Joshi, A. Y. (2013). Vibration Analysis of Double Wall Carbon Nanotube based Resonators for Zeptogram level Mass Recognition. *Computational Materials Science, Elsevier Publishers*, *79*, 230–238. doi:10.1016/j.commatsci.2013.06.022

Patel, A. M., & Joshi, A. Y. (2014c). Investigating the influence of surface deviations in double walled carbon nanotube based nanomechanical sensors. *Computational Materials Science*, *89*, 157–164. doi:10.1016/j.commatsci.2014.03.034

Patel, A. M., & Joshi, A. Y. (2014d). Investigation of Double walled Carbon Nanotubes for Mass Sensing. *Procedia Technology*, *14*, 290–294. doi:10.1016/j.protcy.2014.08.038

Patel, A. M., & Joshi, A. Y. (2015a). Influence of atomic vacancies on the dynamic characteristics of nanoresonators based on double walled carbon nanotube. *Physica E, Low-Dimensional Systems and Nanostructures*, *70*, 90–100. doi:10.1016/j.physe.2015.02.016

Patel, Ajay M., Anand Y Joshi. (2014a). Computational Investigation of mass sensing using defective Double walled Carbon Nanotubes, Procedia. *Materials Science*, *5*, 482–488.

Patel, Ajay M., Anand Y Joshi. (2015b). Evaluating the Vibrational Characteristics of Double Walled Carbon Nanotubes with Pinhole Defects, *Current NanoScience. Bentham Science Publishers*, *11*(3), 371–378.

Pathak, S. K., Basu, S., Basu, K. K., Banerjee, A., Pathak, S., Bhattacharyya, A., & Basu, J. et al. (2007). Direct extracellular interaction between the early secreted antigen ESAT-6 of Mycobacterium tuberculosis and TLR2 inhibits TLR signaling in macrophages. *Nature Immunology*, *8*(6), 610–618. doi:10.1038/ni1468 PMID:17486091

Patolsky, F., Zheng, G., & Lieber, C. M. (2006). Fabrication of silicon nanowire devices for ultrasensitive, label free, real-time detection of biological and chemical species. *Nature Protocols*, *1*(4), 1711–1724. doi:10.1038/nprot.2006.227 PMID:17487154

Patrat, C., Serres, C., & Jouannet, P. (2000). The acrosome reaction in human spermatozoa. *Biology of the Cell*, *92*(3-4), 255–266. doi:10.1016/S0248-4900(00)01072-8 PMID:11043413

Peng, X.-H., Qian, X., Mao, H., & Wang, A. Y. (2008). Targeted magnetic iron oxide nanoparticles for tumor imaging and therapy. *International Journal of Nanomedicine*, *3*(3), 311. PMID:18990940

Perez, J. M., O'Loughin, T., Simeone, F. J., Weissleder, R., & Josephson, L. (2002). DNA-based magnetic nanoparticle assembly acts as a magnetic relaxation nanoswitch allowing screening of DNA-cleaving agents. *Journal of the American Chemical Society*, *124*(12), 2856–2857. doi:10.1021/ja017773n PMID:11902860

Perinchery. (n. d.). A History of Indian healing traditions: A science and society initiative. Retrieved from http://news.ncbs.res.in/story/history-indian-healing-traditions

Persadmehr, A., Torneck, C. D., Cvitkovitch, D. G., Pinto, V., Talior, I., Kazembe, M., & Kishen, A. et al. (2014). Bioactive chitosan nanoparticles and photodynamic therapy inhibit collagen degradation in vitro. *Journal of Endodontics*, *40*(5), 703–709. doi:10.1016/j.joen.2013.11.004 PMID:24767568

Persson, G. R., Salvi, G. E., Heitz-Mayfield, L. J., & Lang, N. P. (2006). Antimicrobial therapy using a local drug delivery system (Arestin®) in the treatment of peri-implantitis. I: Microbiological outcomes. *Clinical Oral Implants Research*, *17*(4), 386–393. doi:10.1111/j.1600-0501.2006.01269.x PMID:16907769

Petros, R. A., & Desimone, J. M. (2010). Strategies in the Design of Nanoparticles for Therapeutic Applications. *Nature Reviews. Drug Discovery*, *9*(8), 615–627. doi:10.1038/nrd2591 PMID:20616808

Piao, Z.-G., Kim, J.-S., Son, J.-S., Lee, S.-Y., Fang, X.-H., Oh, J.-S., & Kim, S.-G. et al. (2014). Osteogenic evaluation of collagen membrane containing drug-loaded polymeric microparticles in a rat calvarial defect model. *Tissue Engineering. Part A*, *20*(23-24), 3322–3331. doi:10.1089/ten.tea.2013.0717 PMID:24967649

Pilgrimm, H. (2003, October 28). Super-paramagnetic particles with increased R1 relaxivity, process for producing said particles and use thereof. Google Patents.

Pinheiro, E., Gomes, B., Ferraz, C., Sousa, E., Teixeira, F., & Souza-Filho, F. (2003). Microorganisms from canals of root-filled teeth with periapical lesions. *International Endodontic Journal*, *36*(1), 1–11. doi:10.1046/j.1365-2591.2003.00603.x PMID:12656508

Pinho, S. L. C., Pereira, G. A., Voisin, P., Kassem, J., Bouchaud, V., Etienne, L., & Geraldes, C. F. G. C. et al. (2010). Fine tuning of the relaxometry of γ-Fe2O3@ SiO2 nanoparticles by tweaking the silica coating thickness. *ACS Nano*, *4*(9), 5339–5349. doi:10.1021/nn101129r PMID:20795638

Pinon-Segundo, E., Ganem-Quintanar, A., Alonso-Perez, V., & Quintanar-Guerrero, D. (2005). Preparation and characterization of triclosan nanoparticles for periodontal treatment. *International Journal of Pharmaceutics*, *294*(1-2), 217–232. doi:10.1016/j.ijpharm.2004.11.010 PMID:15814246

Pinto, R. C., Neufeld, R. J., Ribeiro, A. J., & Veiga, F. (2006). Nanoencapsulation I. Methods for preparation of drug-loaded polymeric nanoparticles. *Nanomed-Nanotechnology*, *2*(1), 8–21. doi:10.1016/j.nano.2005.12.003 PMID:17292111

Plimpton, S. (1995). Fast Parallel Algorithms for Short-Range Molecular Dynamics. *Journal of Computational Physics*, *117*(1), 1–19. doi:10.1006/jcph.1995.1039

Ploughman, M., Windle, V., MacLellan, C. L., White, N., Dore, J. J., & Corbett, D. (2009). Brain-derived neurotrophic factor contributes to recovery of skilled reaching after focal ischemia in rats. *Stroke*, *40*(4), 1490–1495. doi:10.1161/STROKEAHA.108.531806 PMID:19164786

Pool, H., Quintanar, D., Figueroa, J. D., Bechara, J. E. H., Mc Clements, D. J., & Mendoza, S. (2012). Polymeric Nanoparticles as Oral Delivery Systems for Encapsulation and Release of Polyphenolic Compounds: Impact on Quercetin Antioxidant Activity & Bioaccessibility. *Food Biophysics*, *7*, 276–288. doi:10.1007/s11483-012-9266-z

Portenier, I., Waltimo, T. M., & Haapasalo, M. (2003). Enterococcus faecalis–the root canal survivor and 'star' in post-treatment disease. *Endodontic topics, 6*(1), 135-159.

Pourgholi, F., Hajivalili, M., Farhad, J. N., Kafil, H. S., & Yousefi, M. (2016). Nanoparticles: Novel vehicles in treatment of Glioblastoma. *Biomedicine and Pharmacotherapy*, *77*, 98–107. doi:10.1016/j.biopha.2015.12.014 PMID:26796272

Pravina, K., Ravindra, K. R., Goudar, K. S., Vinod, D. R., Joshua, A. J., Wasim, P., & Amit, A. et al. (2007). Safety evaluation of BacoMind in healthy volunteers: A phase I study. *Phytomedicine*, *14*(5), 301–308. doi:10.1016/j.phymed.2007.03.010 PMID:17442556

Qian, D., Wagner, G. J., Liu, W. K., Yu, M. F., & Ruoff, R. S. (2002). Mechanics of carbon nanotubes. *Applied Mechanics Reviews*, *55*(6), 495–533. doi:10.1115/1.1490129

Qi, H., Moran, M. M., Navarro, B., Chong, J. A., Krapivinsky, G., Krapivinsky, L., & Clapham, D. E. et al. (2007). All four CatSper ion channel proteins are required for male fertility and sperm cell hyperactivated motility. *Proceedings of the National Academy of Sciences of the United States of America*, *104*(4), 1219–1223. doi:10.1073/pnas.0610286104 PMID:17227845

Qin, Y., Chen, H., Zhang, Q., Wang, X., Yuan, W., & Kuai, R. et al. (2011). Liposome formulated with TAT-modified cholesterol for improving brain delivery and therapeutic efficacy on brain glioma in animals. *International Journal of Pharmaceutics*, *420*(2), 304–312. doi:10.1016/j.ijpharm.2011.09.008 PMID:21945185

Quill, T. A., Ren, D., Clapham, D. E., & Garbers, D. L. (2001). A voltage-gated ion channel expressed specifically in spermatozoa. *Proceedings of the National Academy of Sciences of the United States of America*, *98*(22), 12527–12531. doi:10.1073/pnas.221454998 PMID:11675491

Rai, K. S., Murthy, K. D., Karanth, K. S., Nalini, K., Rao, M. S., & Srinivasan, K. K. (2002). *Clitoria ternatea* root extract enhances acetylcholine content in rat hippocampus. *Fitoterapia*, *73*(7-8), 685–689. doi:10.1016/S0367-326X(02)00249-6 PMID:12490229

Rai, K. S., Murthy, K. D., Rao, M. S., & Karanth, K. S. (2005). Altered dendritic arborisation of amygdala neurons in young adult rats orally intubated with *Clitoria ternatea* aqueous root extract. *Phytotherapy Research*, *19*(7), 592–598. doi:10.1002/ptr.1657 PMID:16161034

Rai, M., Yadav, A., & Gade, A. (2009). Silver nanoparticles as a new generation of antimicrobials. *Biotechnology Advances*, *27*(1), 76–83. doi:10.1016/j.biotechadv.2008.09.002 PMID:18854209

Rajendran, R., Radhai, R., Kotresh, T. M., & Csiszar, E. (2013). Development of antimicrobial cotton fabrics using herb loaded nanoparticles. *Carbohydrate Polymers*, *91*(2), 613–617. doi:10.1016/j.carbpol.2012.08.064 PMID:23121954

Ramaswami, V., Varghese, N., & Simon, A. (2011). An investigation on cytotoxic and antioxidant properties of *Clitoria ternatea* L. *International Journal of Drug Discovery.*, *3*(1), 74–77. doi:10.9735/0975-4423.3.1.74-77

Rao, K. V., Reddy, M. K., Horning, J. L., & Labhasetwar, V. (2008). TAT-conjugated nanoparticles for the CNS delivery of anti-HIV drugs. *Biomaterials*, *29*(33), 4429–4438. doi:10.1016/j.biomaterials.2008.08.004 PMID:18760470

Reis, K. R., Bonfante, G., Pegoraro, L. F., Conti, P. C. R., Oliveira, P. C. G., & Kaizer, O. B. (2008). In vitro wear resistance of three types of polymethyl methacrylate denture teeth. *Journal of Applied Oral Science*, *16*(3), 176–180. doi:10.1590/S1678-77572008000300003 PMID:19089214

Ren, D., Navarro, B., Perez, G., Jackson, A. C., Hsu, S., Shi, Q., & Clapham, D. E. et al. (2001). A sperm ion channel required for sperm motility and male fertility. *Nature*, *413*(6856), 603–609. doi:10.1038/35098027 PMID:11595941

Ren, G., Hu, D. W. C., Cheng, E. A., Vargas-Reus, M., Reip, P. P., & Allaker, R. (2009). Characterisation of copper oxide nanoparticles for antimicrobial applications. *International Journal of Antimicrobial Agents, 33*(6), 587–590. doi:10.1016/j.ijantimicag.2008.12.004 PMID:19195845

Renvert, S., Lessem, J., Dahlén, G., Lindahl, C., & Svensson, M. (2006). Topical minocycline microspheres versus topical chlorhexidine gel as an adjunct to mechanical debridement of incipient peri-implant infections: A randomized clinical trial. *Journal of Clinical Periodontology, 33*(5), 362–369. doi:10.1111/j.1600-051X.2006.00919.x PMID:16634959

Renvert, S., Lessem, J., Dahlen, G., Renvert, H., & Lindahl, C. (2008). Mechanical and repeated antimicrobial therapy using a local drug delivery system in the treatment of peri-implantitis: A randomized clinical trial. *Journal of Periodontology, 79*(5), 836–844. doi:10.1902/jop.2008.070347 PMID:18454662

Revell, P. A. (2006). The biological effects of nanoparticles. *Nanotechnology Perceptions, 2*, 283–298.

Rho, J. Y., Kuhn-Spearing, L., & Zioupos, P. (1998). Mechanical properties and the hierarchical structure of bone. *Medical Engineering & Physics, 20*(2), 92–102. doi:10.1016/S1350-4533(98)00007-1 PMID:9679227

Richards, D. A. (2012). Stepped care: A method to deliver increased access to psychological therapies. *Canadian Journal of Psychiatry, 57*(4), 210–215. PMID:22480585

Richelson, E. (2010). Pharmacology of antidepressants. *Mayo Clinic Proceedings, 76*(5), 511–527. doi:10.4065/76.5.511 PMID:11357798

Ringsdorf, H. (1975). Structure and properties of pharmacologically active polymers. *Journal* of *Polymer Science. Polymer Symposia, 51*(1), 135–153. doi:10.1002/polc.5070510111

Rivas, G. A., Rubianes, M. D., Rodriguez, M. C., Ferreyra, N. F., Luque, G. L., Pedano, M. L., & Parrado, C. (2007). Carbon nanotubes for electrochemical biosensing. *Talanta, 74*(3), 291–307. doi:10.1016/j.talanta.2007.10.013 PMID:18371643

Roberts, R. J. (1997). Epizootic ulcerative syndrome (EUS): progress since 1985. In T. W. Flegel & I. H. MacRae (Eds.), *Diseases in Asian aquaculture III*. Manila: Asian Fish Soc.

Roca, a G., Costo, R., Rebolledo, a F., Veintemillas-Verdaguer, S., Tartaj, P., González-Carreño, T., … Serna, C. J. (2009). Progress in the preparation of magnetic nanoparticles for applications in biomedicine. *Journal of Physics D: Applied Physics, 42*(22), 224002. doi:10.1088/0022-3727/42/22/224002

Roca, A. G., Marco, J. F., Morales, M. del P., & Serna, C. J. (2007). Effect of nature and particle size on properties of uniform magnetite and maghemite nanoparticles. *The Journal of Physical Chemistry C, 111*(50), 18577–18584. doi:10.1021/jp075133m

Roca, A. G., Morales, M. P., O'Grady, K., & Serna, C. J. (2006). Structural and magnetic properties of uniform magnetite nanoparticles prepared by high temperature decomposition of organic precursors. *Nanotechnology, 17*(11), 2783–2788. doi:10.1088/0957-4484/17/11/010

Roco, M. C., Williams, R. S., & Alivisatos, P. (2000). Nanotechnology Research Directions: *IWGN Workshop Report: Vision for Nanotechnology in the Next Decade*. Springer Science & Business Media.

Rooijakkers, S. H. M., van Kessel, K. P. M., & van Strijp, J. A. G. (2005). Staphylococcal innate immune evasion. *Trends in Microbiology*, *13*(12), 596–601. doi:10.1016/j.tim.2005.10.002 PMID:16242332

Rosa, V., Bona, A. D., Cavalcanti, B. N., & Nör, J. E. (2012). Tissue engineering: from research to dental clinics. *Dental materials: official publication of the Academy of Dental Materials, 28*(4), 341-348.

Royal Pharmaceutical Society of Great Britain. (2010). British National Formulary (59th ed.). BMJ Group, London.

Ruparelia, J. P., Chatterjee, A. K., Duttagupta, S. P., & Mukherji, S. (2008). Strain specificity in antimicrobial activity of silver and copper nanoparticles. *Acta Biomaterialia*, *4*(3), 707–716. doi:10.1016/j.actbio.2007.11.006 PMID:18248860

Saavedra, G., Valandro, L. F., Leite, F. P., Amaral, R., Ozcan, M., & Bottino, M. A. et al.. (2007). Bond strength of acrylic teeth to denture base resin after various surface conditioning methods before and after thermocycling. *The International Journal of Prosthodontics*, *20*(2), 199–201. PMID:17455445

Sahneh, F. D., Scoglio, C., & Riviere, J. (2013). Dynamics of Nanoparticle-Protein Corona Complex Formation: Analytical Results from Population Balance Equations. *PLoS ONE*, *8*(5), e64690. doi:10.1371/journal.pone.0064690 PMID:23741371

Saifuddin, N., Wong, C.W., Nur Yashumira, A.A. (2009). Rapid Biosynthesis of Silver nanoparticles using culture supernatant of bacteria with Microwave irradiation. *e-J Chem,* 6(1), 61-70.

Saito, R., Matsuo, R., Kimura, T., Dresselhaus, G., & Dresselhaus, M. S. (2001). Anomalous potential barrier of double-wall carbon nanotube. *Chemical Physics Letters*, *348*(3-4), 187–193. doi:10.1016/S0009-2614(01)01127-7

Saleh, I., Ruyter, I., Haapasalo, M., & Ørstavik, D. (2004). Survival of Enterococcus faecalis in infected dentinal tubules after root canal filling with different root canal sealers in vitro. *International Endodontic Journal*, *37*(3), 193–198. doi:10.1111/j.0143-2885.2004.00785.x PMID:15009409

Sămaj, J., Baluška, F., Boris, B., Schlicht, M., Volkmann, D., & Menzel, D. (2004). Endocytosis, Actin Cytoskeleton, and Signaling. *Plant Physiology*, *135*(3), 1150–1161. doi:10.1104/pp.104.040683 PMID:15266049

Samdancioglu, S., Calis, S., Sumnu, M., & Atilla Hincal, A. (2006). Formulation and in vitro evaluation of bisphosphonate loaded microspheres for implantation in osteolysis. *Drug Development and Industrial Pharmacy*, *32*(4), 473–481. doi:10.1080/03639040500528871 PMID:16638686

Sanchez-Portal, D., Artacho, E., Soler, J. M., Rubio, A., & Ordejón, P. (1999). Ab initio structural, elastic and vibrational properties of carbon nanotubes. *Physical Review B: Condensed Matter and Materials Physics*, *59*(19), 12678–12688. doi:10.1103/PhysRevB.59.12678

Sandhu, A., Handa, H., & Abe, M. (2010). Synthesis and applications of magnetic nanoparticles for biorecognition and point of care medical diagnostics. *Nanotechnology*, *21*(44), 442001. doi:10.1088/0957-4484/21/44/442001 PMID:20935358

Sanson, C., Diou, O., Thevenot, J., Ibarboure, E., Soum, A., Brûlet, A., & Brisson, A. et al. (2011). Doxorubicin loaded magnetic polymersomes: Theranostic nanocarriers for MR imaging and magneto-chemotherapy. *ACS Nano*, *5*(2), 1122–1140. doi:10.1021/nn102762f PMID:21218795

Santra, S., Tapec, R., Theodoropoulou, N., Dobson, J., Hebard, A., & Tan, W. (2001). Synthesis and characterization of silica-coated iron oxide nanoparticles in microemulsion: The effect of nonionic surfactants. *Langmuir*, *17*(10), 2900–2906. doi:10.1021/la0008636

Saotome, K., Singh, A. K., Yelshanskaya, M. V., & Sobolevsky, A. I. (2016). Crystal structure of the epithelial calcium channel TRPV6. *Nature*, *534*(7608), 506–511. doi:10.1038/nature17975 PMID:27296226

Sarandol, A., Sarandol, E., Eker, S. S., Erdinc, S., Vatansever, E., & Kirli, S. (2007). Major depressive disorder is accompanied with oxidative stress: Short-term antidepressant treatment does not alter oxidative–antioxidative systems. *Human Psychopharmacology Clinical and Experimental*, *22*(2), 67–73. doi:10.1002/hup.829 PMID:17299810

Sarvestani, A. S., He, X. Z., & Jabbari, E. (2007). Effect of osteonectin-derived peptide on the viscoelasticity of hydrogel/apatite nanocomposite scaffolds. *Biopolymers*, *85*(4), 370–378. doi:10.1002/bip.20659 PMID:17183515

Sarvestani, A. S., He, X., & Jabbari, E. (2008). Osteonectin-derived peptide increases the modulus of a bone-mimetic nanocomposite. *European Biophysics Journal with Biophysics Letters*, *37*(2), 229–234. doi:10.1007/s00249-007-0198-3 PMID:17609937

Scheifele, E., Studen-Pavlovich, D., & Markovic, N. (2002). Practitioner's guide to fluoride. [xi.]. *Dental Clinics of North America*, *46*(4), 831–846. doi:10.1016/S0011-8532(02)00019-8 PMID:12436834

Scheller, E. L., Krebsbach, P. H., & Kohn, D. H. (2009). Tissue engineering: State of the art in oral rehabilitation. *Journal of Oral Rehabilitation*, *36*(5), 368–389. doi:10.1111/j.1365-2842.2009.01939.x PMID:19228277

Schindler, M., Ahmed, I., Kamal, J., Nur-E-Kamal, A., Grafe, T. H., Chung, H. Y., & Meiners, S. (2005). A synthetic nanofibrillar matrix promotes in vivo-like organization and morphogenesis for cells in culture. *Biomaterials*, *26*(28), 5624–5631. doi:10.1016/j.biomaterials.2005.02.014 PMID:15878367

Schinkel, A. H. (1999). P-glycoprotein, a gatekeeper in the blood–brain barrier. *Advanced Drug Delivery Reviews*, *36*(2-3), 179–194. doi:10.1016/S0169-409X(98)00085-4 PMID:10837715

Schliwa, M., & Woehlke, G. (2003). Molecular motors. *Nature*, *422*(6933), 759–765. doi:10.1038/nature01601 PMID:12700770

Schnorr, J., Wagner, S., Abramjuk, C., Drees, R., Schink, T., Schellenberger, E. A., & Taupitz, M. et al. (2006). Focal Liver Lesions: SPIO-, Gadolinium-, and Ferucarbotran-enhanced Dynamic T1-weighted and Delayed T2-weighted MR Imaging in Rabbits 1. *Radiology*, *240*(1), 90–100. doi:10.1148/radiol.2393040884 PMID:16684917

Schnyder, A., Krähenbühl, S., Drewe, J., & Huwyler, J. (2005). Targeting of daunomycin using biotinylated immunoliposomes: Pharmacokinetics tissue distribution and *in vitro* pharmacological effects. *Journal of Drug Targeting*, *13*(5), 325–335. doi:10.1080/10611860500206674 PMID:16199376

Schreurs, W. J. A., & Rosenberg, H. (1982). Effect of silver ions on transport and retention of phosphate by *Escherichia coli. Journal of Bacteriology*, *152*, 7–13. PMID:6749823

Schroeder, U., Schroeder, H., & Sabel, B. A. (2000). Body distribution of 3H-labelled dalargin bound to poly(butyl cyanoacrylate) nanoparticles after iv injections to mice. *Life Sciences*, *66*(6), 495–502. doi:10.1016/S0024-3205(99)00619-0 PMID:10794066

Schwartz-Arad, D., & Chaushu, G. (1997). Placement of Implants Into Fresh Extraction Sites: 4 to 7 Years Retrospective Evaluation of 95 Immediate Implants. *Journal of Periodontology*, *68*(11), 1110–1116. doi:10.1902/jop.1997.68.11.1110 PMID:9407405

Selwitz, R. H., Ismail, A. I., & Pitts, N. B. (2007). Dental caries. *Lancet*, *369*(9555), 51–59. doi:10.1016/S0140-6736(07)60031-2 PMID:17208642

Seshadri, R., Joseph, S. W., Chopra, A. K., Sha, J., Shaw, J., Graf, J., & Heidelberg, J. F. et al. (2006). Genome Sequence of *Aeromonas hydrophila* ATCC 7966T: Jack of All Trades. *Journal of Bacteriology*, *188*(23), 8272–8282. doi:10.1128/JB.00621-06 PMID:16980456

Sethiya, N. K., & Mishra, S. H. (2010). Review on ethanomedicinal uses and phyto-pharmacology of memory boosting herb Convolvulus pluricaulis Choisy. *The AustralianJournal of Herbal Medicine*, *22*(1), 19–25.

Sethiya, N. K., Patel, M. B., & Mishra, S. H. (2010). Phyto-pharmacologic aspects of *Canscora decussata* Roem and Schult. *Pharmacognosy Reviews*, *4*(7), 49–57. doi:10.4103/0973-7847.65326 PMID:22228941

Seyoum, A., Asres, K., & El-Fiky, F. K. (2006). Structure–radical scavenging activity relationships of flavonoids. *Phytochemistry*, *67*(18), 2058–2070. doi:10.1016/j.phytochem.2006.07.002 PMID:16919302

Shaffiey, S. F., Shaffiey, S. R., Ahmadi, M., & Azari, F. (2014). Synthesis and evaluation of bactericidal properties of CuO nanoparticles against *Aeromonas hydrophila. Nanomecicine Journal*, *1*(2), 199–205.

Shaffiey, S. R., & Shaffiey, S. F. (2016). Surface enamel remineralization by biomimetic nano hydroxyapatite crystals and fluoride ions effects. *Journal of Ceramic Processing Research*, *17*(2), 109–112.

Shaffiey, S. R., Shaffiey, S. F., & Ahmadi, M. (2015). Synthesis and evaluation of bactericidal properties of Ag_2O nanoparticles against *Aeromonas hydrophila*. *Int. J. Nano Dimens*, *6*(3), 263–269.

Shah, A. A., Hasan, F., Hameed, A., & Ahmed, S. (2008). Biological degradation of plastics: A comprehensive review. *Biotechnology Advances*, *26*(3), 246–265. doi:10.1016/j.biotechadv.2007.12.005 PMID:18337047

Sharma, G., Puranik, M. P., & K.R, S. (2015). Approaches to Arresting Dental Caries: An Update. *Journal of Clinical and Diagnostic Research: JCDR*, *9*(5), ZE08-ZE11.

Sharma, R., Chaturvedi, C., & Tewari, P. V. (1987). Efficacy of *Bacopa monnieri* in revitalizing intellectual functions in children. *The Indian Journal of Medical Research*, *6*, 1–10.

Shin, H., Jo, S., & Mikos, A. G. (2003). Biomimetic materials for tissue engineering. *Biomaterials*, *24*(24), 4353–4364. doi:10.1016/S0142-9612(03)00339-9 PMID:12922148

Shinji, S., Yasukazu, T., Hatsue, W., Kazuo, K., Machiko, I., & Naoki, M. (2011). Analysis of brain cell activation by nano sized particles of Ginkgo biloba extract. *International Journal of Plant Physiology and Biochemistry*, *3*(3), 28–33.

Shoshan-Barmatz, V., & Golan, M. (2012). MitochondrialVDAC1: Function in cell life and death and a target for cancer therapy. *Current Medicinal Chemistry*, *19*(5), 714–735. doi:10.2174/092986712798992110 PMID:22204343

Shoshan-Barmatz, V., & Mizrachi, D. (2012). VDAC1: From structure to cancer therapy. *Frontiers in Oncology*, *29*, 164. PMID:23233904

Shrestha, A., Fong, S.-W., Khoo, B.-C., & Kishen, A. (2009). Delivery of Antibacterial Nanoparticles into Dentinal Tubules Using High-intensity Focused Ultrasound. *Journal of Endodontics*, *35*(7), 1028–1033. doi:10.1016/j.joen.2009.04.015 PMID:19567328

Shrestha, A., Friedman, S., & Kishen, A. (2011). Photodynamically Crosslinked and Chitosan-incorporated Dentin Collagen. *Journal of Dental Research*, *90*(11), 1346–1351. doi:10.1177/0022034511421928 PMID:21911787

Shrestha, A., Hamblin, M. R., & Kishen, A. (2014). Photoactivated rose bengal functionalized chitosan nanoparticles produce antibacterial/biofilm activity and stabilize dentin-collagen. *Nanomedicine; Nanotechnology, Biology, and Medicine*, *10*(3), 491–501. doi:10.1016/j.nano.2013.10.010 PMID:24200522

Siddique, Y. H., Khan, W., Singh, B. R., & Naqvi, A. H. (2013). Synthesis of Alginate-Curcumin Nanocomposite and Its Protective Role in Transgenic *Drosophila* Model of Parkinson's Disease. *ISRN Pharmacology*, *2*, 1–8. doi:10.1155/2013/794582 PMID:24171120

Silva, G. A., Czeisler, C., Niece, K. L., Beniash, E., Harrington, D. A., Kessler, J. A., & Stupp, S. I. (2004). Selective differentiation of neural progenitor cells by high-epitope density nanofibers. *Science*, *303*(5662), 1352–1355. doi:10.1126/science.1093783 PMID:14739465

Singh, R. H., & Udupa, K. N. (1993). *Clinical and experimental studies on rasayana drugs and rasayana therapy. Special Research Monograph, Central Council for Research in Ayurveda and Siddha (CCRAS)*. New Delhi: Ministry of Health and Family Welfare.

Sinico, C., De Logu, A., Lai, F., Valenti, D., Manconi, M., Loy, G., & Fadda, A. M. et al. (2005). Liposomal incorporation of Artemisia arborescens L. essential oil and in vitro antiviral activity. *European Journal of Pharmaceutics and Biopharmaceutics*, *59*(1), 161–168. doi:10.1016/j.ejpb.2004.06.005 PMID:15567314

Sivaranjan V.V. & Indira Balachandran. (1994). *Ayuruedic drugs and their plant sources.* Oxford and IBH Publishing Company, New Delhi, India.

Skrlj, N., Drevenšek, G., Hudoklin, S., Romih, R., Šerbec, V. C., & Dolinar, M. (2013). Recombinant single-chain antibody with the Trojan peptide penetratin positioned in the linker region enables cargo transfer across the blood–brain barrier. *Applied Biochemistry and Biotechnology*, *169*(1), 159–169. doi:10.1007/s12010-012-9962-7 PMID:23160949

Smalley, R. E. (2003). *Carbon nanotubes: synthesis, structure, properties, and applications.* Springer Science & Business Media.

Smeets, R., Gerhards, F., Stein, J., Pereira Paz, R. M., Vogt, S., Pautke, C., & Kolk, A. et al. (2011). A novel hemostatic delivery device for thrombin: Biodegradable poly(D,L-lactide-co-glycolide) 50:50 microspheres. *Journal of Biomedical Materials Research. Part A*, *96A*(1), 177–185. doi:10.1002/jbm.a.32970 PMID:21105166

Socransky, S. S., & Haffajee, A. D. (1994). Evidence of bacterial etiology: A historical perspective. *Periodontology 2000*, *5*(1), 7–25. doi:10.1111/j.1600-0757.1994.tb00016.x PMID:9673160

Sohlberg, K., Sumpter, B. G., Tuzun, R. E., & Noid, D. W. (1998). Continuum methods of mechanics as a simplified approach to structural engineering of nanostructures. *Nanotechnology*, *9*(1), 30–36. doi:10.1088/0957-4484/9/1/004

Solhi, L., Atai, M., Nodehi, A., & Imani, M. (2012). A novel dentin bonding system containing poly(methacrylic acid) grafted nanoclay: Synthesis, characterization and properties. *Dental Materials*, *28*(10), 1041–1050. doi:10.1016/j.dental.2012.06.004 PMID:22795491

Solhi, L., Atai, M., Nodehi, A., Imani, M., Ghaemi, A., & Khosravi, K. (2012). Poly (acrylic acid) grafted montmorillonite as novel fillers for dental adhesives: Synthesis, characterization and properties of the adhesive. *Dental Materials*, *28*(4), 369–377. doi:10.1016/j.dental.2011.11.010 PMID:22169675

Soltani, M., Ghodratnema, M., & Ahari, H., Ebrahimzadeh mousavi, H.A., Atee, M., Dastmalchi, F., Rahmanya, J. (2009). The inhibitory effect of silver nanoparticles on the bacterial fish pathogens, *Streptococcus iniae, Lactococcus garvieae, Yersinia ruckeri* and *Aeromonas hydrophila. J Vet Res*, *3*(2), 137–142.

Sondi, I., & Salopek-Sondi, B. (2004). Silver nanoparticles as an antimicrobial agent: A case study on *E. coli* as a model for gram-negative bacteria. *Journal of Colloid and Interface Science*, *275*(1), 177–182. doi:10.1016/j.jcis.2004.02.012 PMID:15158396

Son, J. S., Choi, Y.-A., Park, E.-K., Kwon, T.-Y., Kim, K.-H., & Lee, K.-B. (2013). Drug delivery from hydroxyapatite-coated titanium surfaces using biodegradable particle carriers. *Journal of Biomedical Materials Research. Part B, Applied Biomaterials*, *101B*(2), 247–257. doi:10.1002/jbm.b.32834 PMID:23143817

Sørensen, M. H., Mortensen, N. A., & Brandbyge, M. (2007). Screening model for nanowire surface-charge sensors in liquid. *Applied Physics Letters*, *91*, 102105. doi:10.1063/1.2779930

Sousa, F. F., Luzardo-Alvarez, A., Perez-Estevez, A., Seoane-Prado, R., & Blanco-Mendez, J. (2010). Development of a novel AMX-loaded PLGA/zein microsphere for root canal disinfection. *Biomedical Materials (Bristol, England)*, *5*(5), 055008. doi:10.1088/1748-6041/5/5/055008 PMID:20844319

Spudich, A., Kilic, E., Xing, H., Kilic, U., Rentsch, K. M., & Allenspach, H. W. et al. (2006). Inhibition of multidrug resistance transporter-1 facilitates neuroprotective therapies after focal cerebral ischemia. *Nature Neuroscience*, *9*(4), 487–488. doi:10.1038/nn1676 PMID:16565717

Srirangarajan, S., Mundargi, R. C., Ravindra, S., Setty, S. B., Aminabhavi, T., & Thakur, S. (2011). Randomized, controlled, single-masked, clinical study to compare and evaluate the efficacy of microspheres and gel in periodontal pocket therapy. *Journal of Periodontology*, *82*(1), 114–121. doi:10.1902/jop.2010.100324 PMID:20681818

Steiniger, S. C., Kreuter, J., Khalansky, A. S., Skidan, I. N., Bobruskin, A. I., & Smirnova, Z. S. et al. (2004). Chemotherapy of glioblastoma in rats using doxorubicin-loaded nanoparticles. *International Journal of Cancer*, *109*(5), 759–767. doi:10.1002/ijc.20048 PMID:14999786

Stevens, B., Yang, Y. Z., Mohanda, S. A., Stucker, B., & Nguyen, K. T. (2008). A review of materials, fabrication to enhance bone regeneration in methods, and strategies used engineered bone tissues. *Journal of Biomedical Materials Research. Part B, Applied Biomaterials*, *85B*(2), 573–582. doi:10.1002/jbm.b.30962 PMID:17937408

Strauss, E. J., & Egol, K. A. (2007). The management of ankle fractures in the elderly. *Injury*, *38*(3), 52–59. PMID:17723786

Strunker, T., Goodwin, N., Brenker, C., Kashikar, N. D., Weyand, I., Seifert, R., & Kaupp, U. B. (2011). The CatSper channel mediates progesterone-induced Ca2+ influx in human sperm. *Nature*, *471*(7338), 382–386. doi:10.1038/nature09769 PMID:21412338

Stuart, C. H., Schwartz, S. A., Beeson, T. J., & Owatz, C. B. (2006). Enterococcus faecalis: Its role in root canal treatment failure and current concepts in retreatment. *Journal of Endodontics*, *32*(2), 93–98. doi:10.1016/j.joen.2005.10.049 PMID:16427453

Suaifan, G. A., Esseghaier, C., Ng, A., & Zourob, M. (2013). Ultra-rapid colorimetric assay for protease detection using magnetic nanoparticle-based biosensors. *Analyst (London)*, *138*(13), 3735–3739. doi:10.1039/c3an36881e PMID:23676852

Suarez, S. S., & Ho, H. C. (2003). Hyperactivation of mammalian sperm. *Cellular and Molecular Biology*, *49*, 351–356. PMID:12887087

Suarez, S. S., & Pacey, A. A. (2006). Sperm transport in the female reproductive tract. *Human Reproduction Update*, *12*(1), 23–37. doi:10.1093/humupd/dmi047 PMID:16272225

Subramani, K., Hosseinkhani, H., Khraisat, A., Hosseinkhani, M., & Pathak, Y. (2009). Targeting nanoparticles as drug delivery systems for cancer treatment. *Current Nanoscience*, *5*(2), 135–140. doi:10.2174/157341309788185406

Suh, W. H., Suslick, K. S., Stucky, G. D., & Suh, Y.-H. (2009). Nanotechnology, nanotoxicology, and neuroscience. *Progress in Neurobiology*, *87*(3), 133–170. doi:10.1016/j.pneurobio.2008.09.009 PMID:18926873

Sunderland, C. J., Steiert, M., Talmadge, J. E., Derfus, A. M., & Barry, S. E. (2006). Targeted nanoparticles for detecting and treating cancer. *Drug Development Research*, *67*(1), 70–93. doi:10.1002/ddr.20069

Sun, E. Y., Josephson, L., Kelly, K. A., & Weissleder, R. (2006). Development of nanoparticle libraries for biosensing. *Bioconjugate Chemistry*, *17*(1), 109–113. doi:10.1021/bc050290e PMID:16417258

Sun, E. Y., Weissleder, R., & Josephson, L. (2006). Continuous analyte sensing with magnetic nanoswitches. *Small*, *2*(10), 1144–1147. doi:10.1002/smll.200600204 PMID:17193579

Sun, J. X., Wang, Y. L., Qian, Z. Y., & Hu, C. B. (2011). An approach to architecture 3D scaffold with interconnective microchannel networks inducing angiogenesis for tissue engineering. *Journal of Materials Science. Materials in Medicine*, *22*(11), 2565–2571. doi:10.1007/s10856-011-4426-0 PMID:21861076

Sun, X. Z., Wang, N., Cao, D., Hu, Z. Y., Mao, P., Zhang, H. B., & Zhou, X. H. (2011). The antimicrobial activities of a series of bis-quaternary ammonium compounds. *Chinese Chemical Letters*, *22*(8), 887–890. doi:10.1016/j.cclet.2011.01.030

Taboada, E., Solanas, R., Rodriguez, E., Weissleder, R., & Roig, A. (2009). Supercritical-Fluid-Assisted One-Pot Synthesis of Biocompatible Core (γ-Fe2O3)/Shell (SiO2) Nanoparticles as High Relaxivity T2-Contrast Agents for Magnetic Resonance Imaging. *Advanced Functional Materials*, *19*(14), 2319–2324. doi:10.1002/adfm.200801681

Tajes, M., Fernández, E. R., Jiang, X. W., Morató, M. B., Guivernau, B., & Pichot, A. E. et al. (2014). The blood-brain barrier: Structure, function and therapeutic approaches to cross it. *Molecular Membrane Biology*, *31*(5), 152–167. doi:10.3109/09687688.2014.937468 PMID:25046533

Takafuji, M., Ide, S., Ihara, H., & Xu, Z. (2004). Preparation of poly (1-vinylimidazole)-grafted magnetic nanoparticles and their application for removal of metal ions. *Chemistry of Materials*, *16*(10), 1977–1983. doi:10.1021/cm030334y

Taktak, S., Sosnovik, D., Cima, M. J., Weissleder, R., & Josephson, L. (2007). Multiparameter magnetic relaxation switch assays. *Analytical Chemistry*, *79*(23), 8863–8869. doi:10.1021/ac701976p PMID:17983206

Talukdar, K., Bhushan, M., Viswanathc, A. K., & Mitrad, A. K. (2013). Simulation study of the performance of a biologically sensitive field effect transistor. *International Journal of Nano Dimension*, *4*, 85–89.

Talukdar, K., & Shantappa, A. (2016a). Molecular dynamics study of nerve ion channel. *Materials Today: Proceedings*, *3*(6), 1828–1834. doi:10.1016/j.matpr.2016.04.081

Talukdar, K., & Shantappa, A. (2016b). Computational study of a calcium release-activated calcium channel. *AIP Conference Proceedings*, *1731*, 050130. doi:10.1063/1.4947784

Tanaka, T., Sakai, R., Kobayashi, R., Hatakeyama, K., & Matsunaga, T. (2009). Contributions of phosphate to DNA adsorption/desorption behaviors on aminosilane-modified magnetic nanoparticles. *Langmuir*, *25*(5), 2956–2961. doi:10.1021/la8032397 PMID:19437706

Tang, J., Wang, Y., Li, J., Da, P., Geng, J., & Zheng, G. (2014). Sensitive enzymatic glucose detection by TiO2 nanowire photoelectrochemical biosensors. *Journal of Materials Chemistry A*, *2*(17), 6153–6157. doi:10.1039/C3TA14173J

Tanimoto, A., & Kuribayashi, S. (2006). Application of superparamagnetic iron oxide to imaging of hepatocellular carcinoma. *European Journal of Radiology*, *58*(2), 200–216. doi:10.1016/j.ejrad.2005.11.040 PMID:16414230

Tartaj, J., Zärate, J., Tartaj, P., & Lachowski, E. E. (2002). Sol-gel Cyclic Self-Production of α-Al2O3 Nanoseeds as a Convenient Route for the Low Cost Preparation of Dense Submicronic Alumina Sintered Monoliths. *Advanced Engineering Materials*, *4*(1-2), 17–21. doi:10.1002/1527-2648(20020212)4:1/2<17::AID-ADEM17>3.0.CO;2-R

Tartaj, P., & Serna, C. J. (2002). Microemulsion-assisted synthesis of tunable superparamagnetic composites. *Chemistry of Materials*, *14*(10), 4396–4402. doi:10.1021/cm021214d

Tartaj, P., & Tartaj, J. (2002). Preparation, characterization and sintering behavior of spherical iron oxide doped alumina particles. *Acta Materialia*, *50*(1), 5–12. doi:10.1016/S1359-6454(01)00336-6

Taupitz, M., Schnorr, J., Abramjuk, C., Wagner, S., Pilgrimm, H., Hünigen, H., & Hamm, B. (2000). New generation of monomer-stabilized very small superparamagnetic iron oxide particles (VSOP) as contrast medium for MR angiography: Preclinical results in rats and rabbits. *Journal of Magnetic Resonance Imaging*, *12*(6), 905–911. doi:10.1002/1522-2586(200012)12:6<905::AID-JMRI14>3.0.CO;2-5 PMID:11105029

Taur, Y., & Ning, T. H. (2009). *Fundamentals of modern VLSI devices* (2nd ed.). Cambridge University Press.

Tenzer, S., Docter, D., Kuharev, J., Musyanovych, A., Fetz, V., Hecht, R., & Stauber, R. H. et al. (2013). Rapid formation of plasma protein corona critically affects nanoparticle pathophysiology. *Nature Nanotechnology*, *8*(10), 772–781. doi:10.1038/nnano.2013.181 PMID:24056901

Thomas, L. A., Dekker, L., Kallumadil, M., Southern, P., Wilson, M., Nair, S. P., & Parkin, I. P. (2009). Carboxylic acid-stabilised iron oxide nanoparticles for use in magnetic hyperthermia. *Journal of Materials Chemistry*, *19*(36), 6529–6535. doi:10.1039/b908187a

Thomson, R. C., Yaszemski, M. J., Powers, J. M., & Mikos, A. G. (1998). Hydroxyapatite fiber reinforced poly(alpha-hydroxy ester) foams for bone regeneration. *Biomaterials*, *19*(21), 1935–1943. doi:10.1016/S0142-9612(98)00097-0 PMID:9863527

Tian, X. H., Wei, F., Wang, T. X., Wang, P., Lin, X. N., & Wang, J. et al. (2012). In vitro and in vivo studies on gelatin-siloxane nanoparticles conjugated with SynB peptide to increase drug delivery to the brain. *International Journal of Nanomedicine*, *7*, 1031–1041. PMID:22403486

Torchilin, V. P. (2008). Tat peptide-mediated intracellular delivery of pharmaceutical nanocarriers. *Advanced Drug Delivery Reviews*, *60*(4–5), 548–558. doi:10.1016/j.addr.2007.10.008 PMID:18053612

Tripathi, K. D. (2008). Essentials of medical Pharmacology (6th ed.). New Delhi, India: Medical Publishers Ltd. doi:10.5005/jp/books/10282

Tsang, K. (2003). Using biofunctional magnetic nanoparticles to capture Gram-negative bacteria at an ultra-low concentration. *Chemical Communications*, (15): 1966–1967. PMID:12932056

Tsang, S. C., Caps, V., Paraskevas, I., Chadwick, D., & Thompsett, D. (2004). Magnetically Separable, Carbon-Supported Nanocatalysts for the Manufacture of Fine Chemicals. *Angewandte Chemie*, *116*(42), 5763–5767. doi:10.1002/ange.200460552 PMID:15495206

Tserpes, K. I., & Papanikos, P. (2005). Finite element modeling of single-walled carbon nanotubes. *Composites. Part B, Engineering*, *36*(5), 468–477. doi:10.1016/j.compositesb.2004.10.003

Tsuji, A., & Tamai, I. (1997). Blood–brain barrier function of P-glycoprotein. *Advanced Drug Delivery Reviews*, *25*(2-3), 287–298. doi:10.1016/S0169-409X(97)00504-8

Tsukada, K., Goan, K., & Furusawa, N. (2008). Fluorescent semiconductor microparticle, method for production of the microparticle, fluorescent labeling agent for biological substance using the microparticle, and bioimaging method using the microparticle: Google Patents.

Tuinstra, H. M., Aviles, M. O., Shin, S., Holland, S. J., Zelivyanskaya, M. L., & Fast, A. G. et al. (2012). Multifunctional multichannel bridges that deliver neurotrophin encoding lentivirus for regeneration following spinal cord injury. *Biomaterials*, *33*(5), 1618–1626. doi:10.1016/j.biomaterials.2011.11.002 PMID:22130565

Ulbrich, K., Hekmatara, T., Herbert, E., & Kreuter, J. (2009). Transferrin-and transferrin-receptor-antibody-modified nanoparticles enable drug delivery across the blood–brain barrier (BBB). *European Journal of Pharmaceutics and Biopharmaceutics*, *71*(2), 251–256. doi:10.1016/j.ejpb.2008.08.021 PMID:18805484

Ulbrich, K., Knobloch, T., & Kreuter, J. (2011). Targeting the insulin receptor: Nanoparticles for drug delivery across the blood–brain barrier (BBB). *Journal of Drug Targeting*, *19*(2), 125–132. doi:10.3109/10611861003734001 PMID:20387992

Upadya, M., Shrestha, A., & Kishen, A. (2011). Role of Efflux Pump Inhibitors on the Antibiofilm Efficacy of Calcium Hydroxide, Chitosan Nanoparticles, and Light-activated Disinfection. *Journal of Endodontics*, *37*(10), 1422–1426. doi:10.1016/j.joen.2011.06.017 PMID:21924195

Utsunomiya, S., & Ewing, R. C. (2003). Application of high-angle annular dark field scanning transmission electron microscopy, scanning transmission electron microscopy-energy dispersive X-ray spectrometry, and energy-filtered transmission electron microscopy to the characterization of nanoparticles in the environment. *Environmental Science & Technology*, *37*(4), 786–791. doi:10.1021/es026053t PMID:12636280

Uttara, B., Singh, A. V., Zamboni, P., & Mahajan, R. (2009). Oxidative stress and neurodegenerative diseases: A review of upstream and downstream antioxidant therapeutic options. *Current Neuropharmacology*, *7*(1), 65–74. doi:10.2174/157015909787602823 PMID:19721819

Vale, R. D. (2003). The Molecular Motor Toolbox for Intracellular Transport. *Cell*, *112*(4), 467–480. doi:10.1016/S0092-8674(03)00111-9 PMID:12600311

Valiathan, S. (2006). Ayurveda – putting the house in order (Guest Editorial). *Current Science*, *90*, 5–6.

van den Ent, F., Amos, L. A., & Löwe, J. (2001). Prokaryotic origin of the actin cytoskeleton. *Nature*, *413*(6851), 39–44. doi:10.1038/35092500 PMID:11544518

Van den Vreken, N. M. F., Pieters, I. Y., Declercq, H. A., Cornelissen, M. J., & Verbeeck, R. M. H. (2010). Characterization of calcium phosphate cements modified by addition of amorphous calcium phosphate. *Acta Biomaterialia*, *6*(2), 617–625. doi:10.1016/j.actbio.2009.07.038 PMID:19654057

van der Meel, R., Vehmeijer, L. J., Kok, R. J., Storm, G., & van Gaal, E. V. (2013). Ligand-targeted particulate nanomedicines undergoing clinical evaluation: Current status. *Advanced Drug Delivery Reviews*, *65*(10), 1284–1298. doi:10.1016/j.addr.2013.08.012 PMID:24018362 van der Pol, E., Böing, A. N., Harrison, P., Sturk, A., & Nieuwland, R. (2012). Classification, Functions, and Clinical Relevance of Extracellular Vesicles. *Pharmacological Reviews*, *64*(3), 676–705. doi:10.1124/pr.112.005983 PMID:22722893

van Oers, R. F. M., Ruimerman, R., van Rietbergen, B., Hilbers, P. A. J., & Huiskes, R. (2008). Relating osteon diameter to strain. *Bone*, *43*(3), 476–482. doi:10.1016/j.bone.2008.05.015 PMID:18619937

Van Rijkom, H., Truin, G., & Van't Hof, M. (1998). A meta-analysis of clinical studies on the caries-inhibiting effect of fluoride gel treatment. *Caries Research*, *32*(2), 83–92. doi:10.1159/000016436 PMID:9544855

van Strijp, A., Buijs, M., & Ten Cate, J. (1999). In situ fluoride retention in enamel and dentine after the use of an amine fluoride dentifrice and amine fluoride/sodium fluoride mouthrinse. *Caries Research*, *33*(1), 61–65. doi:10.1159/000016496 PMID:9831781

Vasita, R., & Katti, D. S. (2006). Nanofibers and their applications in tissue engineering. *International Journal of Nanomedicine*, *1*(1), 15–30. doi:10.2147/nano.2006.1.1.15 PMID:17722259

Venugopal, J., Low, S., Choon, A. T., & Ramakrishna, S. (2008). Interaction of cells and nanofiber scaffolds in tissue engineering. *Journal of Biomedical Materials Research. Part B, Applied Biomaterials*, *84B*(1), 34–48. doi:10.1002/jbm.b.30841 PMID:17477388

Vergoni, A. V., Tosi, G., Tacchi, R., Vandelli, M. A., Bertolini, A., & Costantino, L. (2009). Nanoparticles as drug delivery agents specific for CNS: In vivo biodistribution. *Nanomedicine: NBM*, *5*, 369–377. PMID:19341816

Vijender, S., & Bhavuk, G. (2015). Neuroplasticity and its relevance to psychiatric disorders. *Current Psychopharmacology*, *3*(3), 151–157. doi:10.2174/2211556004666150313234631

Vinogradov, S. V., Bronich, T. K., & Kabanov, A. V. (2002). Nanosized cationic hydrogels for drug delivery: Preparation, properties and interactions with cells. *Advanced Drug Delivery Reviews*, *54*(1), 135–147. doi:10.1016/S0169-409X(01)00245-9 PMID:11755709

Virlan, M. J. R., Miricescu, D., Totan, A., Greabu, M., Tanase, C., Sabliov, C. M., ... & Calenic, B. (2015). Current Uses of Poly(lactic-co-glycolic acid) in the Dental Field: A Comprehensive. *Review Journal of Chemistry*, 2015.

Vuong, Q. L., Berret, J., Fresnais, J., Gossuin, Y., & Sandre, O. (2012). A Universal Scaling Law to Predict the Efficiency of Magnetic Nanoparticles as MRI T2-Contrast Agents. *Advanced Healthcare Materials*, *1*(4), 502–512. doi:10.1002/adhm.201200078 PMID:23184784

Wagner, S., Schnorr, J., Pilgrimm, H., Hamm, B., & Taupitz, M. (2002). Monomer-coated very small superparamagnetic iron oxide particles as contrast medium for magnetic resonance imaging: Preclinical in vivo characterization. *Investigative Radiology*, *37*(4), 167–177. doi:10.1097/00004424-200204000-00002 PMID:11923639

Wang, C. X., Huang, L. S., Hou, L. B., Jiang, L., Yan, Z. T., Wang, Y. L., & Chen, Z.-L. (2009). Antitumor effects of polysorbate-80 coated gemcitabine polybutylcyanoacrylate nanoparticles *in vitro* and its pharmacodynamics in vivo on C6 glioma cells of a brain tumor model. *Brain Research, 1261*, 91–99. doi:10.1016/j.brainres.2009.01.011 PMID:19401168

Wang, C.-Z., Fu, Y.-C., Jian, S.-C., Wang, Y.-H., Liu, P.-L., Ho, M.-L., & Wang, C.-K. (2014). Synthesis and characterization of cationic polymeric nanoparticles as simvastatin carriers for enhancing the osteogenesis of bone marrow mesenchymal stem cells. *Journal of Colloid and Interface Science, 432*, 190–199. doi:10.1016/j.jcis.2014.06.037 PMID:25086394

Wang, F., Song, Y. L., Li, C. X., Li, D. H., Zhang, H. P., Ma, A. J., & Zhou, W. et al. (2010). Sustained release of insulin-like growth factor-1 from poly(lactide-co-glycolide) microspheres improves osseointegration of dental implants in type 2 diabetic rats. *European Journal of Pharmacology, 640*(1-3), 226–232. doi:10.1016/j.ejphar.2010.04.024 PMID:20438725

Wang, H., Liu, J., Cho, K. H., & Ren, D. (2009). A novel, single, transmembrane protein CATSPERG is associated with CATSPER1 channel protein. *Biology of Reproduction, 81*(3), 539–544. doi:10.1095/biolreprod.109.077107 PMID:19516020

Wang, I., Politi, A., Tania, N., Bai, Y., Sanderson, M. J., & Sneyd, J. (2008). A mathematical model of airway and pulmonary arteriole smooth muscle. *Biophysical Journal, 94*(6), 2053–2064. doi:10.1529/biophysj.107.113977 PMID:18065464

Wang, J. (2005). Carbon-nanotube based electrochemical biosensors: A review. *Electroanalysis, 17*(1), 7–14. doi:10.1002/elan.200403113

Wang, J. J., Zeng, Z. W., Xiao, R. Z., Xie, T., Zhou, G. L., & Zhan, X. R. et al.. (2011). Recent advances of chitosan nanoparticles as drug carriers. *International Journal of Nanomedicine, 6*, 765–774. PMID:21589644

Wang, L. C., Wu, H., Chen, X. G., De Li, L., Ji, Q. X., & Liu, C. S. et al.. (2008). Biological evaluation of a novel chitosan-PVA-based local delivery system for treatment of periodontitis. *Journal of Biomedical Materials Research. Part A, 91*(4), 1065–1076. PMID:19107793

Wang, W., Liao, S., Zhu, Y., Liu, M., Zhao, Q., & Fu, Y. (2015). Recent applications of nanomaterials in prosthodontics. *Journal of Nanomaterials, 2015*, 3. doi:10.3390/nano6010003

Wang, X. Y., & Wang, X. (2004). Numerical simulation for bending modulus of carbon nanotubes and some explanations for experiment. *Composites. Part B, Engineering, 35*(2), 79–86. doi:10.1016/S1359-8368(03)00084-2

Wang, Y., Dostalek, J., & Knoll, W. (2011). Magnetic Nanoparticle-enhanced biosensor based on Grating-coupled surface Plasmon resonance. *Analytical Chemistry, 83*(16), 6202–6207. doi:10.1021/ac200751s PMID:21711037

Watson, P. S., Pontefract, H. A., Devine, D. A., Shore, R. C., Nattress, B. R., Kirkham, J., & Robinson, C. (2005). Penetration of fluoride into natural plaque biofilms. *Journal of Dental Research, 84*(5), 451–455. doi:10.1177/154405910508400510 PMID:15840782

Wegner, G., & Rujescu, D. (2013). The current development of CNS drug research. *The International Journal of Neuropsychopharmacology*, *16*(07), 1687–1693. doi:10.1017/S1461145713000345 PMID:23651558

Wei, F., Lillehoj, P. B., & Ho, C. M. (2010). DNA Diagnostics: Nanotechnology-enhanced electrochemical detection of nucleic acids. *Pediatric Research*, *67*(5), 458–468. doi:10.1203/PDR.0b013e3181d361c3 PMID:20075759

Weissleder, R., Lee, A. S., Fischman, A. J., Reimer, P., Shen, T., Wilkinson, R., & Brady, T. J. (1991). Polyclonal human immunoglobulin G labeled with polymeric iron oxide: Antibody MR imaging. *Radiology*, *181*(1), 245–249. doi:10.1148/radiology.181.1.1887040 PMID:1887040

Welch, R. D., Zhang, H., & Bronson, D. G. (2003). Experimental tibial plateau fractures augmented with calcium phosphate cement or autologous bone graft. *J. Bone Joint Surg.*, *85*, 222–231. PMID:12571298

Wetterskog, E., Agthe, M., Mayence, A., Grins, J., Wang, D., Rana, S., & Bergström, L. (2014). Precise control over shape and size of iron oxide nanocrystals suitable for assembly into ordered particle arrays. *Science and Technology of Advanced Materials*, *15*(5), 55010. doi:10.1088/1468-6996/15/5/055010

Whited, B. M., Whitney, J. R., Hofmann, M. C., Xu, Y., & Rylander, M. N. (2011). Pre-osteoblast infiltration and differentiation in highly porous apatite-coated PLLA electrospun scaffolds. *Biomaterials*, *32*(9), 2294–2304. doi:10.1016/j.biomaterials.2010.12.003 PMID:21195474

Whitesides, G. M., & Grzybowski, B. (2002). Self-assembly at all scales. *Science*, *295*(5564), 2418–2421. doi:10.1126/science.1070821 PMID:11923529

Whitman, D., McKinney, J., Hinman, R., Hesby, R., & Pelleu, G. Jr. (1987). In vitro wear rates of three types of commercial denture tooth materials. *The Journal of Prosthetic Dentistry*, *57*(2), 243–246. doi:10.1016/0022-3913(87)90154-5 PMID:3470516

WHO. (n. d.). Depression fact sheet. Retrieved from http://www.who.int/mediacentre/factsheets/fs369/en/index.html

Widera, A., Norouziyan, F., & Shen, W. C. (2003). Mechanisms of TfR-mediated transcytosis and sorting in epithelial cells and applications toward drug delivery. *Advanced Drug Delivery Reviews*, *55*(11), 39–66. doi:10.1016/j.addr.2003.07.004 PMID:14597140

Wikipedia. (2015). Togaviridae. Retrieved from https://en.wikipedia.org/wiki/Togaviridae

Williams, B., Watanabe, C. M., Schultz, P. G., Rimbach, G., & Krucker, T. (2004). Age-related effects of *Ginkgo biloba* extract on synaptic plasticity and excitability. *Neurobiology of Aging*, *25*(7), 955–962. doi:10.1016/j.neurobiolaging.2003.10.008 PMID:15212849

Williams, R. C. (1990). Periodontal Disease. *The New England Journal of Medicine*, *322*(6), 373–382. doi:10.1056/NEJM199002083220606 PMID:2405268

Wilson, B., Samanta, M. K., Santhi, K., Kumar, K. P., Paramakrishnan, N., & Suresh, B. (2008). Poly(n-butylcyanoacrylate) nanoparticles coated with polysorbate 80 for the targeted delivery of rivastigmine into the brain to treat Alzheimer's disease. *Brain Research*, *1200*, 159–168. doi:10.1016/j.brainres.2008.01.039 PMID:18291351

Wilson, B., Samanta, M. K., Santhi, K., Kumar, K. P., Paramakrishnan, N., & Suresh, B. (2008). Targeted delivery of tacrine into the brain with polysorbate 80-coated poly(n-butylcyanoacrylate) nanoparticles. *European Journal of Pharmaceutics and Biopharmaceutics*, *70*(1), 75–84. doi:10.1016/j.ejpb.2008.03.009 PMID:18472255

Wilson, M. (1993). Photolysis of oral bacteria and its potential use in the treatment of caries and periodontal disease. *The Journal of Applied Bacteriology*, *75*(4), 299–306. doi:10.1111/j.1365-2672.1993.tb02780.x PMID:8226389

Wohlfart, S., Gelperina, S., & Kreuter, J. (2012). Transport of drugs across the blood–brain barrier by nanoparticles. *Journal of Controlled Release*, *161*(2), 264–273. doi:10.1016/j.jconrel.2011.08.017 PMID:21872624

Wong, S. S., Joselevich, E., Woolley, A. T., Cheung, C. L., & Lieber, C. M. (1998). Covalently functionalized nanotubes as nanometre-sized probes in chemistry and biology. *Nature*, *394*(6688), 52–55. doi:10.1038/27873 PMID:9665127

Woo, E., Ponvel, K. M., Ahn, I.-S., & Lee, C.-H. (2010). Synthesis of magnetic/silica nanoparticles with a core of magnetic clusters and their application for the immobilization of His-tagged enzymes. *Journal of Materials Chemistry*, *20*(8), 1511–1515. doi:10.1039/B918682D

Wu, S. Z., Hossainy, S. F., Harish, S., Sanders-Millare, D., & Mirzaee, D. (2003). Microparticle coated medical device: Google Patents.

Wu, H., Zhang, Z., Zhao, H., Wu, D., Wu, B., & Cong, R. (2004). Preparation of sodium fluoride-loaded gelatin microspheres, characterization and cariostatic studies. *Journal of Microencapsulation*, *21*(8), 889–903. doi:10.1080/02652040400015445 PMID:15799544

Wu, T. Z., Yen, F. L., Lin, L. T., Tsai, T. R., Lin, C. C., & Cham, T. M. (2008). Preparation, physicochemical characterization, and antioxidant effects of quercetin nanoparticles. *International Journal of Pharmaceutics*, *346*(1–2), 160–168. doi:10.1016/j.ijpharm.2007.06.036 PMID:17689897

Wu, W., Palaniapan, M., & Wong, W. K. (2008). Multiwall carbon nanotube resonator for ultra-sensitive mass detection. *Electronics Letters*, *44*(18), 1932–1933. doi:10.1049/el:20081932

Xia, H., Gao, X., Gu, G., Liu, Z., Hu, Q., & Tu, Y. et al. (2012). Penetratin-functionalized PEG-PLA nanoparticles for brain drug delivery. *International Journal of Pharmacology*, *436*(2), 840–850. doi:10.1016/j.ijpharm.2012.07.029 PMID:22841849

Xie, J., Xu, C., Xu, Z., Hou, Y., Young, K. L., Wang, S. X., & Sun, S. (2006). Linking hydrophilic macromolecules to monodisperse magnetite (Fe3O4) nanoparticles via trichloro-s-triazine. *Chemistry of Materials*, *18*(23), 5401–5403. doi:10.1021/cm061793c PMID:18176627

Xing, Y., Wen, C., Li, S., & Xia, Z. (2016). Non-viral liposome-mediated transfer of brain-derived neurotrophic factor across the blood-brain barrier. *Neural Regeneration Research*, *11*(4), 617–622. doi:10.4103/1673-5374.180747 PMID:27212923

Xu, C. Y., Inai, R., Kotaki, M., & Ramakrishna, S. (2004). Aligned biodegradable nanotibrous structure: A potential scaffold for blood vessel engineering. *Biomaterials*, *25*(5), 877–886. doi:10.1016/S0142-9612(03)00593-3 PMID:14609676

Xu, C., Xu, K., Gu, H., Zhong, X., Guo, Z., Zheng, R., & Xu, B. (2004). Nitrilotriacetic acid-modified magnetic nanoparticles as a general agent to bind histidine-tagged proteins. *Journal of the American Chemical Society*, *126*(11), 3392–3393. doi:10.1021/ja031776d PMID:15025444

Xue, Y. H., Wang, H. X., Yu, D. S., Feng, L. F., Dai, L. M., Wang, X. G., & Lin, T. (2009). Superhydrophobic electrospun POSS-PMMA copolymer fibres with highly ordered nanofibrillar and surface structures. *Chemical Communications*, (42): 6418–6420. doi:10.1039/b911509a PMID:19841795

Xu, K. Y., Aifantis, E. C., & Yan, Y. H. (2008). Vibrations of double-walled carbon nanotubes with different boundary conditions between inner and outer tubes. *Journal of Applied Mechanics*, *75*(2), 021013. doi:10.1115/1.2793133

Xu, S. L., Liu, Y., Tai, H. C., Zhu, J., Ding, H., & Lee, R. J. (2011). Synthesis of transferrin (Tf) conjugated liposomes via Staudinger ligation. *International Journal of Pharmaceutics*, *404*(1-2), 205–210. doi:10.1016/j.ijpharm.2010.10.053 PMID:21056642

Yadava, R. N., & Verma, V. (2003). Antimicrobial activity of a novel flavonol glycosideisolated from the roots of *Clitoria ternatea* Linn. *Asian Journal of Chemistry*, *15*, 842–846.

Yakobson, B., Brabec, C., & Bernholc, J. (1996). Nanomechanics of carbon tubes: Instabilities beyond linear response. *Physical Review Letters*, *76*(14), 2511–2514. doi:10.1103/PhysRevLett.76.2511 PMID:10060718

Yang, J., Lee, C., Ko, H., Suh, J., Yoon, H., & Lee, K. Haam, S. (2007). Multifunctional magneto-polymeric nanohybrids for targeted detection and synergistic therapeutic effects on breast cancer. Angewandte Chemie, 46(46), 8836–8839.

Yang, C.-M., Wang, J.-C., Chiang, T. W., Lin, Y.-T., Juan, T.-W., & Chen, T.-C. et al. (2014). Hydrogen ion sensing characteristics of IGZO/Si electrode in EGFET. *International Journal of Nanotechnology*, *11*(1/2/3/4), 15–26. doi:10.1504/IJNT.2014.059806

Yang, Q., Troczynski, T., & Liu, D. M. (2002). Influence of apatite seeds on the synthesis of calcium phosphate cement. *Biomaterials*, *23*(13), 2751–2760. doi:10.1016/S0142-9612(02)00010-8 PMID:12059025

Yang, X., Koh, C. G., Liu, S., Pan, X., Santhanam, R., & Yu, B. et al. (2009). Transferrin receptor-targeted lipid nanoparticles for delivery of an antisense oligodeoxyribonucleotide against Bcl-2. *Molecular Pharmaceutics*, *6*(1), 221–230. doi:10.1021/mp800149s PMID:19183107

Yang, Y., Callegari, C., Feng, X., Ekinci, K., & Roukes, M. (2006). Zeptogram scale nanomechanical mass sensing. *Nano Letters*, *6*(4), 583–586. doi:10.1021/nl052134m PMID:16608248

Yang, Y., & Rosenberg, G. A. (2011). Blood–brain barrier breakdown in acute and chronic cerebrovascular disease. *Stroke*, *42*(11), 3323–3328. doi:10.1161/STROKEAHA.110.608257 PMID:21940972

Yan, J., Mo, S., Nie, J., Chen, W., Shen, X., Hu, J., & Tong, H. (2009). Hydrothermal synthesis of monodisperse Fe 3 O 4 nanoparticles based on modulation of tartaric acid. *Colloids and Surfaces. A, Physicochemical and Engineering Aspects*, *340*(1), 109–114. doi:10.1016/j.colsurfa.2009.03.016

Yanong, R. P. F. (2003). Fungal diseases of fish. In M.P. Jones (Ed.), Fungal Diseases, veterinary clinics of north America, exotic animal practice (Vol. 6). Philadelphia: W.B. Saunders Co.

Yao, W., Xu, P., Pang, Z., Zhao, J., Chai, Z., & Li, X. et al.. (2014). Local delivery of minocycline-loaded PEG-PLA nanoparticles for the enhanced treatment of periodontitis in dogs. *International Journal of Nanomedicine*, *9*, 3963–3970. PMID:25170266

Yee, N. S. (2016). Chapter four-TRPM8 Ion channels as potential cancer biomarker and target in pancreatic cancer. *Advances in Protein Chemistry and Structural Biology*, *104*, 127–155. doi:10.1016/bs.apcsb.2016.01.001 PMID:27038374

Yee, N. S., Chan, A. S., Yee, J. D., & Yee, R. K. (2012). TRPM7 and TRPM8 Ion Channels in Pancreatic Adenocarcinoma: Potential Roles as Cancer Biomarkers and Targets. *Scientifica.*

Yen, F. L., Wu, T. Z., Lin, L. T., Cham, T. M., & Lin, C. C. (2008). Nanoparticles formulation of *Cuscuta chinensis* prevents acetaminophen-induced hepatotoxicity in rats. *Food and Chemical Toxicology*, *46*(5), 1771–1777. doi:10.1016/j.fct.2008.01.021 PMID:18308443

Yonamine, Y., Matsuyama, T., Sonomura, T., Takeuchi, H., Furuichi, Y., Uemura, M., & Noguchi, K. et al. (2010). Effectable application of vascular endothelial growth factor to critical sized rat calvaria defects. *Oral Surgery, Oral Medicine, Oral Pathology, Oral Radiology, and Endodontology*, *109*(2), 225–231. doi:10.1016/j.tripleo.2009.09.010 PMID:20034820

Yong, C. W. (2014). Study of interactions between polymer nanoparticles and cell membranes at atomistic levels. *Philosophical Transactions of the Royal Society of London B: Biological Sciences, 370*(1661). Doi:10.1098/rstb.2014.0036

Yoon, J., Ru, C. Q., & Mioduchowski, A. (2003). Vibration of an embedded multiwall carbon nanotube. *Composites Science and Technology*, *63*(11), 1533–1542. doi:10.1016/S0266-3538(03)00058-7

Yoshikawa, T., & Pardridge, W. M. (2001). Delivery of peptides and proteins through the blood–brain barrier. *Advanced Drug Delivery Reviews*, *46*(1-3), 247–279. doi:10.1016/S0169-409X(00)00139-3 PMID:11259843

Yoshimoto, H., Shin, Y. M., Terai, H., & Vacanti, J. P. (2003). A biodegradable nanofiber scaffold by electrospinning and its potential for bone tissue engineering. *Biomaterials*, *24*(12), 2077–2082. doi:10.1016/S0142-9612(02)00635-X PMID:12628828

Yue, I. C., Poff, J., Cortes, M. E., Sinisterra, R. D., Faris, C. B., Hildgen, P., & Shastri, V. P. et al. (2004). A novel polymeric chlorhexidine delivery device for the treatment of periodontal disease. *Biomaterials*, *25*(17), 3743–3750. doi:10.1016/j.biomaterials.2003.09.113 PMID:15020150

Yun, Y., Dong, Z., Shanov, V., Heineman, W. R., Halsall, H. B., Bhattacharya, A., & Schulz, M. J. et al. (2007). Nanotube electrodes and biosensors. *Nano Today*, *2*(6), 30–37. doi:10.1016/S1748-0132(07)70171-8

Zeren, K. J. (2006). Minimally invasive extraction and immediate implant placement: The preservation of esthetics. *The International Journal of Periodontics & Restorative Dentistry*, *26*(2), 171–181. PMID:16642906

Zhang, J. Y., Beckman, E. J., Piesco, N. P., & Agarwal, S. (2000). A new peptide-based urethane polymer: Synthesis, biodegradation, and potential to support cell growth in vitro. *Biomaterials*, *21*(12), 1247–1258. doi:10.1016/S0142-9612(00)00005-3 PMID:10811306

Zhang, J., Stevens, M. F., & Bradshaw, T. D. (2012). Temozolomide: Mechanisms of Action, Repair and Resistance. *Current Molecular Pharmacology*, *5*, 102–114. doi:10.2174/1874467211205010102 PMID:22122467

Zhang, L., & Barritt, G. J. (2006). TRPM8 in prostate cancer cells: A potential diagnostic and prognostic marker with a secretory function? *Endocrine-Related Cancer*, *13*(1), 27–38. doi:10.1677/erc.1.01093 PMID:16601277

Zhang, L., Xu, T. Y., Wang, S., Yu, L. Q., Liu, D. X., Zhan, R. Z., & Yu, S. Y. (2012). Curcumin produces antidepressant effects via activating MAPK/ERK-dependent brain-derived neurotrophic factor expression in the amygdala of mice. *Behavioural Brain Research*, *235*(1), 67–72. doi:10.1016/j.bbr.2012.07.019 PMID:22820234

Zhang, M.-L., Fan, X., Zhou, H.-W., Shao, M.-W., Zapien, J. A., Wong, N.-B., & Lee, S.-T. (2010). High efficiency surface-enhanced Raman scattering substrate based silicon nanowires array decorated with silver nanoparticles. *Journal of Physical Chemistry*, *114*, 1969–1975.

Zhang, R. Y., & Ma, P. X. (1999). Poly(alpha-hydroxyl acids) hydroxyapatite porous composites for bone-tissue engineering. I. Preparation and morphology. *Journal of Biomedical Materials Research*, *44*(4), 446–455. doi:10.1002/(SICI)1097-4636(19990315)44:4<446::AID-JBM11>3.0.CO;2-F PMID:10397949

Zhang, Y. Z., Reddy, V. J., Wong, S. Y., Li, X., Su, B., Ramakrishna, S., & Lim, C. T. (2010). Enhanced Biomineralization in Osteoblasts on a Novel Electrospun Biocomposite Nanofibrous Substrate of Hydroxyapatite/Collagen/Chitosan. *Tissue Engineering Part A*, *16*(6), 1949–1960. doi:10.1089/ten.tea.2009.0221 PMID:20088700

Zhang, Y., Chen, R., Xu, L., Ning, Y., Xie, S., & Zhang, G. J. (2015). Silicon Nanowire Boisensor for highly Sensitive and Multiplexed detection of oral squamous cell carcinoma biomarkers in saliva. *Analytical Sciences*, *31*(2), 73–78. doi:10.2116/analsci.31.73 PMID:25746803

Zhao, H., Wu, B., Wu, H., Su, L., Pang, J., Yang, T., & Liu, Y. (2006). Protective immunity in rats by intranasal immunization with Streptococcus mutans glucan-binding protein D encapsulated into chitosan-coated poly(lactic-co-glycolic acid) microspheres. *Biotechnology Letters*, *28*(16), 1299–1304. doi:10.1007/s10529-006-9086-7 PMID:16804764

Zheng, G., Gao, X. P. A., & Leiber, C. M. (2010). Frequency domain detection of biomolecules using silicon nanowire biosensors. *Nano Letters*, *10*(8), 3179–3183. doi:10.1021/nl1020975 PMID:20698634

Zheng, G., Patolsky, F., Cui, Y., Wang, W. U., & Lieber, C. M. (2005). Multiplexed electrical detection of cancer markers with nanowire sensor arrays. *Nature Biotechnology*, *23*(10), 1294–1301. doi:10.1038/nbt1138 PMID:16170313

Zheng, X., Kan, B., Gou, M., Fu, S. Z., Zhang, J., Men, K., & Zhao, X. (2010). Preparation of MPEG-PLA nanoparticle for honokiol delivery in vitro. *International Journal of Pharmaceutics*, *386*(1–2), 262–267. doi:10.1016/j.ijpharm.2009.11.014 PMID:19932160

Zheng, Y., Chen, Y., Jin, L. W., Ye, H. Y., & Liu, G. (2016). Cytotoxicity and Genotoxicity in Human Embryonic Kidney Cells Exposed to Surface Modify Chitosan Nanoparticles Loaded with Curcumin. *AAPS PharmSciTech*. PMID:26718819

Zhu, S., Du, C., & Fu, Y. (2009). Fabrication and characterization of rhombic silver nanoparticles for biosensing. *Optical Materials*, *31*(6), 769–774. doi:10.1016/j.optmat.2008.07.014

Zone, V. (2015). ExPASy. Retrieved from https://en.wikipedia.org/wiki/Caliciviridae

Zou, G.-K., Song, Y.-L., Zhou, W., Yu, M., Liang, L.-H., Sun, D.-C., et al. (2012). Effects of local delivery of bFGF from PLGA microspheres on osseointegration around implants in diabetic rats. *Oral surgery, oral medicine, oral pathology and oral radiology, 114*(3), 284-289.

About the Contributors

Keka Talukdar has a B.Sc. in Physics. Got M.Sc. Degree from the University of Burdwan with specialization of Nuclear and Particle Physics. Received Ph.D. degree on computational nanoscience from NIT Durgapur on 2011. Still working with collaboration with NIT Durgapur. Also worked as Research Associate in the Institute of Genomics and Bio-medical Research. Research interests: Modeling and simulation of CNTs and graphene properties, nanobiophysics, biosensor and ion channel simulation. Publications: Published 23 papers in reputed international journals. Also published 4 book chapters with international publishers including one with Wiley. Award and Scholarship: Got National Scholarship from Education Directorate of West Bengal for M.Sc. study. Got one poster award and one best paper award in the international conferences. Have three conference papers outside India. Other achievements: Reviewer of two reputed Elsevier Journals: Materials Science and Engineering B and Mechanics of Materials, Springer journal: Nanoscale Research Letters. Also reviewer of ANTJ Journal and Merit Research Journal of Medicine and Medical Sciences. Member of editorial board of Advanced Nanoscience and Technology: An International Journal. Got call for Invited Talk from the International conferences outside India.

Mayank Bhushan obtained his B.Tech degree in Biotechnology in 2010 from SRM University, Chennai, India. He has completed his M.Tech. degree in Nanoscience and Technology in 2012 from Pondicherry University, Puducherry, India. Subsequently, in August 2012 he has joined PhD program in Nanoscience and Technology at Centre for Nanoscience and Technology, Pondicherry University and final degree is yet to be awarded. Currently he is working as guest lecturer at Department of Nanotechnology, North Eastern Hill University, Shillong, India. His current research interests include development of inorganic nanomaterials based new generation antibiotics, bioimaging probes and nano carriers for targeted drug delivery.

Anil Shantappa Malipatil (M.Sc., M.Phil., PhD) did his M.Sc in Nuclear Physics (2006), M.Phil in Spectroscopy (2007) and completed his Doctor of Philosophy in Radiation Physics (2015) from Gulbarga University, Kalaburgi, and Karnataka, INDIA. He was awarded UGC-BSR (RFSMS) fellowship during 2013-2014. His biography is published in the 33rd Edition of Marquis Who's Who in the World-2016. He Published 11 International Journal Publication and presented his research work at International (6) and National Conferences (8) including Japan and Nepal Conferences and also participated in college on Multiscale Computational Modeling of materials for Energy application held at ICTP, Trieste, ITALY. He has 7 years of teaching experience (UG+PG). He is presently working as Assistant Professor in Physics at Guru Nanak Dev Engineering College, Bidar, Karnataka, India. He is Reviewer of reputed Springer Journal (Radiation and Environmental Biophysics). He is presently working under the supervision of Dr Keka Talukdar on Nano-Bio, CNT, Biosensor, Ion channel simulation and CNT heterojunction properties.

* * *

George A Avtandilov is a Dentist prosthodontist who studied at MSMSU 2000, passed all stages from student to graduate students, and in 2013 he defended his thesis on "Biodegradation dentures of polymer materials (experimental research)". He continues to research activities with the selected direction of the master's thesis on the biodestruction of the materials used in the oral cavity.

Dimple Sethi Chopra is presently working as Senior Assistant Professor (Pharmaceutics) in Department of Pharmaceutical Sciences and Drug Research, Punjabi University, Patiala. Her present areas of interest are molecular nanotechnology, targeted drug delivery, nanobiotechnology and nanomedicine. She has expertise in teaching wide variety of subjects like mathematics, statistics, computer applications, cosmeticology, physical pharmacy, dispensing pharmacy, unit operations, advanced pharmacokinetics and biopharmaceutics at undergraduate and postgraduate level. She has supervised 17 M.Pharm students for their thesis work and presently 2 Ph.D students are enrolled under her. Her research interest includes exploring different types of nanoparticles for treatment of dementia related to various neurodegenerative disorders like Alzheimer's disease. She is also actively engaged in developing nanocarrier systems for treatment of rheumatoid arthritis and osteoarthritis. She has also worked on other particulate systems like microspheres and niosomes for development of controlled drug delivery systems for oral as well as topical use. She has published several research papers in national and international journals of repute. She has presented her work at several national and international conferences. She won the best presentation award at National Seminar on current trends and Future challenges in Pharmaceutics held at UIPS, PU, Chandigarh. She has filed in two Indian patents on brain permeable nanoparticles. She is an invited manuscript reviewer of many international journals like Drug Discovery today, Journal of Nanoparticle Research, drug delivery, seizure, Open Access Rheumatology: Research and Reviews, International Journal of Nanomedicine, Drug Design, Development and Therapy. She is a member of American Nano Society. She has filed in two Indian Patents on Brain Permeable Nanoparticles. They have been successfully examined by the Indian Patent office. She has successfully published four chapters with publishing houses like INTECH OPEN, Studium Press LLC, USA, McGraw Hill Education (India) Private Limited, New Delhi. All the book chapters are on nanoparticles. The chapter with INTECH OPEN has reached 3000 downloads.

Roberta Curia started to take interest in the multifaceted behavior of Staphylococcus aureus and its interactions with plastic materials during her thesis work in Biological Sciences at University of Milano-Bicocca, Milan (Italy). In the frame of a collaboration between the Department of Materials Science (University of Milano-Bicocca, Milan, Italy) with the Laboratory of Microorganisms' anatomy at Gamaleya Research Institute for Epidemiology and Microbiology, Moscow (Russian Federation) she investigates the aspects of bacteria as both nanoparticles producers and vectors influenced by the nanoparticles themselves via electron microscopy, highlighting the novelties from the toxicological and medical standpoints.

Nileshkumar Dubey's research is in the interface of bone tissue engineering, biomaterials, cellular & molecular biology, materials engineering, and analytical chemistry. Specifically, his research involves the development of nano-biomaterials for tissue engineering applications. Long term implications of his research are directed towards the next-generation of materials-based tissue engineering approaches for the rational design of biomedical implants and coatings, cellular-support biomaterials.

Anand Y. Joshi is currently working as a Professor in Mechatronics Engineering Department at G.H. Patel College of Engineering & Technology, Vallabh Vidyanagar, Gujarat, India. He completed his Ph.D. in the year 2011 from Mechanical and Industrial Engineering Department, Indian Institute of Technology Roorkee. His research areas include Areas of vibration, nano biosensors, nanomechanics, fracture mechanics, computational mechanics, functionally graded material, carbon nanotubes, Optimization, Finite Element Analysis etc. He has to his credit more than 45 publications in various Journals and conferences of high repute including one of his articles published by Nature India and an article published in a Journal, Physica E: Low dimensional systems and Nanostructures, Elsevier Publishers, being Cited as one of the Top 25 hottest articles of the Journal and the Most cited article of the

Journal since 2008. He has been the recipient of a number of Research Grants of more than Rs. 25 Lacs, including the startup research grant which is awarded to Young Scientists below the age of 35 years by Department of Science and Technology, Govt. Of India, New Delhi for his project on Computational Investigation of Dynamic Characteristics of Carbon Nanotubes for Atomic Scale sensing of Objects. His overseas presence includes (i) International Technical Programme Committee member of the 2nd International Conference on Sensors and Electronic Instrumental Advances (SEIA' 2016) 22-23 September 2016, Barcelona, Castelldefefs, Spain, (ii) Scientific Program Committee member of the International Academy of Industrial, Mechanical and Aeronautical Engineering., IAIMAE, (iii) International Technical Programme Committee member for the International Conference on Advances in Sensors, Actuators, Metering and Sensing - ALLSENSORS 2016, April 24 - 28, 2016 - Venice, Italy, (iv) International Conference Scientific Committee member for the International Conference on Food and Biosystems Engineering (I.C. FaBE) 2015, 28-31 May 2015, Mykonos island, GREECE, (v) Scientific Committee member for the International Conference on Nonlocal Mechanics of Composites (NONMECH2013), (vi) 23-25 October 2013 Istanbul, Turkey, (vii) Associate Editor for the Journal of Nanotechnology and Nanoscience, Bioinfo Publications etc.

Ozan Karaman has received his B.S. degree in Bioengineering, M.S. and PhD. degree in Biomedical Engineering from University of South Carolina, USA. He has been working as an assistant professor at İzmir Katip Çelebi University Faculty of Engineering and Architecture, Biomedical Engineering Department since August 2014. He has been working on bone and cartilage tissue engineering using surface modified biodegradable nanofibrous composites by biomimetic peptide units and investigating the osteoconductivity and vascularization capacity of designed materials in order to develop functional scaffolds for bone regeneration.

Anita Margret is the author is working as an assistant Professor in the department of Biotechnology and has pursued her degrees from Bharathidasan university Tiruchirappali. The contemporary research is in the fields of Nanoscience and Neurobehavioural science (Post Natal Depression).

Marziale Milani is Professor of Physics at the University of Milano-Bicocca Department of Materials Science.

Ajay M Patel is currently working as an Associate Professor in Mechatronics Engineering Department at G.H. Patel College of Engineering & Technology, Vallabh Vidyanagar, Gujarat, India. He has been associated with the research in the area of nano sensors, nanomechanics, non- traditional manufacturing, double walled carbon nanotube, Robotics, Finite Element Method, Simulation and Modeling, CAD, advance manufacturing, Automation etc. his research is focused on the analysis of double walled carbon nanotube base mass sensor. His main research contribution includes detection of biological object using double walled carbon nanotubes, use of Double walled carbon nanotube as biosensors, application and analysis of bio and virus nanomaterials etc. Moreover, he has to his credit a number of papers published in reputed international journal and international conferences. He has been serving as a Reviewer in various international journals of high repute.

Balasankar Meera Priyadarshini completed her Bachelors in B.Tech Biotechnology, SRM University India after which she pursued her Masters in Biomedical Engineering, NTU, Singapore in 2009.

Sayedeh Fatemeh Shaffiey is a managing director and researcher at Novin Pak Nanotech Co. Ltd. Company that focuses on nano materials. Her research interest is bactericidal nano particles.

Sayed Reza Shaffiey is president and CEO of Novin Pak Nanotech Co. Ltd. and his interests are in topics such as sensors, fuel cells, nano materials and bio materials. The research center of this company is focused on the development of new products based on nano and bio materials for drug delivery.

Natalia V Shevlyagina is a Senior researcher in Gamaleya research institute for epidemiology and microbiology in Moscow, Russia.

Francesco Tatti is an SEM and Dual Beam Application Specialist at European Nanoport – FEI.

Irshad Ahmad Wani was born and raised in state, Jammu and Kashmir, India. He received his M.Sc in Physical chemistry from Jamia Millia Islamia, New Delhi in 2007. He received his Ph.D in 2012 from the same university under the tender supervision of Dr. Tokeer Ahmad. His Ph.D mainly focused on synthesis of noble metal nanoparticles and their oxides by employing various chemical-biological routes and studying their application in biomedical area. He has been able to publish a good number of research papers in various esteemed international journals. His research work has been cited by a number of scientific journals. Currently he is working as lecturer in chemistry serving J&K govt. education services. He is also a peer reviewer of various international journals and a member of various scientific societies. His current research interests include Nanoparticle-Cancer, Nanoparticle-Protein and carbohydrate interactions, conjugation of inorganic nanoparticles with various biologically significant molecules and role of nanoparticles in sensor applications.

Index

R

S

T

V

W

www.ingramcontent.com/pod-product-compliance
Lightning Source LLC
LaVergne TN
LVHW080600300926
849162LV00005B/1012